I0007299

Vivian W. Y. Tam, Khoa N. Le and Liyin Shen (Eds.)

Life Cycle Assessment on Green Building Implementation

This book is a reprint of the Special Issue that appeared in the online, open access journal, *Sustainability* (ISSN 2071-1050) from 2015–2016 (available at: http://www.mdpi.com/journal/sustainability/special_issues/GBI).

Guest Editors
Vivian W. Y. Tam
School of Computing, Engineering and Mathematics
Western Sydney University
Australia
College of Civil Engineering
Shenzhen University
China

Khoa N. Le
School of Computing, Engineering and Mathematics
Western Sydney University
Australia

Liyin Shen
Faculty of Construction Management and Real Estate
Chongqing University
China

Editorial Office
MDPI AG
St. Alban-Anlage 66
Basel, Switzerland

Publisher
Shu-Kun Lin

Managing Editor
Guoshui Liu

1. Edition 2016

MDPI • Basel • Beijing • Wuhan • Barcelona

ISBN 978-3-03842-256-3 (Hbk)
ISBN 978-3-03842-257-0 (PDF)

Table of Contents

List of Contributors

Antonio Aguado Department of Civil and Environmental Engineering, UPC, c Jordi Girona 1, Barcelona 08034, Spain.

Cheonghoon Baek Korea Institute of Civil Engineering and Building Technology, 283, Goyangdae-ro, Goyang-si, Gyeonggi-do 10223, Korea.

Ginevra Balletto Department of Civil and Environmental Engineering and Architecture, DICAAR, University of Cagliari, via Marengo, 3, Cagliari 09123, Italy.

Myeongsoo Chae Department of Architectural Engineering, Yonsei University, Seoul 03722, Korea.

Chang U. Chae Building and Urban Research Institute, Korea Institute of Civil Engineering and Building Technology, Daehwa-dong 283, Goyandae-Ro, ILsanseo-Gu, Goyang-Si 10223, Korea.

Albert de la Fuente Department of Civil and Environmental Engineering, UPC, C. Jordi Girona 1, Barcelona 08034, Spain.

Gokhan Egilmez Department of Civil, Mechanical and Environmental Engineering, University of New Haven, West Haven, CT 06516, USA.

Lin Gan School of Construction Management and Real Estate, and International Research Center for Sustainable Built Environment, Chongqing University, Chongqing 400044, China.

Chiara Garau Department of Civil and Environmental Engineering and Architecture, DICAAR, University of Cagliari, via Marengo, 3, Cagliari 09123, Italy.

Taehoon Hong Department of Architectural Engineering, Yonsei University, Seoul 03722, Korea.

Jaemin Jeong Department of Architectural Engineering, Yonsei University, Seoul 03722, Korea.

Jimin Kim Department of Architectural Engineering, Yonsei University, Seoul 03722, Korea.

Rakhyun Kim Department of Architectural Engineering, Hanyang University, 55 Hanyangdaehak-ro, Sangrok-gu, Ansan-si, Gyeonggi-do 15588, Korea.

Tae Hyoung Kim Building and Urban Research Institute, Korea Institute of Civil Engineering and Building Technology, Daehwa-dong 283, Goyandae-Ro, ILsanseo-Gu, Goyang-Si 10223, Korea.

Murat Kucukvar Department of Industrial Engineering, Istanbul Sehir University, Uskudar, Istanbul 34662, Turkey.

Han-Seung Lee Department of Architectural Engineering, Hanyang University, Ansan 426-791, Korea.

Kanghee Lee Department of Architectural Engineering, Andong National University, 1375, Gyeongdong-Ro, Andong-Si 36729, Korea.

Minhyun Lee Department of Architectural Engineering, Yonsei University, Seoul 03722, Korea.

Giovanni Mei Department of Civil and Environmental Engineering and Architecture, DICAAR, University of Cagliari, via Marengo, 3, Cagliari 09123, Italy.

Joonho Park Department of Architectural Engineering, Yonsei University, Seoul 03722, Korea.

Oriol Pons Department of Architectural Technology, UPC, av. Diagonal 649, Barcelona 08028, Spain.

Payam Rahnamayiezekavat School of Built Environment, Curtin University, Bentley WA 6102, Australia.

Seungjun Roh Innovative Durable Building and Infrastructure Research Center, Hanyang University, 55 Hanyangdaehak-ro, Sangnok-gu, Ansan 426-791, Korea.

Liyin Shen School of Construction Management and Real Estate, and International Research Center for Sustainable Built Environment, Chongqing University, Chongqing 400044, China.

Sungwoo Shin School of Architecture & Architectural Engineering, Hanyang University, 55 Hanyangdaehak-ro, Sangrok-gu, Ansan-si, Gyeonggi-do 15588, Korea.

Monty Sutrisna School of Built Environment, Curtin University, Bentley WA 6102, Australia.

Sungho Tae School of Architecture & Architectural Engineering, Hanyang University, 55 Hanyangdaehak-ro, Sangrok-gu, Ansan-si, Gyeonggi-do 15588, Korea.

Vivian W.Y. Tam School of Computing, Engineering and Mathematics, Western Sydney University, Australia; College of Civil Engineering, Shenzhen University, China.

Nicole Tassicker School of Built Environment, Curtin University, Bentley WA 6102, Australia.

Omer Tatari Department of Civil, Environmental, and Construction Engineering, University of Central Florida, Orlando, FL 32816, USA.

Xiao-Yong Wang Department of Architectural Engineering, Kangwon National University, Chuncheon 200-701, Korea.

Kunhui Ye School of Construction Management and Real Estate, and International Research Center for Sustainable Built Environment, Chongqing University, Chongqing 400044, China.

Zongnan Zhao School of Construction Management and Real Estate, and International Research Center for Sustainable Built Environment, Chongqing University, Chongqing 400044, China.

About the Guest Editors

Vivian W.Y. Tam is an Associate Professor at the School of Computing, Engineering and Mathematics, Western Sydney University, Australia and an Adjunct Professor at the College of Civil Engineering, Shenzhen University, China. She received her Ph.D. in sustainable construction from the Department of Building and Construction at the City University of Hong Kong in 2005. Her research interests are in the areas of environmental management in construction and sustainable development. She is currently the Associate Editor of the *International Journal of Construction Management* and the Research Group Leader for Sustainable Construction Management and Education Research Group through the School of Computing. She has published two books, 20 book chapters, 180 referred journal articles and 73 referred conference articles.

Khoa N. Le received his Ph.D. in October 2002 from Monash University, Melbourne, Australia. From 2003–2009, he was a lecturer at Griffith University, Gold Coast Campus, Griffith School of Engineering. From January to July 2008, he was a visiting professor at the Intelligence Signal Processing Laboratory, Korea University, Seoul, Korea. In 2009, he was a visiting professor at the Wireless Communication Centre, University of Technology, Johor Bahru, Malaysia. He is currently a Senior Lecturer at the School of Computing, Engineering and Mathematics, University of Western Sydney. His research interests are in wireless communications with applications to structural problems, image processing and wavelet theory. Dr. Le is the Editor in Chief of the *International Journal of Ad hoc, Sensor & Ubiquitous Computing (IJASUC)*. He has also been on the Editorial Board of the *International Research Magazine of Computer Science* and other wireless communications journals.

Liyin Shen was a Professor in the Department of Building & Real Estate at the Hong Kong Polytechnic University before he joined Chongqing University of China in 2012 as a Distinguished Professor in the Faculty of Construction Management and Real Estate. He received his first class BSc in Mechanical Engineering in Construction from Chongqing University, his PgD in Management Engineering from Harbin Institute of Technology in China, and his PhD in Construction Management from the University of Reading in the United Kingdom. His major research interests are in the areas of sustainable built environment; risk management for built environment; and sustainable competitiveness for construction business. Shen has widely published his research in international journals. He is the Editor-in-Chief of the *International Journal of Construction Management* and the Founding President of the Chinese Research Institute of Construction Management. He is the Director of the International

Research Centre for Sustainable Built Environment based in Chongqing University. Shen has a number of external appointments in the capacity of Honorary Professor and Visiting Professor.

Improving Sustainability Performance for Public-Private-Partnership (PPP) Projects

Liyin Shen, Vivian W.Y. Tam, Lin Gan, Kunhui Ye and Zongnan Zhao

Abstract: Improving sustainability performance in developing infrastructure projects is an important strategy for pursuing the mission of sustainable development. In recent years, the business model of public-private-partnership (PPP) is promoted as an effective approach in developing infrastructure projects. It is considered that the distribution of the contribution on project investment between private and public sectors is one of the key variables affecting sustainability performance of PPP-type projects. This paper examines the impacts of the contribution distribution between public and private sectors on project sustainability performance. A model named the sustainability performance-based evaluation model (SPbEM) is developed for assisting the assessment of the level of sustainability performance of PPP projects. The study examines the possibility of achieving better sustainability through proper arrangement of the investment distribution between the two primary sectors in developing PPP-type infrastructure projects.

Reprinted from *Sustainability*. Cite as: Shen, L.; Tam, V.W.Y.; Gan, L.; Ye, K.; Zhao, Z. Improving Sustainability Performance for Public-Private-Partnership (PPP) Projects. *Sustainability* **2016**, *8*, 289.

1. Introduction

The role of infrastructure projects is significant in socio-economic development due to their contributions to national competitiveness and social welfare in both developing and developed countries [1–4]. Traditionally, infrastructure projects are developed by governments using public budgets. However, the limitations of public capital and the shortage of management expertise in government departments have led to the development of an alternative procurement mechanism for developing and running infrastructure projects. This procurement mechanism is named public-private-partnership (PPP) approach. PPP mechanism serves for promoting the engagement of private sectors in the process of developing infrastructure projects, including project design, financing, construction, maintenance, and project operation. In recent years, the popularity of PPP-type infrastructure (PTI) projects has been rising particular in developing countries. Ke noted that the demand for PPP infrastructures is increasing in China whilst there are increasing number of PPP projects in the country [2]. According to the PPP Demonstration Project List of Second Batch announced by Ministry of Finance of China, there are 206 projects with national financial support in 2015, which is seven times that of 2014 [5]. In fact, most developing countries have been actively applying PPP policy as a contractual arrangement and condition on loans from International Organizations [3,6]. On one hand, PPP mechanism mitigates the burden of public fiscal shortage and ensures the timely provision of infrastructures in need [2,7]. On the other hand, the development of PPP-type projects presents private sectors with good business opportunities where they can apply innovative technological and the advanced management skills they have [8,9]. Considering the very important role and significance of PPP type infrastructures in the built environment sector, it can be seen that improving sustainability performance for PPP projects can contribute significantly to the mission of sustainable construction.

Research on examining the applications of PPP approach has been conducted extensively from many perspectives including technology innovation [10], risk management [11], critical success factors [12], cooperate governance [13], concessionaire determinants [14,15], and sustainability performance [16,17]. An increasing number of researchers have appealed for sustainability performance appraisal on PPP-type projects as an instrument for striking the trade-off among economic, social and environmental performance criteria in examining project feasibility [16,18,19]. This has led to the development of several models for project sustainability performance appraisal. Ogwu *et al.* proposed a sustainability appraisal in Infrastructure projects (SUSAIP) model for evaluating sustainability of infrastructure projects in Hong Kong, and shed light on various dimensions (e.g., knowledge, problem analysis, and application) of designing and constructing for better infrastructure sustainability [16,17]. Dasgupta and Tam synthesized sustainability indicators of civil infrastructure projects by using a multi-objectives decision approach to facilitate the choice of practical alternatives for better sustainability performance [20]. Koo and Ariaratnam employed a sustainability assessment model to determine the option for implementing a water main replacement project towards better sustainability performance [21]. The study by Shen *et al.* suggested that a concession time period in operating infrastructure projects must incorporate the benefits, authorities and responsibilities among various project parties for the interests of better sustainability [14]. Shen and Wu investigated a risk-based concession time period in developing build-operate-transfer-type infrastructure projects in ensuring the sustainability interests that benefit to various project parties [15].

Whilst existing studies present various effective appraisal models for assessing the sustainability performance of various types of infrastructure projects, it appears that there is little existing research on examining the ways of improving the sustainability performance of PPP-type projects. The difficulty of sustainability appraisal for PPP-type projects is well appreciated [14,15]. Previous studies have addressed the selection of sustainability indicators for PPP-type projects [16].

Nonetheless, many of the selected indicators are qualitative and how to quantify them has not been addressed. In particular, it is noted that the proportion of investment or contribution between public and private sectors has not been taken into account in the establishment of sustainability performance indicators. In fact, investment distribution from the two primary parties has major influence to project performance across various aspects including economic, social and environmental performance. These dimensional performances are the determinants to the sustainability performance of the concerned PPP projects. Furthermore, public and private sectors have different consideration for involving PPP projects. According to the study by Ling *et al.* [21], the typical factor considered by public clients include suitability to PPP, difficulties in evaluating PPP deals, political fallout, affordability and profitability, value creation and changing needs. The main factors considered by private sectors include PPP champion, speed of implementation, capability of private clients to use PPP procurement route, tendering cost and risk allocation. Koppenjan and Enserink identified governance practices that help or hinder the reconciliation of private sector participation in infrastructure projects with the objective to increase the sustainability of the urban environment. They further appreciated that private sector participation in urban infrastructure does not automatically contribute to sustainability as private sector is expected to focus on short-term financial return on investment whilst the sustainability performance of project can only be obtained from long-term perspective [22].

Although the two primary sectors have different perceptions on the performance of a concerned PPP-type project, for example, private sector's perception is usually based on short-term performance of project, as appreciated by Koppenjan and Enserink [22], the main decision at inception stage for a PPP-type project is the investment distribution between public and private sectors. The investment distribution is the key variable that determines the interests of both public and private sectors from the development of a PPP-type project. It can be appreciated that the investment contributed from private sector is motivated by the expectation to earn certain level of profits. However, the investment level by private sector will affect the public interests which can only be realized from long-term perspective. For example, the public interests of environmentally friendly implementation of an infrastructure may not be considered as a major criterion if the public sector assumes a very small proportion of project investment, where the private sector's profit level is one of the dominant factors for the consideration in project feasibility study. Robert *et al.* identified four major factors which affect the levels of the public and private sectors, including benefits to local economic development, access to the public sector market, tax exemptions and reduction, and incentives to new market penetration [3]. These factors are considered differently when the investment proportions changes between public and private sectors. The different attention given into the factors will in turn affect project sustainability performance. Therefore, a proper investment distribution in implementing a PPP scheme is the key to gaining better project sustainability performance which aims to balance social, economic and environmental interests. The aim of this study is thus to use the investment distribution between public and private sectors as the key variable in developing a sustainability performance-based evaluation model (SPbEM). The appraisal results from using the model SPbEM can help decision-makers know what distribution level of investment between public and private sectors will enable better sustainability performance of PPP-type projects.

2. Research Method

This paper starts with conducting literature reviews for gaining proper understanding on the principle of the mechanism for developing PPP-type projects. The level of investment distribution between public and private sectors is based on the types of PPP projects and the perceptions possessed by the two primary sectors. Based on this understanding, the research team develops a sustainability performance-based evaluation model (SPbEM) to assist in decision making on the distribution level that can lead to better sustainability performance of an infrastructure project. In further stage of the study, the research demonstrates the application of the model SPbEM through a selected practical case. The case study demonstrates the applicability of the model. In the final stage of the study, discussion is conducted to address the implications of using the model SPbEM and the ways of improving project sustainability performance by adjusting investment distributions between the public and private sectors. The results from analyzing various scenarios are then presented.

3. Development of the Sustainability Performance Based Evaluation Model (SPbEM)

The involvement of private and public sectors in PPP-type projects is characterized with many dimensions such as contractual arrangements, structure of management organization, investment distribution, and others [14]. Nevertheless, the investment distribution between the two sectors is the key variable to be considered in the process of contract negotiation. The level of the distribution between the two sectors will affect contractual arrangements and the organization structure of implementing the project under the principle of PPP. According to previous studies [1,4,23,24], there are sixteen contractual arrangements in line with the principle of PPP methods, as shown in Table 1.

Table 1. Contractual arrangements under the principle of PPP methods.

Synonym	Stage \ Full Name	Private Sponsor								Public Agencies		Impact on Sustainability Indicators		
		D	B	O_p	L	T	F	M	O_w	O_p	O_w	E_n	E_c	S
DBT	Design-build-transfer	√	√							√	√	low	low	high
BLT [1]	Build-lease-transfer		√		√					√	√			
DOT [25]	Design-operate-transfer	√		√						√	√			
BOT	Build-operate-transfer		√	√						√	√			
BOR [1]	Build-operate-renewal		√	√						√	√	↓	↓	↑
ROT [23]	Refurbish-operate-transfer			√				√		√	√			
DBOM [4]	Design-build-operate-maintain	√	√	√				√		√	√			
DBFO [4]	Design-build-finance-operate	√	√	√			√			√	√			
DBO [4]	Design-build-operate	√	√	√						√	√			
BTO [1]	Build-transfer-operate		√	√						√	√			
BOOTT [23]	Build-own-operate-train-transfer		√	√		√			√	√	√			
BOOT [24]	Build-own-operate-transfer		√	√					√	√	√			
BLO [1]	Build-lease-own		√	√					√	√				
BOOM [23]	Build-own-operate-maintain		√	√				√	√					
ROO [25]	Rehabilitate-own-operate			√				√	√					
BOO [1]	Build-own-operate		√	√					√					

Note: B—Build; D—Design; E_c—Economic; E_n—Environmental; F—Finance; L—Lease; M—Maintain; O_p—Operate; O_w—Own; P_r—Private; P_u—Public; S—Social; T—Train.

These approaches can be organized to a list in the order of ascending extent to which private sectors invest in developing infrastructure projects. Private businesses are by nature profit-orientated and are keen to invest public facilities provided they can make profits particularly in short-term. However, public sectors need to take into account public interests pertaining to non-economic aspects such as social and environmental interests. It is reasonable to assume that the economic performance of PPP-type projects will be one of the dominant factors considered in the decision making for considering project development if the project is fully invested by private investors. In contrast, the economic performance would not be a major consideration if a project is wholly invested within public budgets. Therefore, there exists a proper proportion of project investment between private and public sectors in order to gain better project sustainability performance which is a balanced integration among economic, social and environmental performance. In line with this analogy, the proposed model SPbEM should be enabled to assess the impact of the investment distribution between the two sectors on economic, social and environmental performance of the concerned projects.

3.1. The Primary SPbEM Model

The model SPbEM synthesizes the triple–bottom-line (TBL) indicators of project sustainability performance, namely, economical performance indicator E_c, social performance indicator S, and environmental performance indicator E_n. The contribution of these TBL indicators to project sustainability can be expressed in Equation (1).

$$\begin{aligned} IfSV &= W_1 E_c + W_2 S + W_3 E_n \\ \sum_{i=1}^{3} W_i &= 1, and\, 0 \leqslant W_i \leqslant 1 \end{aligned} \tag{1}$$

where $IfSV$ refers to the sustainability index value for a concerned PPP-type project; E_c, S and $E_n$are the performance values for economic, social and environmental performance indicators respectively; and $W_i(i = 1, 2, \text{and } 3)$ denotes the weighting values of the TBL indicators (economic, social and environmental performance respectively).

3.2. Investment Distribution Coefficients (α, β)

As discussed above, the investment distribution between private and public sectors in developing PPP-type projects has major impact on project sustainability performance. The investment distribution between the two sectors can be described by two coefficients, namely, α, denoting for the proportion of public investment; and β, denoting the proportion of private investment. The two coefficients can be described in the following model in Equation (2):

$$\begin{aligned} \alpha &= \frac{P_u}{P_u + P_r} \\ \beta &= \frac{P_r}{P_u + P_r} \end{aligned} \quad (2)$$

where P_u denotes the investment level (or volume) by public sector; and P_r refers to the investment level (volume) by private sector.

The relationships between α and β are exhibited graphically in Figure 1. The two coefficients interacted reversely whilst both the coefficients can assume any value between 0 and 1.

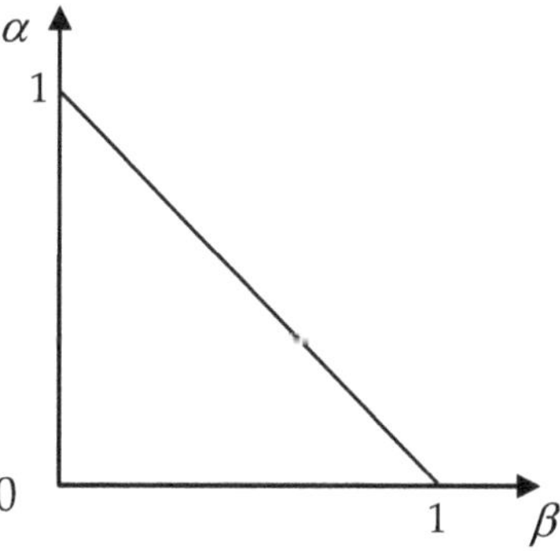

Figure 1. Relationships between public and private investment coefficients.

There are infinite number of possible combinations of values between α and β. Among these combinations, there are three special scenarios:

(1) $\alpha = 0$ and $\beta = 1$, indicating that no financial investment is contributed by public sector;

(2) $\alpha = 0.5$ and $\beta = 0.5$, indicating that the project investment is equally contributed between public and private sectors; and

(3) $\alpha = 1$ and $\beta = 0$, indicating that no financial investment is contributed by private sector.

These three special scenarios can be described in the following matrix model in Equation (3):

$$(\alpha, \beta) = \begin{pmatrix} 0, & 1 \\ 0.5, & 0.5 \\ 1, & 0 \end{pmatrix} \quad (3)$$

3.3. Weighting W_1, W_2, and W_3

For a given PPP-type project, the values of the investment distribution coefficients α or β will be changed under different contractual terms or arrangements between private and public sectors. On the other hand, when different types of PPP-type projects are considered, the investment distribution between the two sectors will also be changed, namely, the values of α and β will also be changed. Therefore, α and β change with different types of contractual arrangements and different types of PPP-type projects. For example, in applying Building-Lease-Transfer (BLT) type contractual system, private sector will contribute 100% finance investment [1]. In adopting Design-Building-Transfer (DBT) type project, public sector, namely, government will normally contribute to most part of the investment [9].

The discussion in the previous section suggests that the investment distribution between the two investment sectors determines the project sustainability performance. In other words, the change of the investment distribution between the two sectors will also lead to the changes of project sustainability performance. In this context, when α and β change with different emphasis placed on economic, social and environmental aspects, the weighting values for the TBL indicators will change as well. Consequently, there are certain relationships among α, β and the weighting values W_i. For example, if $0 \leqslant \alpha \leqslant 0.5$ or $0.5 \leqslant \beta \leqslant 1$, suggesting that the private party dominates the project finance investment, economic indicator E_c will be given a higher weighting value than that to E_n or S indicators. Likewise, if $0.5 \leqslant \alpha \leqslant 1$ or $0 \leqslant \beta \leqslant 0.5$, suggesting the dominance of public party over project investment, E_n and S indicators should deserve more weights in this case. In line with these arguments, the relationships between W_i and $\alpha(\beta)$ can be described in Equation (4).

$$\begin{aligned} W_1 &= f_1(\alpha/\beta) \\ W_1 &= f_2(\alpha/\beta) \\ W_1 &= f_3(\alpha/\beta) \end{aligned} \quad (4)$$

(1) When $0 \leqslant \alpha \leqslant 0.5$ (or $0.5 \leqslant \beta \leqslant 1$)

$$W_1 \geqslant W_2$$
$$W_1 \geqslant W_3$$

(2) When $0.5 \leqslant \alpha \leqslant 1$ (or $0 \leqslant \beta \leqslant 0.5$)

$$W_1 \leqslant W_2$$
$$W_1 \leqslant W_3$$

3.4. Functional Relationships between $\alpha(\beta)$ and W_i

The above discussions suggests that α and β determine the weighting values of W_1, W_2 and W_3. In other words, certain functional relationships exist between $\alpha(\beta)$ and W_i. By considering the discussion in the previous sections, the following five assumptions are adopted to demonstrate the relationships between $\alpha(\beta)$ and W_i.

(1) When $\alpha = 0$ (or $\beta = 1$), indicating no public investment involved, E_c indicator shall become the most important factor to be considered in making contractual arrangements, E_n has less importance while S has the least importance. In this circumstance, the reasonable assumptions on the weighting values can be considered as: $W_1 = 0.65$, $W_2 = 0.15$, and $W_3 = 0.20$ respectively.
(2) When $\alpha = 0.25$ (or $\beta = 0.75$), indicating less investment from public sector and more from private sector, thus economic aspect still assumes more important role that of environmental and social aspect. In line with this, reasonable weighting values are allocated as: $W_1 = 0.45$, $W_2 = 0.28$, and $W_3 = 0.27$.
(3) When $\alpha = 0.5$ (or $\beta = 0.5$), indicating an equal amount of investment from private and public sectors. In this circumstance, all TBL indicators are considered equally important, thus the assumption of $W_1 = 0.34$, $W_2 = 0.33$, and $W_3 = 0.33$ is considered reasonable.
(4) When $\alpha = 0.75$ (or $\beta = 0.25$), indicating that public sector contributes more than that by private sector, the environmental and social dimensions are considered more important than economic dimension. Nevertheless, as private sector in this case still makes significant part of investment, economic performance has to be given reasonable attention. Thus, the assumptions on the weighting values are $W_1 = 0.30$, $W_2 = 0.35$, and $W_3 = 0.35$.
(5) When $\alpha = 1$ (or $\beta = 0$), indicating no private finance investment involved in this. The private sector may provide technical service and management skill in the process of construction and operation of the concerned project. In this case, the indicators of S and E_n become significant for consideration in the process of contract negotiation, and E_c has less importance. Accordingly, the assumptions on the weighting values between three dimensional indicators are given as $W_1 = 0.15$, $W_2 = 0.45$, and $W_3 = 0.40$.

The above five assumptions are summarized in Table 2.

Table 2. Assumptions on the data for W_i and $\alpha(\beta)$.

α	β	E_c Indicator	S Indicator	E_n Indicator
		W_1	W_2	W_3
0	1	0.65	0.15	0.20
0.25	0.75	0.45	0.28	0.27
0.50	0.50	0.34	0.33	0.33
0.75	0.25	0.30	0.35	0.35
1	0	0.15	0.45	0.40

Furthermore, it is assumed that there is a linear relationship between W_i and $\alpha(\beta)$:

$$W = a\alpha + b$$

or

$$W = a\beta + b$$

These functions are determined by the values of constants a and b, which have been listed in Table 3.

Table 3. Parameters for the relationship functions between W_i and $\alpha(\beta)$.

α	W_1		W_2		W_3	
	a	**b**	**a**	**b**	**a**	**b**
(0,0.25)	−0.80	0.65	0.52	0.15	0.28	0.20
(0.25,0.5)	−0.44	0.56	0.20	0.14	0.24	0.22
(0.5, 0.75)	−0.16	0.42	0.08	0.12	0.08	0.29
(0.75, 1)	−0.60	0.75	0.40	−0.21	0.20	0.05
β	W_1		W_2		W_3	
	a	**b**	**a**	**b**	**a**	**b**
(0,0.25)	0.60	0.15	−0.40	0.45	−0.20	0.40
(0.25,0.5)	0.16	0.26	−0.08	0.29	−0.08	0.37
(0.5, 0.75)	0.44	0.12	−0.20	0.27	−0.24	0.43
(0.75, 1)	0.80	−0.15	−0.52	0.39	−0.28	0.66

By applying the values in Table 3, the following functional relationships are obtained accordingly:

$$W_1 = \begin{cases} -0.80\alpha + 0.65, & (0 < \alpha < 0.25) \\ -0.44\alpha + 0.56, & (0.25 < \alpha < 0.5) \\ -0.16\alpha + 0.42, & (0.5 < \alpha < 0.75) \\ -0.60\alpha + 0.75, & (0.75 < \alpha < 1) \end{cases}$$

or

$$W_1 = \begin{cases} 0.60\beta + 0.15, & (0 < \beta < 0.25) \\ 0.16\beta + 0.26, & (0.25 < \beta < 0.5) \\ 0.44\beta + 0.12, & (0.5 < \beta < 0.75) \\ 0.80\beta - 0.15, & (0.75 < \beta < 1) \end{cases}$$

$$W_2 = \begin{cases} 0.52\alpha + 0.15, & (0 < \alpha < 0.25) \\ 0.20\alpha + 0.14, & (0.25 < \alpha < 0.5) \\ 0.08\alpha + 0.12, & (0.5 < \alpha < 0.75) \\ 0.40\alpha - 0.21, & (0.75 < \alpha < 1) \end{cases}$$

or

$$W_2 = \begin{cases} -0.40\beta + 0.45, & (0 < \beta < 0.25) \\ -0.08\beta + 0.29, & (0.25 < \beta < 0.5) \\ -0.20\beta + 0.27, & (0.5 < \beta < 0.75) \\ -0.52\beta + 0.39, & (0.75 < \beta < 1) \end{cases}$$

$$W_3 = \begin{cases} 0.28\alpha + 0.20, & (0 < \alpha < 0.25) \\ 0.24\alpha + 0.22, & (0.25 < \alpha < 0.5) \\ 0.08\alpha + 0.29, & (0.5 < \alpha < 0.75) \\ 0.20\alpha + 0.05, & (0.75 < \alpha < 1) \end{cases}$$

or

$$W_3 = \begin{cases} -0.20\beta + 0.40, & (0 < \beta < 0.25) \\ -0.08\beta + 0.37, & (0.25 < \beta < 0.5) \\ -0.24\beta + 0.43, & (0.5 < \beta < 0.75) \\ -0.28\beta + 0.66, & (0.75 < \beta < 1) \end{cases}$$

Based on the above relationship functions, the relationships among $\alpha(\beta)$ and W_i can be presented graphically in Figure 2.

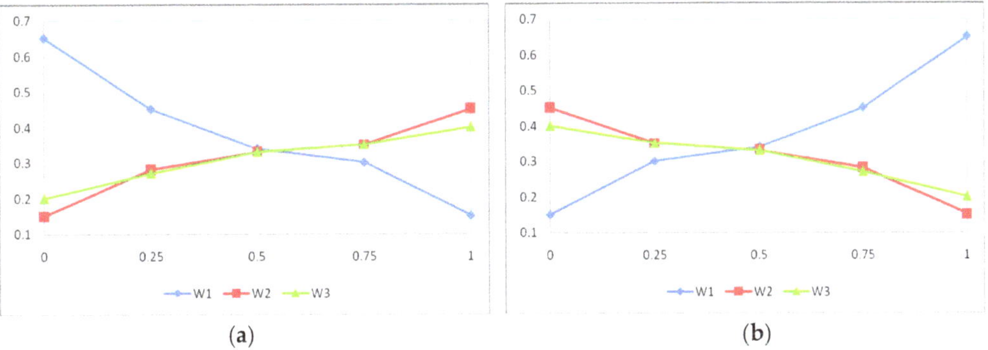

(**a**) (**b**)

Figure 2. Relationships between W_i and α (β). (**a**) W_i and α; (**b**) W_i and β.

3.5. Measuring the TBL Indicators (E_c, S, and E_n)

The study by Ogwu *et al.* showed the effectiveness of quantifying the variables of project sustainability by adopting several methods, including credit-based scoring system, scaled scoring, comparison with benchmark or other available options, credit system, and subjective judgments [17]. Scoring method is employed in this study to measure the contribution of the TBL indicators (E_c, S, and E_n) for reflecting sustainability performance of a particular PPP-type project. In scoring the three TBL indicators, experts are invited to judge the contribution of a concerned project to each dimensional sustainability indicator on a nine-level Likert scale (9—best, and 1—no contribution). For ensuring the effectiveness of engaging expert participation, it is suggested that at least 10 experts should be invited to participate in judging the sustainability performance of a given PPP project [25].

There are many sub-indicators under each of the TBL indicators for measuring sustainability performance of PPP projects. The significance of individual indicators varies widely from project to project [17,20,21,26]. The framework proposed by Gan *et al.* for appraising sustainability performance of PPP projects in China is adopted for illustration [26]. The framework is composed of a set of sub-TBL indicators as shown in Table 4.

As the backgrounds and experience are different among the invited experts in study, the scores allocated by the experts are considered in probability distributions. The triangular probability distribution (TPD) is considered effective when small sample size is used in defining probability distribution [15], which is adopted in this study. In using TPD, there are three parameters to be defined, namely, the minimum (M_i), the maximum (M_a) and the mode (M_o). In line with this argument, the TPDs for defining the score performance for dimensional sustainability indicators E_c, S and E_n are shown in Figures 3–5 respectively. In these distributions, parameters

M_i and M_a determine the interval of triangle distribution, whilst the parameter M_o determines whether a distribution is a left (negative) or right (positive) skewed.

Table 4. Effective TBL indicators.

	TBL Indicators	Marking Grade								
		1	**2**	**3**	**4**	**5**	**6**	**7**	**8**	**9**
E_c	Economic prosperity and development policy(E_{c1})	○	○	○	○	○	○	○	○	○
	Complexity of project construction technique (E_{c2})	○	○	○	○	○	○	○	○	○
	Channel and cost of project financing (E_{c3})	○	○	○	○	○	○	○	○	○
	Project investment schedule (E_{c4})	○	○	○	○	○	○	○	○	○
	Project life-cycle cost (E_{c5})	○	○	○	○	○	○	○	○	○
	Internal return rate(E_{c6})	○	○	○	○	○	○	○	○	○
S	Contribution to local socio-economy in terms of household income and life quality (S_1)	○	○	○	○	○	○	○	○	○
	Provision of quality services to local economic activities (S_2)	○	○	○	○	○	○	○	○	○
	Promotion of public health and sanitation (S_3)	○	○	○	○	○	○	○	○	○
	Consumption of land and impact on land utility (S_4)	○	○	○	○	○	○	○	○	○
	Creation of employment (S_5)	○	○	○	○	○	○	○	○	○
E_n	Impact on geographic condition (E_{n1})	○	○	○	○	○	○	○	○	○
	Air pollution (E_{n2})	○	○	○	○	○	○	○	○	○
	Water pollution (E_{n3})	○	○	○	○	○	○	○	○	○
	Noise pollution (E_{n4})	○	○	○	○	○	○	○	○	○
	Environment protection measures (E_{n5})	○	○	○	○	○	○	○	○	○
	Energy saving (E_{n6})	○	○	○	○	○	○	○	○	○

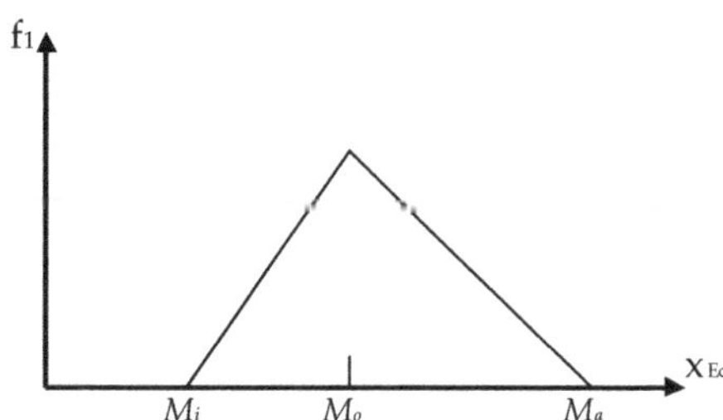

Figure 3. TPD Distribution of expert's perceived performance score for economic parameter.

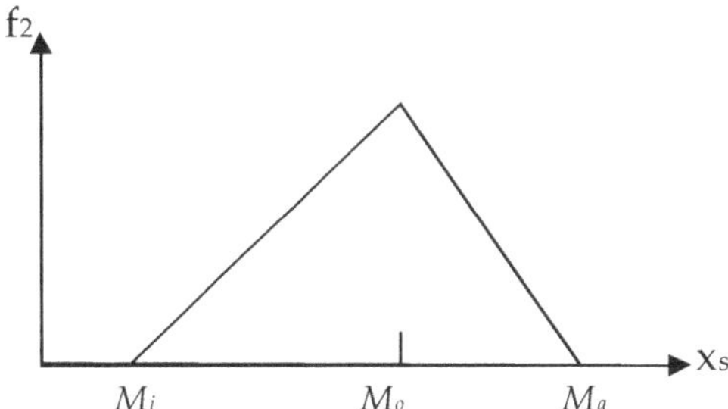

Figure 4. TPD Distribution of expert's perceived performance score for social parameter.

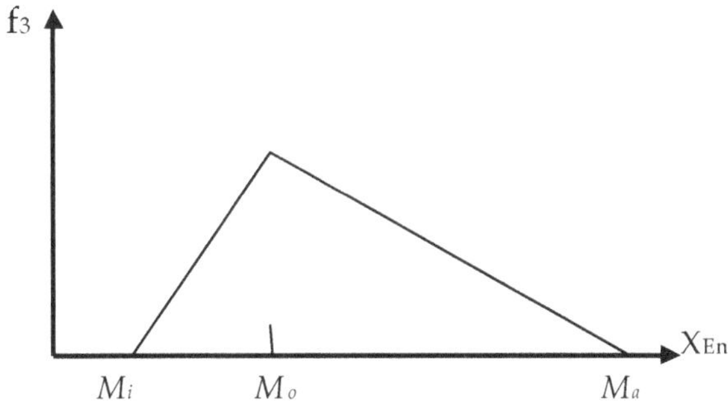

Figure 5. TPD Distribution of expert's perceived performance score for environmental parameter.

For further discussion, a general TPD is referred, as shown in Figure 6. According to the principle of probability density function that the total probability value is 1, the probability density function in referring to the triangle distribution Figure 6 can be derived by Equation (5).

$$f(x\,|M_i\ , M_o, M_a) = \begin{cases} \dfrac{2(x - M_i)}{(M_a - M_i)(M_o - M_i)} & (M_i \leqslant x \leqslant M_o) \\ \dfrac{2(M_a - x)}{(M_a - M_i)(M_a - M_o)} & (M_o \leqslant x \leqslant M) \end{cases} \quad (5)$$

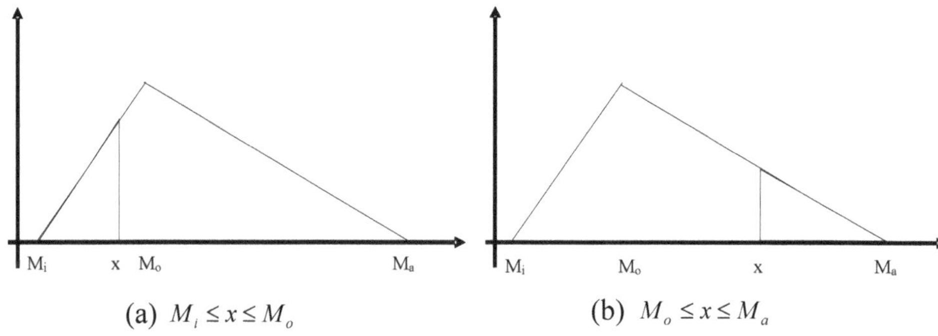

(a) $M_i \le x \le M_o$ (b) $M_o \le x \le M_a$

Figure 6. A general triangle probability distribution.

The generic Equation (5) can be applied to describing the probability density functions for the triangle probability distributions as shown in Figures 3–5 which describe the distribution or experts' perception on the sustainability performance score for economic, social and environmental dimensions. The sustainability performance values of E_c, S and E_n can be calculated by Equation (6).

$$\begin{aligned} E_c &= \int_{E_c(M_i)}^{E_c(M_a)} f(x_{E_c}) x_{E_c} dx_{E_c} \\ S &= \int_{S(M_i)}^{S(M_a)} f(x_S) x_S dx_S \\ E_n &= \int_{E_n(M_i)}^{E_n(M_a)} f(x_{E_n}) x_{E_n} dx_{E_n} \end{aligned} \quad (6)$$

By further referring to the sustainability index value for a concerned PPP-type project, $IfSV$ can be rewritten as Equation (7):

$$IfSV = f_1(\alpha)\int_{E_c(M_i)}^{E_c(M_a)} f(x_{E_c}) x_{E_c} dx_{E_c} + f_2(\alpha)\int_{S(M_i)}^{S(M_a)} f(x_S) x_S dx_S + f_3(\alpha)\int_{E_n(M_i)}^{E_n(M_a)} f(x_{E_n}) x_{E_n} dx_{E_n} \quad (7)$$

The application of the SPbEM model (Equation (7)) will be illustrated through a case study in the next section.

4. Case Study

Researchers have identified many types of PPP-type infrastructure projects, including highways, railways, ports, tunnels, bridges, power plants, hydraulic structures, mass transit, and municipal facilities [1,26]. A specific PPP project is selected to demonstrate the applicability of the model SPbEM (in Equation (7)) is an expressway connecting Anxi County to Xiamen City in China's Fujian Province. The expressway is 6.82 kilometers long with six large tunnels crossing many mountains with the total investment cost of RMB388.5 million (about US$52 million). This project was procured by adopting PPP mechanism with the concession period of ten years. There was an extensive discussions between the two parties on the contract terms and investment distributions between the public sector and private sector. The public sector of the project was the Anxi County Government. The private sponsor was Xiamen Hengxing Industrial Co. Ltd., which paid around 75% of the total project investment. The main contractor was Changsha Central-South Construction Engineering Group Corporation of Nuclear Industry. The project construction commenced in October 2004, finished in December 2009 and would be transferred to Anxi County Government without any charges upon the end of the concession period.

By using the data collection framework in Table 4, responses were obtained from fifteen professionals who were invited to assess the significance of the TBL indicators in referring to this particular PPP project. The data obtained from the respondents are used to develop the triangular probability distribution for each TBL indicator.

Take the indicator E_{c1} for example, the minimum score is 3 (M_i = 3), the most likely score is 5 (M_o = 5), and the maximum score is 9 (M_a = 9). By applying these values to model (5), the probability density function of E_{c1} can be obtained by Equation (8):

$$f(x\,|m_i\,, m_o, m_a) = \begin{cases} \dfrac{2(x-3)}{(9-3)(5-3)} & 3 \leqslant x \leqslant 5 \\ \dfrac{2(9-x)}{(9-3)(5-3)} & 5 \leqslant x \leqslant 9 \end{cases} \tag{8}$$

This function can be expressed graphically in Figure 7.

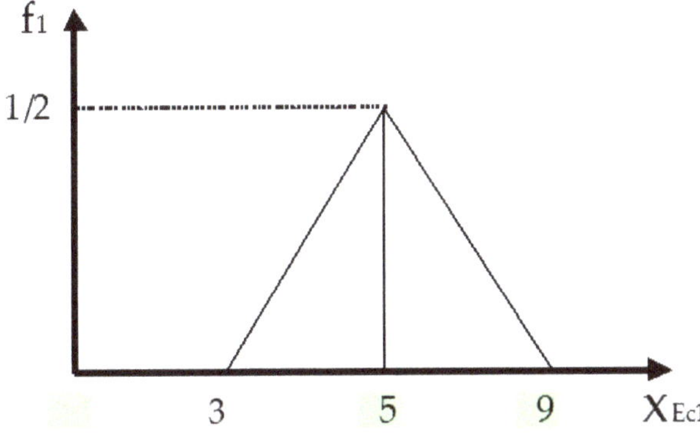

Figure 7. Distribution of the experts' perceived score on the performance of the indicator E_{c1}.

Thus, the overall performance grade of the indicator Ec_1 can be calculated by using Equation (9):

$$E(\mathrm{E}_{c1}) = \int_3^5 \frac{1}{6}(x-3)xdx + \int_5^9 \frac{1}{12}(9-x)xdx = 8.67 \quad (9)$$

Similar to the analysis for the indicator E_{c1}, the functions for all other TBL indicators listed in Table 4 can be derived, and the performance values of all the indicators are derived and presented in Table 5.

Table 5. The performance values of the TBL indicators for the example project.

TBL Indicators	Performance Grade	TBL Indicators	Performance Grade	TBL Indicators	Performance Grade
E_{c1}	8.67	S_1	5.3	E_{n1}	8.86
E_{c2}	6.43	S_2	8.21	E_{n2}	7.59
E_{c3}	8	S_3	5.89	E_{n3}	6.28
E_{c4}	6.72	S_4	6.63	E_{n4}	6.86
E_{c5}	5.5	S_5	6.25	E_{n5}	6.96
E_{c6}	5.45			E_{n6}	6.47

By using the data in Table 5, the performance values of the three dimensional parameters in this case study project, namely, E_c, S and E_n, can be calculated by using Equation (10).

$$\begin{aligned} E(E_c) &= \sum_i^6 E_{c_i}/6 = 6.80 \\ E(S) &= \sum_j^5 S_j/5 = 6.46 \\ E(E_n) &= \sum_k^6 E_{n_k}/6 = 7.17 \end{aligned} \quad (10)$$

As regards weighting values between the three sustainability dimensions, the project information provided shows that the private investment in the case study project accounts for about 75% of the total project investment, namely, $\alpha = 0.25$ and $\beta = 0.75$. According to this, the weighting values for the three dimensional parameters can be established by referring to the assumptions in Table 2, namely,

$$W_1 = 0.45, W_2 = 0.28, \text{and } W_3 = 0.27 \quad (11)$$

By applying the data in Equations (10) and (11) to the Equation (1), the total contribution of TBL indicators to the sustainability performance of the case study project can be calculated as follows:

$$IfSV = 0.45 \times 6.80 + 0.28 \times 6.46 + 0.27 \times 7.17 = 6.80 \quad (12)$$

This value of 6.80 demonstrates a positive project sustainability performance of this project, indicating that the 75% contribution from private sector to this PPP project can provide positive contribution to the project sustainability performance.

5. Findings and Discussions

Results of the case study in the above section demonstrate that the level of sustainability performance for a PPP-type project is affected by the investment distributions between public and private sectors, and this performance can be assessed if the investment distributions are defined. This shows the possibility of achieving an expected level of project sustainability performance ($IfSV$) by adjusting investment distributions, denoted by the coefficients α and β In other words, sensitivity analysis can be conducted to find out the optimal level of investment distribution between the two sectors in a PPP project towards better level of sustainability. For example, by referring to the case study discussed in the previous section, when the coefficient α assumes the value 0, 0.25, 0.5, 0.75 and 1 respectively, the sensitivity analysis results on the value of $IfSV$ for this case study PPP project will be obtained, as shown in Table 6.

Table 6. Sensitivity analysis results on $IfSV$ values for the case study project.

α	$IfSV$
0	6.783
0.25	6.801
0.5	6.807
0.75	6.808
1	6.792

In further analysis, a curve line can be drawn based on the data in Table 6, as shown in Figure 8. The curve line indicates the relationship between $IfSV$ and the coefficient α. The curve shows that an investment wholly contributed either by private or public sector cannot lead to better project sustainability performance. There is a point where the investment distributions between public and private sectors can contribute to the best level of project sustainability performance. In other words, better sustainability performance can be obtained by arranging proper investment distributions between the two sectors in a PPP-type project. In the current practice, the determination of the investment distribution in a PPP-type scheme is a complicated decision process, affected by various factors including economic performance of the given project, private sector's expected return on project investment, and the concession period [23,27]. The application of the proposed model SPbEM to the case study suggests that the effective investment distributions between the public and private sectors should be arranged in the way that public sector contributes around 25% investment ($\alpha = 0.25$), and private sector shares around 75% investment ($\beta = 0.75$). This distribution arrangement can contribute to reasonable level of sustainability performance in implementing this project. Nevertheless, such investment arrangement does not contribute to the best level sustainability, as can be observed in Figure 8.

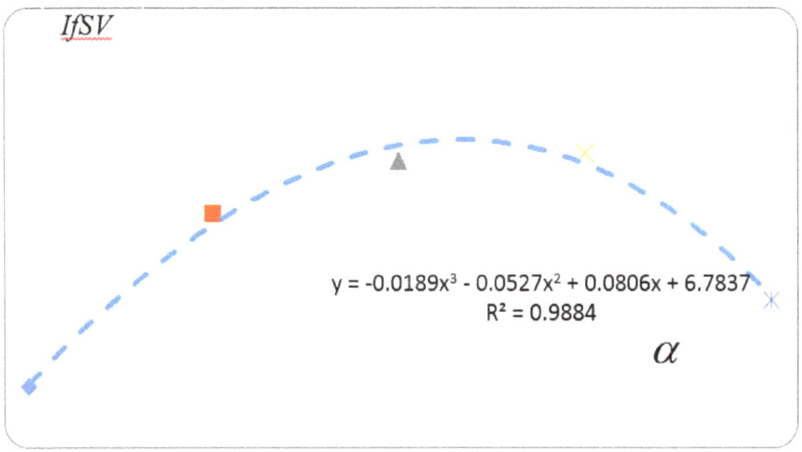

Figure 8. Relationship between investment distribution coefficient α and *IfSV*.

The relationship between the investment distributions and project sustainability performance, exampled by the case study in Figure 8, provides a valuable mechanism for determining investment distributions in a PPP-type project towards better project sustainability performance. Taking the concerned case as an example, the distribution involving about 60% public investment will lead to the best level of project sustainability performance.

6. Conclusions

The performance of PPP-type infrastructure projects should be appraised in line with the sustainable development principles considering the significant impacts of infrastructure projects to economic, social and environment aspects. In other words, the appraisal should consider project performance from the perspective of project sustainability, which is measured by economic, social and environmental indicators.

The key in implementing a PPP-type project is the investment distribution between public and private sectors. This study showed that there is a major impact of the investment distribution on the project sustainability performance, and this impact can be assessed. A sustainability performance-based evaluation model (SPbEM) is therefore developed to facilitate this impact assessment. The model developed in this study can present the relationship between the investment distribution and the project sustainability performance, from which an optimal level of investment distribution between two PPP sectors can be identified for contributing best level of project sustainability. In addition, this has been demonstrated effectively from the case study in this paper.

The findings from this study not only promotes the application of PPP approach in the area of infrastructure development, but also provide an effective method for supporting decision making on the choice for a proper alpha (α), namely, a proper level of investment distribution between public and private sectors. The value α can be determined when a certain level of project sustainability performance is defined. The understanding on this mechanism can help public and private sectors find a better contractual arrangement to work towards better sustainability performance during the process of implementing PPP-type projects.

By seeking for collaboration with governmental and industry sectors, it is planned for future study in this research team to utilize the developed model SPbEM for different types of prospective infrastructure projects based on $IfSV$ values.

Author Contributions: All authors have contributed equally to the designed research, researched and analyzed the data, and wrote up the paper. All authors have read and approved the final manuscript.

Conflicts of Interest: All authors have no conflict of interests to any other parties.

References

1. Algarni, A.M.; Arditi, D.; Polat, G. Build-operate-transfer in infrastructure projects in the United States. *J. Constr. Eng. Manag.* **2007**, *133*, 728–735.
2. Ke, Y. Is public-private partnership a panacea for infrastructure development? The case of Beijing National Stadium. *Int. J. Constr. Manag.* **2014**, *14*, 90–100.
3. Robert, O.K.; Dansoh, A.; Ofori-Kuragu, J.K. Reasons for adopting Public–Private Partnership (PPP) for construction projects in Ghana. *Int. J. Constr. Manag.* **2014**, *14*, 227–238.
4. Buertey, J.I.T.; Asare, S.K. Public Private Partnership in Ghana: A Panacea to the Infrastructural Deficit? *Int. J. Constr. Eng. Manag.* **2014**, *3*, 135–143.
5. Sohu Website. 2015 Inventory of PPP Projects Signed. Available online: http://mt.sohu.com/20151211/n430932858.shtml (accessed on 2 July 2015).
6. Appuhami, R.; Perera, S.; Perera, H. Coercive policy diffusion in a developing country: The case of Public-Private Partnerships in Sri Lanka. *J. Contemp. Asia* **2015**, *41*, 431–451.

7. Chowdhury, A.N.; Chen, P.H.; Tiong, R.L. Analysing the structure of public–private partnership projects using network theory. *Constr. Manag. Econ.* **2011**, *29*, 247–260.
8. Hwang, B.G.; Zhao, X.; Gay, M.J.S. Public private partnership projects in Singapore: Factors, critical risks and preferred risk allocation from the perspective of contractors. *Int. J. Proj. Manag.* **2013**, *31*, 424–433.
9. Tang, L.; Shen, Q.; Cheng, E.W.L. A review of studies on Public–Private Partnership projects in the construction industry. *Int. J. Proj. Manag.* **2010**, *28*, 683–694.
10. Leiringer, R. Technological innovation in PPPs: Incentives, opportunities and actions. *Constr. Manag. Econ.* **2006**, *24*, 301–308.
11. Bing, L.; Akintoye, A.; Edwards, P.J.; Hardcastle, C. The allocation of risk in PPP/PFI construction projects in the UK. *Int. J. Proj. Manag.* **2005**, *23*, 25–35.
12. Abraham, G.L.; Chinowsky, P. Critical success factors for the construction industry. In Proceedings of the Construction Research Congress, Honolulu, Hawaii, USA, 19–21 March 2003; pp. 19–21.
13. Aras, G.; Crowther, D. Governance and sustainability: An investigation into the relationship between corporate governance and corporate sustainability. *Manag. Decis.* **2008**, *46*, 433–448.
14. Shen, L.Y.; Li, H.; Li, Q.M. Alternative concession model for build operate transfer contract projects. *J. Constr. Eng. Manag.* **2002**, *128*, 326–330.
15. Shen, L.Y.; Wu, Y.Z. Risk concession model for build/operate/transfer contract projects. *J. Constr. Eng. Manag.* **2005**, *131*, 211–220.
16. Ugwu, O.O.; Kumaraswamy, M.M.; Wong, A.; Ng, S.T. Sustainability appraisal in infrastructure projects (SUSAIP): Part 1. Development of indicators and computational methods. *Autom. Constr.* **2006**, *15*, 239–251.
17. Ugwu, O.O.; Kumaraswamy, M.M.; Wong, A.; Ng, S.T. Sustainability appraisal in infrastructure projects (SUSAIP): Part 2: A case study in bridge design. *Autom. Constr.* **2006**, *15*, 229–238.
18. Shen, L.Y.; Wu, M.; Wang, J.Y. A model for assessing the feasibility of construction project in contributing to the attainment of sustainable development. *J. Constr. Res.* **2002**, *3*, 255–269.
19. Plessis, C.D. Sustainability and sustainable construction: The African context. *Build. Res. Inf.* **2001**, *29*, 374–380.
20. Dasgupta, S.; Tam, E.K. Indicators and framework for assessing sustainable infrastructure. *Can. J. Civ. Eng.* **2005**, *32*, 30–44.
21. Koo, D.H.; Ariaratnam, S.T. Application of a sustainability model for assessing water main replacement options. *J. Constr. Eng. Manag.* **2008**, *134*, 563–574.
22. Koppenjan, J.F.M.; Enserink, B. Public-Private Partnerships in Urban Infrastructures: Reconciling Private Sector Participation and Sustainability. *Public Admin. Rev.* **2009**, *69*, 284–296.
23. Ling, F.Y.Y.; Kumaraswamy, M.M.; Dulaimi, M.F.; Khalfan, M. Public private partnerships: Overcoming pre-contract problems faced by public clients and private providers in infrastructure projects in Singapore. *Int. J. Constr. Manag.* **2011**, *11*, 63–77.

24. Jnr, E.Q. Development projects through build-operate schemes: Their role and place in developing countries. *Int. J. Proj. Manag.* **1996**, *14*, 47–52.
25. Zhuang, C.; He, C. *Applied Mathematical Statistics*; The Press South China University of Technology: Guangzhou, China, 2013.
26. Gan, L.; Shen, L.Y.; Fu, H. Study of the infrastructure project evaluation factors based on sustainable development principles. *China Civ. Eng. J.* **2009**, *42*, 133–138.
27. Chen, B.; Mao, C.K.; Hu, J.L. The optimal debt ratio of public–private partnership projects. *Int. J. Constr. Manag.* **2015**, *15*, 239–253.

Relationship between Quarry Activity and Municipal Spatial Planning: A Possible Mediation for the Case of Sardinia, Italy

Ginevra Balletto, Giovanni Mei and Chiara Garau

Abstract: Despite its economic importance, quarrying activity for the production of natural aggregates (sand, gravel, and crushed stone) can result in overexploitation of the natural environment. This paper investigates the current state of natural and recycled aggregates in Sardinia Italy and how to limit the production of natural aggregates (NA) and increase the use of recycled aggregates (RA). The municipalities of Cagliari, Sant'Antioco and Tortolì of Sardinia, Italy, were chosen as case studies because they fall within a particular territorial context. Owing to its geographic condition, the island of Sardinia must produce its own raw materials. The results of this research show how the combined use of NA and RA can help meet local and regional demand for aggregates. This proposal is derived from a needs assessment of NA based on urban masterplans for each municipality. Possible strategies for limiting the consumption of NA, as well as the use of RA, are also described.

Reprinted from *Sustainability*. Cite as: Balletto, G.; Mei, G.; Garau, C. Relationship between Quarry Activity and Municipal Spatial Planning: A Possible Mediation for the Case of Sardinia, Italy. *Sustainability* **2015**, *7*, 16148–16163.

1. Introduction

Designing with industrial and recyclable materials leads to more sustainable buildings. Most certificates for green building recommend such practices; two of the best-known certification systems are the US Green Building Council's Leadership in Energy and Environmental Design (LEED) rating system and the Green Globes Green Building Initiative. The introduction of these systems encourages the incorporation of many environmentally friendly programs in urban masterplans (UMPs). With this work, we intend to introduce a methodology processes of transition that addresses both certification systems.

Natural resources, which are inherently non-renewable, have been overexploited for centuries. Mining and quarrying are industrial activities that can cause irreversible changes to the earth's surface, leading to the degradation of the environment [1–3]. Improper management of these activities, particularly in the industrial processing and waste management stages, can result in dangerous consequences for the environment. Despite its economic importance, quarrying activity for the production of natural aggregates (NA) (sand, gravel and crushed stone) is a significant source of environmental degradation.

Aggregates are used primarily in the construction sector, both without being laboriously processed (road or railway ballast), and in the production of other high-quality materials such as concrete, asphalt or pre-cast products. The European Union produces approximately two billion cubic meters of aggregates per year in order to meet the demand for this material [4]. Since 2007, the per capita production of construction aggregates in Europe increased from approximately 6 tons/inhabitant [5]. One possible solution to the environmental costs of aggregate production is to satisfy a part of the demand for construction aggregates with construction and demolition waste (CDW), as determined by Directive 2008/98/EC of the European Union (19 November 2008) and international research [6–8].

Recycled aggregates (RA) can be produced from CDW and reused in the construction sector. This practice tends to be common in those countries where high residential density and a shortage of raw materials reduce the opportunity for new quarry sites. European regulation also strongly recommends using RA to meet natural resource demand [8,9]. This regulation states that members states should reach a waste recovery rate of 70% by weight by 2020. Since the regulation was established, the recycling of aggregates increased in some European countries. In fact, some States, including Denmark, the Netherlands, and Germany, have been able to recycle more than 70% of their CDW since the mid-nineties [8]. Germany, for example, had a recycling rate of 89.2% in 2007 [10], and Denmark recently adopted a landfill tax to further encourage the recycling of CDW. However, in some other countries, such as Italy, this practice is still so little diffused that it is very difficult to determine the amount of RA obtained from the recycling of CDW. Furthermore, the use of RA in concrete is virtually non-existent.

This paper analyses the use of CDW as RA in the Sardinia region, where CDW amounted to approximately 1,127,644 tons in 2008 [11]. The Sardinia region must supply the demand for aggregates with the extraction of its own raw materials because of its particular insular geographic condition. Therefore, the Sardinia region can be considered a "closed system" in regard to the supply of NA and RA.

This work is a part of an extensive research project on the recovery of inert waste [12], in which the use of recycled aggregates is integrated into the urban masterplans (UMPs) [13], inspired by international literature, in particular a new approach to waste management demolition in China [14]. In addition, this paper attempts to innovate the urban masterplan (UMP) [13] to create more sustainable development strategies that utilize the measurement and evaluation of the use of RA.

This study aims to investigate the production and recycling of natural aggregates (NA) using the UMPs [13] of three different urban areas in Sardinia, Italy (Cagliari, Sant'Antioco and Tortolì) as a tool for evaluating demand. These three case studies were chosen because they represent the three major urban typologies of Sardinia. Cagliari is the regional capital of Sardinia and has the densest urban configuration. Tortolì and Sant'Antioco are two different costal municipalities with widespread urban configurations. The authors compared the results of this study with similar municipalities in other countries, but found difficulty comparing data, usingthe diversity evaluation method [15–18]. The proposed methodology provides valuable insights, useful at the local, national, and international levels. In fact, this methodology identifies a correlation between the materials demand and urban planning processes on an international level, through the main planning tool of urban government, the UMP [13]. This relationship is especially important in urban contexts with population and economic growth because optimizes the flow of materials in a sustainable environmental lens.

This paper is divided into two parts. In the first part, the CDW streams that can be reused in Sardinia are estimated through a census of recycling facilities in Sardinia and compared by interpreting the demand for aggregates from each municipality's UMP [13]. In the second, the authors focus on the definition of the amount of natural aggregates (NA)that can be used in combination with RA to meet the municipal demand for aggregates. This ratio is also derived from each UMP. In all three case studies, the UMP is valid for 10 years. The paper concludes by analysing the study's findings.

2. Estimation and Comparison of Natural and Recycled Aggregates in Sardinia, Italy

According to the Regional Waste Management Plan (*Piano Regionale Gestione Rifiuti* (PRGR)), Sardinia has a per capita CDW production rate of approximately 670 kg/inhabitant/year. This production rate is significantly lower than the national average of approximately 900 kg/inhabitant/year [19]. The region of Sardinia has proven to be sensitive to the problem of proper management of the materials produced in CDW by adopting the PRGR. However, this policy still does not cover the recycling of materials. Despite lacking means for the import and export of CDW in Sardinia, the use of certified RA for the construction of public and private works, is possible. The Regional Plan of Extractive Activities (*Piano Regionale Attività Estrattive* (PRAE)) [20] does not identify any initiative aimed at limiting the extraction of NA. Although these plans are closely related, there are no initiatives with a unified and sustainable vision for the territory. As such, the authors intend to outline a proposal for orienting initiatives towards the concept: less quarrying, less landfill. In particular, this proposal optimizes the flow of materials derived from construction activities (RA) to their use as a substitute for NA.

2.1. *Natural Aggregates and Recycled Aggregates in the Construction Sector. A Comparison between Italy and Other European States*

Inert materials [21] are of primary importance for the construction industry. In industrialized countries, inert materials represent 10% of the gross national product (*prodotto interno lordo* [PIL]). The analysis conducted at the European level reveals that the average quantity of extracted aggregates amounted to 2.95 billion tons/year, which corresponds to approximately 7.9 tons/year per capita [22].

In Italy, 62.2% of mining is inert. Gravel, sand, and limestone for cement make up 27% of the industry. This confirms the close and long-established correlation between mining and construction activity, which from the second post-war characterized the building of the historic Italian city [23].

There was a substantial decline in aggregate production from 142 million cubic meters in 2009, to 89 million cubic meters in 2010, and to approximately 80 million cubic meters in 2012. This decline is likely linked to the economic crisis in the housing market, which has affected Italy in particular. Nevertheless, Italy remains the third largest European producer of aggregates, after Germany and France [24].

In addition, the construction industry and urban architecture are based, as in all national and international cities, on the principle of availability of construction materials: aggregates, ornamental rocks and concrete [25]. In Italy, these materials are strictly inert, ornamental rocks and cement.

In 2012, Italy has held the record for the greatest cement consumption per capita, at 4322 kg against the EU average of 314 kg [26].

Although the economic crisis drastically reduced production and consumption (Table 1), demand remained high. In fact, all urban renovation projects generate a strong demand for materials and provide the opportunity to utilize recycled materials in the building process [27]. The "extraordinary maintenance" sector plunged in 2011 (Table 1) due to crisis in the international real estate market, then grew to an estimated +21% in 2015. The "extraordinary maintenance" sector is the only sector with a positive trend compared to others in Table 1. In addition, Table 1 shows how the construction sector in Italy is particularly oriented to urban renewal through "extraordinary maintenance" (recovery of existing buildings), which overshadows all other types of building intervention.

Meanwhile the European Commission [9] required the recovery of inert materials from CDW to reach 70% by 2020. Governments could take the following initiatives concerning the reuse of products and preparing for reuse of waste to uphold this directive (Legislative Decree of 3 December 2010, n. 205, article 6):

(i) Use of economic instruments;
(ii) Logistical measures, such as the establishment and support of accredited centres and networks of repair/reuse; or

(iii) Adoption of the framework of procedures for granting public contracts.

Table 1. Investments in the construction sector [26,28].

	2014 Millions (€)	% Variations in Quantity									
		2008	2009	2010	2011	2012	2013	2014	2015	2008–2014	2008–2015
Buildings	**135,332**	**−2.4**	**−8.6**	**−4.7**	**−4.2**	**−7.6**	**−6.9**	**−3.5**	**−2.4**	**−32.0**	**−33.6**
Houses	66,482	−0.4	−8.1	−0.1	−2.9	−6.4	−5.7	−2.4	−1.3	−28.7	−29.7
• *New*	20,565	−3.7	−18.7	−6.1	−7.5	−17.0	−19.0	−10.2	−8.8	−62.3	−65.6
• *Extraordinary Maintenance*	**45,917**	**3.5**	**3.1**	**4.8**	**0.6**	**0.8**	**2.9**	**1.5**	**2.0**	**18.5**	**20.9**
Non-Residential	68,850	−4.4	−9.1	−9.4	−5.7	−9.1	−8.0	−4.6	−3.5	−35.0	−37.2
• *Private*	43,357	−2.2	−10.7	−6.9	−2.1	−8.0	−7.2	−4.3	−3.0	−23.6	−25.9
• *Public*	25,493	−7.2	−7.0	−12.6	−10.5	−10.6	−9.3	−5.1	−4.3	−48.1	−50.3

Some member states are close to, or have already met, the minimum target for the recovery of waste from construction and demolition imposed by the CE directive [9]. For Italy, however, this is a particularly ambitious target. The trend in Italy is to consider the business of mining and digging easier and more profitable than the business of recovery and reuse.

Lacking a unified building code under the Law of 11 November 2014, n. 164 (also called Unlock Italy—simplification of planning rules), the authors note that many municipalities have introduced specific measures for the use of CDW through zoning laws. Although scattered and uncoordinated, the initiatives suggested by the CE directive listed above assist in the reduction of land consumption [29]. In fact, use of recycled aggregates can reduce the amount of land occupied by levy mining and landfills.

In this sense, evaluating the annual demand forNA and RA, asreported to the civil and public construction sector in reference to the UMP, is the first step to identifying possible strategies for an environmental compromise.

2.2. CDW Management in Sardinia, Italy

The tool used to analyse CDW management in Sardinia is the Regional Special Waste Plan (*il Piano Regionale dei Rifiuti Speciali* (PRRS)), approved on 21 December 2012 [30]. Thisannual report considers allwaste produced, processed, transported or sent for disposal in the region. According to these data, CDW production amounted to a total of 658,676,965 kg, which accounts for 9% of the total regional production of hazardous waste, and corresponds to 670 kg per capita. It should be noted, however, that this production is considered only part of CDW; this is just the amount that is declared by treatment plants. The production of CDW in Sardinia (approximately 670 kg/individual/year) appears to be under-valuedas compared to the national average (approximately900 kg/inhabitant/year), and by comparison with the data obtained from the Model of the Environmental Declaration (*Modello Unico di Dichiarazione Ambientale* (MUD) [31]) for other territories in the national context.

Table 2 shows that most of this type of waste in Sardinia is subject to recovery (295,092,875 kg, approximately 53% of the total) and the remaining share, 258,736,682 kg, is destined for disposal.

Table 2. CDW production in Sardinia.

Description	Production
Waste mixed construction and demolition waste	352,314,211 kg
Mixture of concrete, bricks, tiles and ceramics	120,371,652 kg
iron and steel	120,371,652 kg
cement	79,337,576 kg
bituminous mixtures	35,051,773 kg

The Regional Special Waste Plan (PRRS) does not include items of import/export for CDW in Sardinia. This plan sets objectives forthe recovery and disposal of this waste as indicated in Table 2. In 2008, material recovery was 53%, 12 percentage points below the 2015 objective goal of 65% (Table 3). The rate of landfilling is still too high to be overlooked. In addition, Table 3 indicates that the percentage of recycling should be distributed, according to the objectives for 2015 and 2020.

Based on Table 3, the authors decided to focus on how to increase the current recovery rates in Sardinia, for material recovery in particular. The authors quantified the demand for aggregate from the UMP, to evaluate the possibility of replacing the use of NA with RA from CDW. However, before doing so, it was necessary to understand Sardinia's situation.

Table 3. Recovery rate for CDW in Sardinia (2008), and Objectives for 2015 and 2020 [32].

Current Recovery Rate in Sardinia		*2015 Objective*	*2020 Objective*
Material recovery	53.00%	65.00%	70.00%
Energy recovery	0.10%	0.00%	0.00%
Treatment	46.70%	35.00%	30.00%
Disposal	46.00%	0.00%	0.00%
Total		100.00%	100.00%

3. Strategies to Reduce Quarrying Activity in Sardinia

The sustainable cycle of development represents an important part of green building materials, in fact establishes a beneficial reutilization of waste resources. In this context, reducing not only mining activity but also its impact on the landscape is urgent. Data from other European countries demonstrate that it is possible to reduce the amount of material extracted through a policy of reuse of waste from the construction industry. This is, currently, the only possible way to give a future to many areas that may otherwise be condemned to increasingly degraded identities and landscape quality. The countries leading in using RA (United Kingdomand Denmark) demonstratethat it is possible to promote innovative new jobs associated with the mining industry, including green jobs in the recovery of aggregates, that further contributes to the protection of the landscape [24].

The following actions are required in order to reduce quarry activity in Sardinia:

(i) Define a maximum threshold of demand.
(ii) Return to legalizing the transfer of inert materials.
(iii) Introduce concession fee for levy mining of aggregates, which is widespread in Sardinia and the main source of aggregates production.

The authors note that the first two points are subject to extensive research and the introduction of a monetary counterpart is completely absent in some Italian regions (Basilicata, Liguria and Sardinia). In this regard, it is important to show the situation in Italy by regions (Table 4).

Table 4. Revenues of royalties and profits from the sale of sand and gravel in Italy [33].

Regions Geographical Distributions	The Annual Revenue from the Royalties (in Euros)	Annual Business Volume from Mining Activities with Sales Prices (in Euros)	Revenues of Royalties Compared to the Selling Price for Sand and Gravel (%)
Piedmont	5,384,980	137,371,962	3.9
Aosta Valley	62,400	2,600,000	2.3
Lombardy	9,728,796	173,728,500	5.6
Trentino-Alto Adige	no data	10,875,000	-
Bolzano-Bozen	471,350	11,783,750	3.9
Veneto	3,786,891	76,040,625	4.9
Friuli Venezia Giulia	420,338	9,553,137	4.4
Liguria	0	0	-
Emilia-Romagna	3,593,716	78,809,562	4.5
Tuscany	1,434,554	37,358,187	3.8
Umbria	229,867	7,662,250	2.9
Marche	811,718	14,290,812	5.6
Lazio	4,494,150	187,256,250	2.4
Abruzzo	2,119,326	20,069,375	10.5
Molise	414,886	5,186,075	7.9
Campania	118,950	1,486,875	7.9
Puglia	827,410	129,282,887	0.7
Basilicata	0	10,051,250	0
Calabria	420,000	14,975,000	2.9
Sicily	208,337	10,416,875	2.1
Sardinia	**0**	**59,625,000**	**0**
ITALY	**34,527,669**	**998,731,372**	

Table 4 shows the different weights that individual regions attributed to royalties, which indicates a lack of unified national vision for aspects of mining, environmental protection and trade.

In the international scenario, similar conditions occur. However, the cases of Denmark and the United Kingdom are interesting [24]. Denmark has been struggling with how to reduce quarry extraction and promote the recovery of waste from construction and demolition for over 20 years, while the United Kingdom presents uniform royalties in single regions that are 5–6 times higher than the Italian average.

The fragmented Italian approach distorts the market for aggregates, encouraging higher production and sales in territories not subject to (or subject to modest) royalties. This will inevitably lead to the enlargement of the rays of action between places of origin and destination.

4. Evaluation of the Demand for Natural and Recycled Aggregates Resulting from Implementation of the Urban Masterplan

Green buildings represent a great market opportunity and, in this sense, that the assessment of the aggregate resulting from the UMP should contribute. In fact, the demand for aggregates in Cagliari, Sant'Antioco and Tortolì was derived from forecasts in the UMP for each municipality by (i) quantitative and qualitative analysis of the CDW flow in the Sardinian Region; and (ii) the amount of RA that may be used to meet the demand.

More specifically, the authors proceeded by following these steps:

- analyze flow of the Regional Special Waste Plan of Sardinia (*il Piano Regionale dei Rifiuti Speciali* (PRRS)) of 2012;
- census Sardinian inert material treatment facilities;
- interview operators and catalogue plants;
- sample the CDW physical and chemical analysis system periodically [34];
- derive the aggregate demand from forecasts in UMP; and
- compare quantities produced and estimated demand, derived from the UMP tool.

4.1. Assessment of the Demand for Inert Materials in Cagliari, Sant'Antioco and Tortolì

In order to meet the demand for aggregates in a local context, a crucial step is to know the amount of material consumed in the territory of interestduring a specified period. Mining plans use this information to estimate the demand for aggregates of a territory. Recently, these plans play a major role inmeeting demand for natural materials by using aggregates in the construction industry. The primary purpose of these plans is to understand the demand for aggregates and facilitate the use of recycled material from CDW in place of NA.

The demand for aggregates within a territory is destined primarily to private construction and public works, since the construction sector is the primary field of application for aggregates. Badino *et al.* [35] identified a number of approaches for assessing the demand for aggregates, which include using local planning tools (UMP), the method that was adopted in this paper . This method is based on the estimation of the possible consumption of minerals. In each case study (Cagliari, Sant'Antioco and Tortolì, Figure 1), the hypothesis that land consumption is equivalent to the demand for aggregates, appears to be supported. In fact, Sardinia can only count on its own resources due to the low market value of aggregates and to the high costs of transport to and from the island. Therefore, the market of inert materials is fully represented at the local level in insular regions and, consequently, land consumption appears to be closely linked to the UMP forecast.

Figure 1. The Municipalities of Cagliari, Sant'Antioco and Tortolì.

In Italy, the UMP usually applies to a period of 10 years, which is also the period used to assess demand for aggregates [36]. According to the adopted methodology, the demand for aggregates is derived from expected aggregate volumes. In accordance with the regional legislation and to Zoppi *et al.* [37] the zoning rules of cities' masterplans categorize urban areas (expressed in square meters) using the following abbreviationsin parentheses:

- Historic centre zone ("A" zone);
- Residential completion zone ("B" zone);
- Residential expansion zone ("C" zone); and
- Tourism zone ("F" zone)

The achievable volumes were calculated for every homogeneous zone (A, B, C and F zones) for each municipality by adding the existing volume to the maximum realized volume (Table 5). Subsequently, the coefficients of use were applied according to Italian and Regional law These coefficients define the relationship betweenhomogeneous areas and the percentage of aggregates required by the corresponding building sector. In this way, the authors deduced the amount of materials required for the execution of works envisaged by the UMP [38]. The estimated demand for aggregates (deduced from the indices of use) for the municipalities of Cagliari, Sant'Antioco, and Tortolì are outlined in Table 5. In particular, Table 5 shows for each homogeneous area of each municipality under studied the urban planning volumes, established by the legislation of the UMP, from which we can deduct the amount of aggregates required to realize them (for instance, new construction and maintenance of construction (construction of buildings, public works, private works, *etc.*) and road infrastructure), The authors determined the results presented in Table 5 using conversion coefficients calibrated to the final function of the building. Because these coefficients are numerous and have a complex application, the table shows only the results in the last two columns [39–41].

Table 5. Estimation of the demand for aggregates in Cagliari, Sant'Antioco and Tortoli based on UMP forecasts (Source: data from UMPs under study).

Cagliari						Demand Estimation (Aggregates for New Construction and Maintenance (Cubic Meters))		Estimation of Required Aggregates For Max Expected Inhabitants and for Homogeneous Zone	
Homogeneous Zone	Existing Volume (Vex)	Max Realizable Volume (Vmax_R)	Volume to be Realized (Vmax_r-Vex)	Inhabitants (Data from UMP)	Max Expected Inhabitants (Data from UMP)	Construction for Homogeneous Zone	Viability	Aggregates Per Capita (Total Estimation of Aggregates Demand/Max Expected Inhabitants (Cubic Meters/Inhabitants))	
A zone	5,522,043	5,522,043	0	18,208	18,654	33,132	6185	0.211	0.554
B zone	26,514,752	28,317,489	1,802,737	141,141	159,168	519,636	27,835	0.343	
C zone	1,222,686	3,116,655	1,893,969	12,659	30,599	386,130	102,291	1.596	1.596
TOTAL	**33,259,481**	**36,956,187**	**3,696,706**	**172,008**	**208,421**	**938,898**	**136,311**	**2.150**	
Sant'Antioco						Demand estimation (Aggregates for new construction and maintenance (cubic meters))		Estimation of required aggregates for max expected inhabitants and for homogeneous zone	
Homogeneous Zone	Existing Volume (Vex)	Max Realizable Volume (Vmax_R)	Volume to be Realized (Vmax_r-Vex)	Inhabitants (data from UMP)	Max Expected Inhabitants (data from UMP)	Construction for Homogeneous Zone	Viability	Aggregates Per Capita (Total Estimation of Aggregates Demand/Max Expected Inhabitants (cubic meters/inhabitants))	
A zone	120,500	135,500	15,000	800	900	3723	414	0.460	1.270
B zone	2,544,259	2,544,259	0	25,442	25,442	15,266	5249	0.810	
C zone	485,032	884,862	399,830	4851	8849	82,876	41,859	1.410	4.059
F zone	174,124	985,080	810,956	2902	16,418	163,236	271,596	2.649	
TOTAL	**3,323,915**	**4,549,701**	**1,225,786**	**33,995**	**51,609**	**265,101**	**319,118**	**5.329**	
Tortoli						Demand estimation (Aggregates for new construction and maintenance (cubic meters))		Estimation of required aggregates for max expected inhabitants and for homogeneous zone	
Homogeneous Zone	Existing Volume (Vex)	Max Realizable Volume (Vmax_R)	Volume to be Realized (Vmax_r-Vex)	Inhabitants (data from UMP)	Max Expected Inhabitants (data from UMP)	Construction for Homogeneous Zone	Viability	Aggregates Per Capita (Total Estimation of Aggregates Demand/Max Expected Inhabitants (cubic meters/inhabitants))	
A zone	548,737	548,737	0		1960	3292	11,469	0.753	0.975
B zone	1,410,815	1,410,815	0		5039	8465	2713	0.222	
C zone	1,388,553	2,119,832	731,278		8231	154,587	483,407	7.751	8.543
F zone	483,001	629,484	224,024		11,749	47,703	45,387	0.792	
TOTAL	**3,831,106**	**4,708,868**	**955,302**	**0**	**26,979**	**214,047**	**542,976**	**9.518**	

In order to focus on the residential completion ("C" zone) and tourism areas ("F" zone), the authors did not consider the enterprise and industrial areas ("D" zone) and recreational and service areas ("G" zone). In this way, we could compare C and F zones with the remaining residential areas that have a predominantly historic fabric (A and B zones). This analysis allowed us to understand how to contain both the consumption and demand for soil resulting from expansion in these zones. The authors identified that the highest territorial indices (TI) [42] correspond to A and B zones, while the remaining zones (C and F) correspond lower values (Table 6). The values in Table 6 were obtained by total area and maximum achievable volume (extracted from the values declared by each UMP). An average value of TI is thus obtained for each homogeneous zone. This average value can express compactness or spread of the urban configuration of each homogeneous zone in the three municipalities studied.

Table 6. TI Distribution for A, B, C and F zones in Cagliari, Sant'Antioco and Tortolì, based on UMP forecasts (Source: data from UMPs under study).

Cagliari			
Homogeneous Zone	**(TA) Total Area**	**Vmax_R**	**TI =Vmax_R/TA (Cubic m/sqm)**
A zone	1,237,007	5,522,043	4.46
B zone	5,567,059	28,317,489	5.09
C zone	2,846,603	3,116,655	1.09
F zone	0	0	0.00
Total	9,650,669	36,956,187	
Sant'Antico			
Homogeneous Zone	**(TA) Total Area**	**Vmax_R**	**TI =Vmax_R/TA (Cubic m/sqm)**
A zone	41,501	135,500	3.26
B zone	1,049,823	2,544,259	2.42
C zone	1,292,324	884,862	0.68
F zone	6,459,541	985,080	0.15
TOTAL	8,843,189	4,549,701	
Tortolì			
Homogeneous Zone	**(TA) Total Area**	**Vmax_R**	**TI =Vmax_R/TA (Cubic m/sqm)**
A zone	131,262	548,737	4.18
B zone	541,352	1,410,815	2.61
C zone	24,446,262	2,119,832	0.09
F zone	1,999,102	629,484	0.31
TOTAL	**27,117,978**	**4,708,868**	

High values of TI correspond to an urban form more compact than those associated with low TI values, which correspond to a dispersed configuration. The building and urban density, which is a measure of building volume per square meter of territorial surface (TS), is determined by the manufacturability of territorial indices (TIs). Land use decisions with compact, dense configurations have greater environmental sustainability in terms of the energy use reduction [43] and building material selection [39]. However, these choices are strongly influenced by national and regional legislation [44] that can affect decisions made by the designer and the planner.

4.2. Valuation Assumptions for the Maximum Limit of the Demand for Aggregates

The above analysis shows that the only way to limit levy mining aggregates is through an urban design that favours compact configurations over dispersed. This is because it is not possible to encourage the use of CDW when the distance between production centres and product destinations exceeds (30 km [45]).

Aligning local planning (UMP) forecasts for the reuse of materials with the targets that the region of Sardinia has established for itself in the period 2015–2020 (Table 3) is not possible. The primary obstacle lies in the distance between the places of production and those of potential destination. Therefore, rather than acting only on reuse it is important also to consider limiting the average requirement per capita per year, making an appropriate reduction of 35%, in line with the guidelines restricting land consumption [46].

To allow for reduction in per capita demand on the implementation of the UMP, the authors argue that a specific parameter associated with the UMP should be introduced. This specific parameter expresses the maximum demand for aggregates, that, for simplicity, we will call "Da-max (UMP)".

A generic municipality (n) with a UMP has a demand for aggregates, Da-n (UMP), that can have the following conditions:

(1) Da-n (UMP) > Da-max (UMP), implies environmental incompatibility
(2) Da-n (UMP) = Da-max (UMP), implies environmental neutrality
(3) Da-n (UMP) < Da-max (UMP), implies environmental compatibility

The best place for discussion about the definition of the parameter "Da-max (UMP)" is within the Strategic Environmental Assessment (SEA), which delegates the definition of policies for environmental sustainability. Pursuant to Legislative Decree no. 152/2006 (Art. 6 and subsequent amendments), all plans that can have significant impacts on the environment, including UMPs and mining activities must perform a SEA.

The SEA is a process that accompanies the development and adoption of the UMP in order to ensure the integration of environmental aspects. The SEA requires that from the earliest stages of UMP development both local and regional environmental externalities must be accounted for.

Three possible scenarios are formulated in the UMPs for Sant'Antioco and Tortolì. Cagliari constitutes a special case because it lacks C zones.

All scenarios include a compact configuration of A and B zones for which no reduction is made in relation to the potential levy. Configurations in the remaining zones were considered based on the reduction of the potential levy. A brief description of the three proposed scenarios is shown below.

1 Scenario 1: Widespread urban form for the residential C zones, and compact urban form for the tourism F zone.
2 Scenario 2: Equivalent urban form between C and F zones.
3 Scenario 3: Compact urban form for C zones and widespread urban form for F zones.

These scenarios represent the unique assets of urban form that are possible in the drafting process of a UMP.

This proposal builds on limiting the use of natural materials.

The additional contribution that the authors intend to introduce is a portion of RA use in the implementation of the UMP. The inter-ministerial decree D.M. 203/2003 states that public offices and companies with a majority public capital should cover the annual demand of manufactured goods, and goods with a portion of products made from recycled material in not less than 30% of the same demand.

In regards to the construction industry, the procedures for implementing these prescriptions are contained in the Circular of the Italian Ministry of the Environment No. 5205 of 15/07/2005 [47]. This Circular defines the technical and performance criteria that recycled materials should possess, including the frequencies of control.Furthermore, the annexes of Circular specify the values of the technical and environmental characteristics of the products, with respect to its destination.

In this sense, the Green Public Procurement (GPP), as defined by the Action Plan for the environmental sustainability of consumption in the field of public administration (Decree of 10 April 2013), plays an important role. The GPP is defined by the European Commission as "[...] the approach by which Public Bodies integrate environmental criteria into all stages of their procurement process, thus encouraging the spread of environmental technologies and the development of environmentally sound products, by seeking and choosing outcomes and solutions that have the least possible impact on the environment throughout their whole life-cycle" [48].

New construction and maintenance viability can be an important test case for two reasons. First, because these are works for which the literature and recycling technologies are widely used and second because these are works for which it is easier to overcome cultural mistrust.

The imposition in the reuse of RA in UMPs through the GPP, referring to transport structures with a margin of 30% compared to the demand, is an important test case for the pursuit of environmental sustainability for public administrations.

Specifically the three case studies as described in Table 7 would occur.

Table 7. Optimizing the use of recycled aggregates.

	Aggregates for New Construction and Maintenance (Cubic Meters)—NA		30% of RA (Cubic Meters)	Total Demand of NA
	Homogeneous Zone	**Viability**	**Viability**	**Viability**
Cagliari	938,898	1,363,311	40,893	95,418
Sant'antioco	265,101	319,118	95,735	223,383
Tortolì	214,047	542,976	162,893	380,083
Total			**299,522**	**698,884**

In other words, a saving of NA equal to about 300,000 cubic meters—that would be replaced by RA, not necessarily coming from a local basin—could be experienced, thereby breaking the insularity that has always characterized aggregates.

5. Conclusions

Resource conservation is a national effort to conserve energy and other resources and reduce greenhouse gas emissions by managing materials more efficiently. Industrial materials recycling (IMR) helps accomplish these goals by conserving natural resources and decreasing energy use and greenhouse gas emissions. The authors estimated and evaluated the demand for natural and recycled aggregates, and focused on the correlation between implementation of the UMP and its demand for building materials, which is little discussed in the literature.

This document reviewed the current literature on the relationship between the aggregate materials demand and urban planning, in order to analyse the Italian situation and, in particular, the Sardinian Region condition, which is a special case due to its insularity. In this regard, this paper shows how the assessment of demand for aggregates, linked to the implementation of the UMP, can provide insight into the definition of the urban form and planning process. In fact, the urban case studies of Cagliari, Sant'Antioco and Tortolì, which are representative of the remaining urban areas in Sardinia, confirmed that the historic areas (A zone) require lower quantities of aggregates, compared to the surroundings residential expansion (C zone) and tourism (F zone). The case study of Cagliari is emblematic of demand in the A zone, equal to about 5522 thousand cubic meters, compared to 31 million cubic meters for the remaining B and C zones.

The authors have shown that compact and dispersed city forms are associated with different per capita demands for aggregates. Low demand per capita is associated with a compact urban form. The relationships identified above show that we have the opportunity to quantitatively orient the urban form, by defining the achievable Da-max (UMP), for every UMP.

Furthermore, the proposed approach allows full control of land use resulting from the municipal development plan, in accordance with international requirements of the Europe 2020 Strategy.

This methodological approach to the evaluation of the demand for natural aggregates associated with the UMP is also consistent in today's national [49] and international [50] debate. The correlation with mining, arising from building activities and the urban planning sector, can no longer be ignored or neglected. Strategic objectives must be pursued to address these issues. Finding the point of balance between the form of the city and the delayed impact—that the same city generates in order to be implemented—is the key to this work. In addition, introducing the compulsory use of RA, at least for minor works, is a realistic possibility.

In addition, the control of the ecological footprint of aggregate mining through urban spatial planning constitutes a new approach to pursuing strategic objectives for environmental sustainability that is repeatable in national and international contexts.

Acknowledgments: This study is supported by the Autonomous Regions of Sardinia (Regione Autonoma della Sardegna, RAS) through a project entitled The recovery of inert waste for the packaging of recycled aggregates to be used in concrete. Experiments and applications in the provinces of Cagliari and Carbonia-Iglesias. This project is financed through the "Sardinia PO FSE 2007–2013" funds and provided according to the L.R. 7/2007 for the "Promotion of the Scientific Research and of the Technological Innovation in Sardinia". We authorize the RAS to reproduce and distribute reprints for Governmental purposes notwithstanding any copyright notation thereon. Any opinions, findings and conclusions or recommendations expressed in this material are those of the authors and do not necessarily reflect the views of the RAS.

Author Contributions: This paper is the result of the joint work of the authors. In particular, Ginevra Balletto wrote Sections 1, 2 and 4.2; Giovanni Mei wrote Section 3; and Chiara Garau wrote Sections 4, 4.1 and 5.

Conflicts of Interest: Conflicts of Interest: The authors declare no conflict of interest.

Glossary

Construction and Demolition Waste (CDW): Unwanted material produced directly or incidentally by the construction industry. This includes building materials such as aggregates, many of which can be recycled.

Gross Domestic Product (GDP) (*prodotto interno lordo (PIL)*): The value of everything a country produces. The size of a country's PIL is very important in assessing the health of an economy.

Green Public Procurement (GPP): An environmental policy tool that aims to encourage voluntary development of a market for products and services with reduced environmental impact through the leverage of public demand. Public authorities that undertake GPP streamline purchasing and consumption increase the environmental quality of their supplies and credit lines (The handbook: Buying Green—http://ec.europa.eu/environment/gpp/pdf/handbook.pdf).

Model of the Environmental Declaration (MUD) (*Modello Unico di Dichiarazione Ambientale*): A collection of statements, presented annually by different actors such as landfills, waste producers and transporters.

Natural aggregate (NA): The component of a composite material that resists compressive stress and provides bulk to the composite material (e.g., the particles of stone used to make concrete typically include both sand and gravel). For efficient filling, aggregate should be much smaller than the finished item, but have a wide variety of sizes.

Regional Plan of Waste Management Special Sardinia (PRRS)(*Piano Regionale dei Rifiuti Speciali*): The document represents a major updating of the document "Section Special waste" approved by resolution No. 13/34 of 30/04/02. It is the result of a thorough analysis of the current situation of the installation and logistics of the regional system of treatment of this category of waste and is aimed, above all, at a further determination of the needs and to plant more incentive for its recovery, with regard to the general guidelines set by the EU and national legislation.

Recycled Aggregate (RA): A broad category of coarse particulate material used in construction, including sand, gravel, crushed stone, slag, recycled concrete and geosynthetic aggregates. Aggregates are the most mined materials in the world. Aggregates are a component of composite materials such as concrete and asphalt concrete; the aggregate serves as reinforcement to add strength to the overall composite material.

Strategic Environmental Assessment (SEA): A fundamental tool that supports decision-making processes that characterizes the urban masterplan document. (Directive 2001/42/CE- D.G.R. n. 8/1563 del 22/12/2005).

Territorial Indices (TI): The ratio between the manufacturable volume, expressed in cubic meters, and the land area, measured in square meters.

Territorial Surface (TS): Surface area including the areas earmarked for development in public use sectors. It considers areas of primary and secondary urbanization, including roads.

Urban Masterplan (UMP): A complex planning tool owned by a city that regulates and protects the urban and territorial processes of transformation, in accordance with the Italian National Law no. 1942/1150. This tool has a relevance of at least one decade. Every Italian municipality, from small village to sprawling municipality, can have an urban masterplan. Small communities will hire a private planning firm to prepare a plan and submit it to the local government for approval. In larger cities or metropolises, the city administrative planning sector prepares the urban masterplan.

References and Notes

1. Pavan, V. *Quarry Architecture*; Faenzaindustrie Grafiche: Faenza (RA), Italy, 2010.
2. Gisotti, G. *Le Cave. Recupero e Pianificazione Ambientale (Quarries. Recovery and environmental* planning); Dario Flaccovio Editore: Palermo, Italy, 2008. (In Italian)
3. Nautiyal, H.; Shree, V.; Khurana, S.; Kumar, N. Recycling Potential of Building Materials: A Review. In *Environmental Implications of Recycling and Recycled Products*; Springer: Singapore, 2015; pp. 31–50.
4. UEPG (European Aggregates Association). *Annual Review 2011–2012*; UEPG Aisbl: Brussels, Belgium, 2012; Available online: http://www.uepg.eu (accessed on 30 January 2015).
5. Furcas, C.; Balletto, G. Effects of quarrying activity and the construction sector on environmental sustainability. A brief report on the rapid growth of emerging Central and Eastern European States. *Diam. Appl. Tecnol.* **2012**, *70*, 74–81.
6. Pani, L.; Francesconi, L. Influence of Replacement Percentage of Recycled Aggregates on Recycled Aggregate Concrete Properties. In Proceedings of the Fib Symposium, Prague, Czech Republic, 8–10 June 2011.
7. Ding, T.; Xiao, J. Estimation of building-related construction and demolition waste in Shanghai. *Waste Manag.* **2014**, *34*, 2327–2334.
8. European Parliament, Policy Department Economy and Science DG Internal Policies. Impact Assessment of Recycling Targets in the Waste Framework Directive (IP/A/ALL/FWC/2006-105/Lot4/C1/SC3). Available online: http://www.pedz.uni-mannheim.de/daten/edz-ma/ep/08/EST21011.pdf (accessed on 9 November 2015).
9. European Parliament and Council. Directive 2008/98/EC on Waste. Available online: http://ec.europa. eu/environment/waste/framework/ (accessed on 30 Novermber 2015).
10. Spies, S. German Technical Cooperation. 3R in Construction and Demolition Waste (CDW), potentials and constraints. In Proceedings of the Inaugural Meeting of the Regional 3R Forum in Asia, Tokyo, Japan, 11–12 November 2009; Available online: www.uncrd.or.jp (accessed on 15 September 2015).
11. Autonomous Region of Sardinia (Regione Autonoma della Sardegna—RAS, 2012). Available online: http://www.regione.sardegna.it (accessed on 30 August 2015).
12. This broader work (called *Recovery of inert waste in recycled aggregates for use in concrete*) is divided into two main lines of research. The first has a technical character and deals with the evaluation of the mechanical, physical and chemical properties of the RA in some urban masterplans of Sardinia. The second aims to investigate the consequences of the containment requirements of aggregates (NA) and of the recycling of CDW for the production of RA in some municipalities of Sardinia. This document will explain the first results of the second line of intervention.
13. The urban masterplan (UMP)is a complex planning tool owned by a city that regulates and protects the urban and territorial processes of transformation, in accordance with the Italian National Law No. 1942/1150. This tool has a valence of at least one decade.
14. Wu, H.; Wang, J.; Duan, H.; Ouyang, L.; Huang, W.; Zuo, J. An innovative approach to managing demolition waste via GIS (geographic information system): A case study in Shenzhen city, China. *J. Clean. Prod.* **2015**, *99*, 1–10.

15. Horvath, A. Construction materials and the environment. *Annu. Rev. Environ. Resour.* **2004**, *29*, 181–204.
16. Chau, C.K.; Yik, F.W.H.; Hui, W.K.; Liu, H.C.; Yu, H.K. Environmental impacts of building materials and building services components for commercial buildings in Hong Kong. *J. Clean. Prod.* **2007**, *15*, 1840–1851.
17. Bribián, I.Z.; Capilla, A.V.; Usón, A.A. Life cycle assessment of building materials: Comparative analysis of energy and environmental impacts and evaluation of the eco-efficiency improvement potential. *Build. Environ.* **2011**, *46*, 1133–1140.
18. Tam, V.W.; Zuo, J.; Zhu, J. Designers' attitude and behaviour towards construction waste minimization by design: A study in Shenzhen, China. *Resour. Conserv. Recycl.* **2015**, *105*, 29–35.
19. RAS (Autonomous Region of Sardinia). *Regional Waste Management Plan (Piano Regionale Gestione Rifiuti)*; Oikosprogetti: Milano, Italy, 2008. (In Italian)
20. Regional Plan of Extractive Activities (Piano Regionale Attività Estrattive). Available online: https://www.regione.sardegna.it/documenti/1_82_20080627104231.pdf (accessed on 30 November 2015).
21. Varghese, P.C. *Building Materials*; PHI Learning Pvt. Ltd.: Delhi, India, 2015.
22. Competitiveness Report on the Productive Sectors (Rapporto sulla competitività dei settori produttivi) ISTAT. 2013. Available online: http://www.istat.it/it/files/2013/02/Rapporto-competitivit%C3%A0.pdf (accessed on 29 September 2015).
23. Balletto, G.; Naitza, S.; Mei, G.; Furcas, C. Compromise between mining activities and reuse of recycled aggregates for development of sustainable local planning. Available online: http://www.seekdl. org/nm.php?id=6135 (accessed on 30 November 2015).
24. The Numbers, the Regulatory Framework, the Point on the Economic and Environmental Impact of Mining in the Italian Territory (Rapporto Cave 2014. I Numeri, il Quadro Normativo, il Punto Sull'impatto Economico e Ambientale Dell'attività Estrattiva nel Territorio Italiano). Available online: http://www. legambiente.it/sites/default/files/docs/rapporto_cave_2014_web_2.pdf (accessed on 9 November 2015). (In Italian).
25. Hesse, M. Cities, material flows and the geography of spatial interaction: Urban places in the system of chains. *Glob. Netw.* **2010**, *10*, 75–91.
26. AITEC (Associazione Italiana Tecnico Economica Cemento). Relazione Annuale 2013. Available online: http://www.aitecweb.com/Portals/0/pub/Repository/Area%20 Economica/Pubblicazioni%20AITEC/Relazione _ annuale_2013.pdf (accessed on 9 November 2015).
27. Endl, A.; Berger, G. Sustainable raw materials management: A perspective on stakeholders roles and policy strategies. In Proceedings of the 2nd Symposium for Urban Mining, Bergamo, Italien, 19–21 May 2014.
28. AITEC (Associazione Italiana Tecnico Economica Cemento). Relazione Annuale 2014. Available online: http://www.aitecweb.com/Portals/0/pub/Repository/Area%20 Economica/Pubblicazioni%20AITEC/ Relazione_ Annuale_2014.pdf(accessed (accessed on 9 November 2015).

29. Jaeger, J.A.G.; Soukup, T.; Madriñán, L.F.; Schwick, C.; Kienast, F. *Landscape Fragmentation in Europe*; European Environment Agency: Copenhagen, Denmark, 2011.
30. Data from this instrument refer to the statement Model of the Environmental Declaration (*Modello singolo di Dichiarazione Ambientale* (MUD)) of 2009, which collects 2008 data (Law 70/94).
31. The Model of the Environmental Declaration (*Modello singolo di Dichiarazione Ambientale* (MUD)) identifies a whole set of statements, presented annually by different actors such as landfills, waste producers and transporters.
32. Regional Special Waste Plan (*il Piano Regionale dei Rifiuti Speciali*). Sardegna, Italy, 2008. Available online: http://www.regione.sardegna.it/documenti/1_106_20120423163058.pdf (accessed on 2 September 2015).
33. Author's elaborations on Data of [24].
34. Pani, L.; Balletto, G.; Naitza, S.; Francesconi, L.; Trulli, N.; Mei, G.; Furcas, C. Evaluation of mechanical, physical and chemical properties of recycled aggregates for structural concrete. In Proceedings of the 14th International Waste Management and Landfill Symposium, Sardinia, Italy, 30 September–4 October 2013; CISA Publisher: Cagliari, Italy, 2013.
35. Badino, V.; Blengini, G.A.; Garbarino, E. Technical analysis—Environmental—Economic aggregates for the construction industry in Italy. Part 2. Needs estimation (Analisi tecnico—Economico—Ambientale degli aggregati per l'industria delle costruzioni in Italia. Parte 2. La stima dei fabbisogni). *Geam* **2006**, *3*, 5–16.
36. In reality this condition does not always come true, and actions by the UMP may be modified or delay and cover much longer periods of time.
37. Zoppi, C.; Argiolas, M.; Lai, S. Factors influencing the value of houses: Estimates for the city of Cagliari, Italy. *Land Use Policy* **2015**, *42*, 367–380.
38. Balletto, G. *A Cura di, La Pianificazione Sostenibile Delle Risorse (Sustainable planning of environmental resources)*; Franco Angeli: Milano, Italy, 2005.
39. Pauleit, S.; Duhme, F. Assessing the environmental performance of land cover types for urban planning. *Landsc. Urban Plan.* **2000**, *52*, 1–20.
40. Drew, L.J.; Langer, W.H.; Sachs, J.S. Environmentalism and natural aggregate mining. *Nat. Resour. Res.* **2002**, *11*, 19–28.
41. Balletto, G.; Naitza, S.; Mei, G.; Furcas, C. Compromise between mining activities and reuse of recycled aggregates for development of sustainable local planning. (Sardinia). In Proceedings of the 3rd Intnational Conference on Advances in Civil, Structural and Mechanical Engineering—CSM, Birmingham, UK, 26–27 April 2015; pp. 136–142.
42. TI (territorial index—Cubic meters/sqm) = It is the ratio of the volume (V) maximum achievable in a given area and the territorial surface (TS) of the zone.
43. Vaccaro, V. Energia nel paesaggio: Sistemi di produzione di energia elettrica da fonte rinnovabile a Pantelleria (Energy in the landscape: Systems of electricity production from renewable sources in Pantelleria). In *A cura di, Atlante Delle Smart City*; Riva Sanseverino, E., Riva Sanseverino, R., Vaccaro, V., Eds.; Franco Angeli: Milano, Italy, 2015; pp. 283–308.

44. Bressi, G.; Volpe, G.; Pavesi, E. La produzione di aggregati riciclati da rifiuti inerti. I manuali di SARMa. In *The Production of Recycled Aggregates from Inert Waste. The Manuals of SARMa*; Centro Stampa Regione Emilia-Romagna: Bologna, Italy, 2011; Available online: http://www.sarmaproject.eu/index.php?id=1964 (accessed on 15 September 2015).
45. It should be specified that the situation observed in the municipalities of Cagliari, Sant'Antioco and Tortolì (but likewise also for all the remaining 362 municipalities of Sardinia) is mainly due to the regional legislation (Decree of 22 December 1983 No. 2266/U—Called also "Floris" Decree), and to the national legislation (ministerial decree April 2, 1968, n. 1444). In particular the aforementioned regional law establishes for C zones a TImax (cubic m/sqm) equal to 1 for municipalities up to 10,000 inhabitants and a TImax (cubic m/sqm) of 1.5 for municipalities with more than 10,000 inhabitants.
46. General Report of the Regional Plan of Extractive Activities (Piano Regionale Attività Estrattive). Available online: http://www.regione.sardegna.it/documenti/1_82_20080110174612.pdf (accessed on 11 November 2015).
47. Ministry of Environment Guidelines for operation in the construction industry, road and environment, according to DM 08/05/2003, 203, OJ of 25.07.2005 n. 171.
48. Bouwer, M.; Jonk, M.; Berman, T.; Bersani, R.; Lusser, H.; Nappa, V. Green Public Procurement in Europe 2006: Conclusions and Recommendations. 2006. Available online: http://ec.europa.eu/environment/gpp/pdf/ take_5.pdf (accessed on 20 September 2015).
49. In particular, we mention the following Draft law emerged in 2014 from the national debate "Containment of land consumption and reuse of soil built" (*Contenimento del consumo del suolo e riuso del suolo edificato*).
50. We refer to BREEAM (BRE Environmental Assessment Method or "methodology of the BRE Environmental Assessment"). It is the first and most widely used environmental assessment protocol in the world. BREEAM was born in London, for the Olympics of 2012 and establishes the evaluation criteria used to represent the environmental performance of a building. Available online: http://www.breeam.org/about.jsp?id=66 (accessed on 15 September 2015).

Life Cycle Assessment and Optimization-Based Decision Analysis of Construction Waste Recycling for a LEED-Certified University Building

Murat Kucukvar, Gokhan Egilmez and Omer Tatari

Abstract: The current waste management literature lacks a comprehensive LCA of the recycling of construction materials that considers both process and supply chain-related impacts as a whole. Furthermore, an optimization-based decision support framework has not been also addressed in any work, which provides a quantifiable understanding about the potential savings and implications associated with recycling of construction materials from a life cycle perspective. The aim of this research is to present a multi-criteria optimization model, which is developed to propose economically-sound and environmentally-benign construction waste management strategies for a LEED-certified university building. First, an economic input-output-based hybrid life cycle assessment model is built to quantify the total environmental impacts of various waste management options: recycling, conventional landfilling and incineration. After quantifying the net environmental pressures associated with these waste treatment alternatives, a compromise programming model is utilized to determine the optimal recycling strategy considering environmental and economic impacts, simultaneously. The analysis results show that recycling of ferrous and non-ferrous metals significantly contributed to reductions in the total carbon footprint of waste management. On the other hand, recycling of asphalt and concrete increased the overall carbon footprint due to high fuel consumption and emissions during the crushing process. Based on the multi-criteria optimization results, 100% recycling of ferrous and non-ferrous metals, cardboard, plastic and glass is suggested to maximize the environmental and economic savings, simultaneously. We believe that the results of this research will facilitate better decision making in treating construction and debris waste for LEED-certified green buildings by combining the results of environmental LCA with multi-objective optimization modeling.

Reprinted from *Sustainability*. Cite as: Kucukvar, M.; Egilmez, G.; Tatari, O. Life Cycle Assessment and Optimization-Based Decision Analysis of Construction Waste Recycling for a LEED-Certified University Building. *Sustainability* **2016**, *8*, 89.

1. Introduction

Residential and commercial buildings generate a significant amount of construction and debris (C&D) waste in the United States. The estimated total amount of building-related C&D materials is approximately 170 million tons [1]. Based on the U.S. Environmental Protection Agency's (EPA) waste report, 39% of these wastes are residential and 61% are from commercial buildings [1]. Recycling or appropriate treatment of these C&D wastes not only reduces the amount of waste land-filled or incinerated, but additionally minimizes the environmental impacts associated with producing new materials from virgin resources. In this context, one of the barriers for effective policy making towards shifting to a more sustainable C&D waste management is that C&D generation statistics are not rigorously collected [2]. Even though the statistics vary significantly, a recent report indicates that recycling could create credible benefits as a sustainable solution [2]. For instance, according to the same report, in 2012, the estimated magnitude of GHG emissions offset corresponded to taking 4.7 million passenger cars off the road for an entire year. The green building movement has adopted several strategies to reduce C&D waste. Among the green building initiatives, the LEED (Leadership in Energy and Environmental Design) rating system, which was established by the U.S. Green Building Council (USGBC), has gained wide acceptance and has been adopted by several federal and state agencies for evaluating their building designs. LEED green building certification systems employ a simplified checklist that is mainly used in the design process [3]. To obtain LEED certification, a building must first satisfy certain prerequisites and then obtain points for credits related to sustainable sites, water efficiency, energy and atmosphere, materials and resources, indoor environmental quality and design process. In general, LEED can be applied to new constructions, major renovations, existing buildings, commercial interiors, core and shell, schools, retail, healthcare, homes and neighborhood development [4]. LEED has two main construction waste material diversion credits, which are as follows [3]:

- Credit 2.1 (one point): recycle and/or salvage at least 50 percent of construction, demolition and land-clearing waste.
- Credit 2.2 (one point): recycle and/or salvage an additional 25 percent (75 percent total) of construction, demolition and land-clearing waste.

The credits are proposed to divert construction, renovation and demolition debris from landfill areas and redirect recyclable materials back to the manufacturing process. To accomplish this goal, it is necessary to develop a detailed waste management plan for recycling of various construction waste materials, such as cardboard, metal, brick, mineral fiber panel, concrete, asphalt, plastic, clean wood, glass, gypsum wallboard, carpet and insulation materials. However, setting a goal, such as "50 percent or 75 percent of construction waste must be recycled" without defining the possible economic and environmental impacts associated with recycling of each C&D waste material may lead to a misrepresented understanding of the comprehensive impact that recycling of that nature may cause. Therefore, recycling goals should be supported by robust decision making models considering environmental and economic impacts, simultaneously [5].

Studying the construction materials from a life cycle point of view is critical for an overall understanding about the sustainability impacts of processes associated with the entire life cycle of buildings. In fact, the literature is abundant with works that focus on the process life cycle of construction materials based on case studies. Several construction materials are analyzed from a life cycle perspective to quantify the environmental impacts; for example, wood and concrete [6], wood and alternative materials [7], generic *vs.* product specific comparisons [8], socio-economic aspects of life cycle impacts [9] and the case of multiple life cycles [10–13]; for recent comprehensive reviews, see [14].

Novelty and Organization of the Research

Even though a significant amount of work related to LCA-based sustainability assessment of construction materials and buildings has been done, the current literature lacks a comprehensive LCA of the recycling of construction materials that considers both process and supply chain-related impacts as a whole. Additionally, a decision support framework has not been also addressed in any work, which provides a quantifiable understanding about the potential savings and implications associated with recycling of such construction materials from the life cycle perspective. The goal of this research is to develop a comprehensive framework that aids in quantifying and minimizing the environmental impact of C&D waste, while maximizing the economic value added to the solid waste industry. To realize this goal, the following tasks were undertaken: (1) assess the net environmental impacts of C&D waste management strategies using the proposed hybrid LCA model; (2) optimize C&D waste recycling strategies using a multi-criteria optimization model considering all direct and indirect environmental and economic impacts of C&D waste management alternatives; and (3) optimize the sustainable waste recycling strategies by taking LEED requirements into consideration. Hence, this research aims at integrating the solid waste requirements of the LEED green building rating system in order to devise the most environmentally-friendly C&D waste treatment strategies. For a general research framework, please see Figure 1.

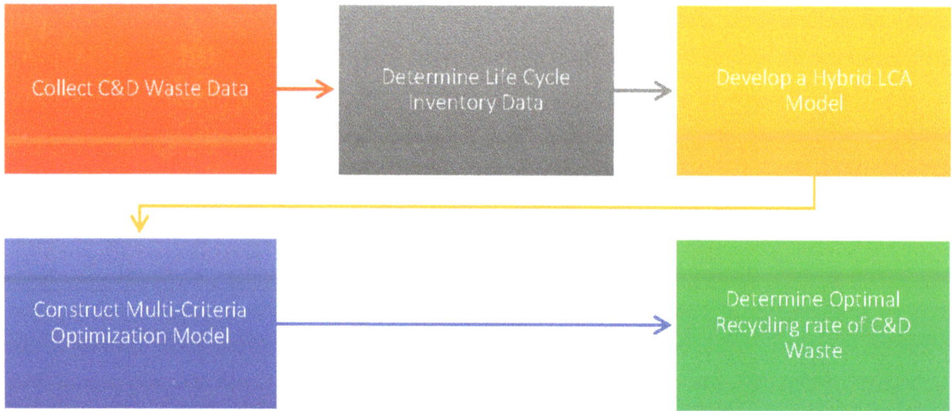

Figure 1. General research framework. C&D, construction and debris.

The rest of the paper is organized as follows: First, the case study is explained. Then, the proposed methodology that consists of hybrid LCA and optimization models is presented. Next, the LCA results of the certified green building and the results of the optimization model are presented. Finally, the conclusion and the future work are pointed out.

2. Case Study

In this paper, the LEED-certified Physical Science Building at the University of Central Florida was chosen as a case study for this research. In this analysis, the amount of C&D waste and its composition data were gathered from the LEED waste documentation, which is publicly available through the University of Central Florida (UCF) Office of Sustainability for the Physical Science Building [15]. The percentage of the waste composition of C&D materials of the building is included, such as asphalt, concrete, wood, non-ferrous and ferrous metals, cardboard, plastic, glass and cardboard. Total waste composition shows that concrete and asphalt have the largest share (over 60% of the total) among the C&D waste materials. The composition of all other building waste materials is presented in Figure 2, which shows the % shares of C&D waste of different materials related to the case study. Although the total composition of these waste materials is important to know, the life cycle impacts related to treatment of these wastes are critical. Hence, the net greenhouse gas (GHG) emissions, energy consumption and water withdrawal associated with recycling, land filling and incineration of C&D materials were quantified using the economic input-output (EIO)-based hybrid LCA methodology. Later, an optimization model is developed to optimize the construction waste recycling strategies.

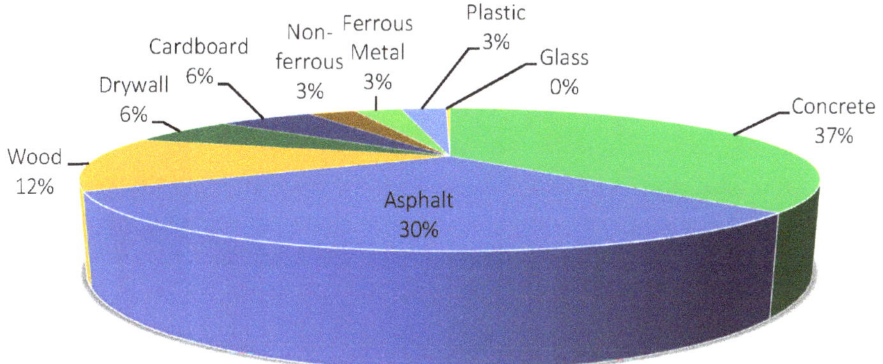

Figure 2. Percentage distribution of C&D wastes for the LEED-certified university building.

3. Methodology

Several important life cycle phases, including production of building materials from virgin and recycled resources, transportation, material recovery, incineration and conventional land filling, were analyzed for nine C&D waste materials, namely: asphalt, concrete, wood, non-ferrous and ferrous metals, cardboard, plastic, glass and cardboard. For the recycling, collection and transportation of wastes, material recovery process and producing new products from recycled materials were holistically investigated under the scope of this research. For the incineration, transportation of waste to the incineration facility, processing of waste and energy recovery from combustible waste are analyzed. In addition, the environmental impacts related to a conventional land filling process are quantified for the transportation of C&D to landfill and land filling of each C&D waste material, respectively.

3.1. Hybrid LCA Model

The proposed hybrid LCA model consists of process-based LCA (P-LCA) and economic input-output LCA (EIO-LCA). P-LCA is a commonly-used method to analyze the life cycle impacts of solid waste management systems. This LCA approach provides a detailed view of the processes and impacts involved in the management of waste. The EIO-LCA model, augmented with sector-level water use, energy consumption and GHG emissions vectors, has been used in this LCA study. In general, EIO analysis tackles the sector-level interdependencies and represents sectoral direct requirements, which are represented by the matrix A [16,17]. This matrix includes the dollar value of inputs required from other sectors to produce one dollar of output. The total output of a sector in this economic model with a final demand of f can be written as [18,19]:

$$x = (I\text{–}A)^{-1} \times f \quad (1)$$

where x is the total output vector, I represents the diagonal identity matrix and f refers to the final demand vector representing the change in the final demand of the desired sector. After the EIO model has been established, the total environmental impacts can be calculated by multiplying the economic output of each industrial sector by the environmental impacts associated with per dollar of output. A vector of environmental outputs can be expressed as [20]:

$$R_i = E_i \times X = E_i \times (I\text{–}A)^{-1} \times f \quad (2)$$

where R_i is the total environmental output vector for the environmental impact category of i and E_i represents a diagonal matrix, which consists primarily of the environmental impacts per dollar of output for each industrial sector. In this research, a hybrid EIO-LCA model is built to consider the environmental impacts associated with different waste management scenarios. This LCA model quantifies the total environmental burdens associated with the waste management system, which is presented in the following equation [20]:

$$K_i = E_i \times (I\text{–}A)^{-1} \times f + Q_i \times e_i \quad (3)$$

where K_i denotes the total environmental impact defined as the summation of environmental burdens associated with the production of resource inputs (by tracing all supply chains) and the direct environmental impacts related to waste treatment processes. Q_i is the total input requirement for a process, and e_i is the unit environmental impact factor associated with the consumption of Q_i. For example, the production of reinforced steel, which is widely used in residential and commercial buildings, has high GHG emissions. During its production process, electricity is consumed as an energy source for steel manufacturing. In our hybrid LCA model, to quantify the direct and indirect GHG emissions considering the whole supply chain of electricity production, we used Equation (2). In addition, Q_i represents the amount of electricity used in the steel production process, and e_i represents the emission factor related to electricity generation. In this way, Equation (3) presents the total carbon emission related to the indirect supply chains of electricity production ($E_i \times (I\text{–}A)^{-1} \times f$) and onsite electricity production processes ($Q_i \times e_i$).

3.2. Multi-Criteria Optimization Model

After quantifying the total environmental impacts of different waste management strategies, the next challenge is selecting the best recycling strategy considering environmental and economic benefits, simultaneously. As mentioned earlier, several environmental impact categories, such as energy consumption, GHG emissions and water withdrawals, are quantified using a hybrid LCA model. In addition, the developed LCA model quantified the direct and indirect economic value added associated with the recycling of the analyzed C&D materials. At this point, a multi-criteria optimization model will be critical for finding a feasible alternative that yields the most preferred amount of recycling for each building waste material. Hence, a compromise programming model was developed. This approach is widely used for optimally solving multi-objective linear, non-linear or integer programming problems [21–24]. The compromise programming model measures the distance based on the L_a metric. The L_a metric defines the distance between two points, such as $Z_k^*(X)$ and $Z_k(X)$. As can been seen from Equation (4), compromise programming uses a distance-based function in order to minimize the difference between ideal and compromise solutions. The formulation of the L_a metric is presented as follows:

$$L_a = \mathrm{Min}(\sum \pi_k{}^a(Z_k^*(X) - Z_k(X))^a)^{\frac{1}{a}} \tag{4}$$

Due to each objective function having different units, normalization is needed before the optimization analysis is performed. The values after normalization will be confined to a given range, such as zero to one. The normalization function Z is presented in Equation (5):

$$Z = \frac{Z_k^*(X) - Z_k(X)}{Z_k^*(X)} \tag{5}$$

After completing the normalization procedure, the distance-based compromise programming formulation can be written as [25]:

$$\mathrm{MinL_a} = \mathrm{Min}(\sum \pi_k{}^a(\frac{Z_k^*(X) - Z_k(X)}{Z_k^*(X)})^a)^{\frac{1}{a}} \tag{6}$$

Subject to:

$$\sum_{k=1}^{p} \pi_k{}^a = 1 \tag{7}$$

$$1 \leqslant a \leqslant \infty \tag{8}$$

In this formulation, Z_k^* represents the ideal solution for objective k. The parameter p represents the total number of objectives, and $\pi_k{}^a$ refers to the corresponding weight associated with each objective. Since we give an equal importance for our economic and environmental objectives, $\pi_k{}^a$ is assumed to be equal for each objective function. In general, three points of the compromise set are calculated for decision analysis, such as a = 1, 2 and ∞. After presenting the theoretical background of the compromise programming, this model has been used for selecting the best recycling strategy. As mentioned earlier, we have the following four primary objectives: maximizing economic value added, maximizing GHG savings, maximizing the net reductions in energy and water consumption. Based on these objectives, the following equations are solved using a multi-objective optimization approach, which is presented as follows:

Notation:

Index:

i: Material index

Parameters:

C_i: economic value added per ton recycled waste for material *i*

GHG_i: GHG emission savings per ton recycled waste for material *i*

W_i: water savings per ton recycled waste for material *i*

E_i: energy savings per ton recycled waste for material *i*

Q_i: total amount of waste generated by the LEED-certified building for material *i*

$LEED_{rf}$: recycling factor

Decision variable:

X_i: optimal amount of recycled waste allocated for material *i*

Objective function:

$$Max\ Z_1(X_i) = \sum_{i=1}^{M} (C_i \times X_i) \quad (9)$$

$$Max\ Z_2(X_i) = \sum_{i=1}^{M} (GHG_i \times X_i) \quad (10)$$

$$Max\ Z_3(X_i) = \sum_{i=1}^{M} (W_i \times X_i) \quad (11)$$

$$Max\ Z_4(X_i) = \sum_{i=1}^{M} (E_i \times X_i) \quad (12)$$

Subject to:

$$\sum_{i=1}^{M} X_i \leqslant LEED_{rf} \times \sum_{i=1}^{M} Q_i \quad (13)$$

$$X_i \leqslant Q_i\ for\ i = 1, 2, \ldots, M \quad (14)$$

$$\forall X_i \geqslant 0 \quad (15)$$

The first objective is to maximize the total economic value added (Equation (10)). The second objective maximizes the GHG emission-based savings (Equation (10)). The water savings are addressed by the third objective, as shown in Equation (11). The energy savings objective function is also represented in Equation (12). The total of the optimal waste for each material *(i)* is less than or equal to the total recycled waste multiplied by the LEED recycling factor ($LEED_{rf}$) (see Equation (13)). Subsequently, the decision variable (X_i) must be less than or equal to the recycled waste (Q_i), as shown in Equation (14). Finally, all decision variables are greater than or equal to zero (see Equation (15)).

Since our goal is to select the best combination of C&D materials for 50% recycling of overall construction waste, the total waste amount multiplied by 0.5, which is known as the $LEED_{rf}$, is used. Due to being one of the most robust optimization software in the applied optimization field, the LINGO© software package is used for solving the multi-objective optimization model [26]. Since we have four objective functions (Z_1, Z_2, Z_3 and Z_4) and three compromise programming functions (a = 1, a = 2 and a = ∞), the LINGO© program has been run to solve each mathematical model. By using the mathematical optimization model, the optimal recycling amount of each construction material has been calculated for the overall 50% recycling goal for single and multiple objectives.

3.3. Data Collection

In this paper, several sources have been used to collect the life cycle inventory data for different C&D waste management alternatives. First, the process data for producing each building material from virgin resources and recycled C&D waste are gathered from Christensen [27]. This detailed data included all electricity, fuel and other resource inputs, as well as atmospheric GHG emissions associated with producing building materials from recycled or virgin materials. Additionally, electricity, fuel consumption and GHG emissions data for the material recovery process were compiled from the LCA study of Denison [28]. The waste reduction model (WARM), which was developed by the U.S. EPA, is utilized for quantifying the emissions related to incineration and landfilling of each C&D waste [29]. The energy production efficiency and electricity generation associated with incineration of cardboard, paper, plastic and wood waste are obtained from the Waste Analysis Software Tool for Environmental Decisions (WASTED) model developed by Diaz and Warith [30].

For transportation of C&D materials from the building construction site to both the material recovery facility and final disposal area, a 50-km transportation distance is assumed for each transfer process. Diesel fuel consumption and emission factors data are also provided by the National Renewable Energy Laboratory (NREL) life cycle inventory database for a diesel-powered single-unit truck [31]. After quantifying the life cycle inventory data for each waste management alternative, the producer prices of each energy and material input are obtained, and the Carnegie Mellon's EIO-LCA software was used for calculating direct plus indirect environmental impacts related to C&D waste management [32].

4. Results

4.1. LCA Results

In this part, environmental impacts analysis results are presented in terms of energy, GHG and water savings.

4.1.1. Energy Savings

The amount of fossil fuel consumption has been quantified in terms of terajoules (TJ) for recycling, landfilling and incineration of per ton building-related C&D waste. Results indicated that recycling of non-ferrous metals and plastics resulted in considerable reductions for the energy consumption among the alternative waste management approaches (see Table 1). This is due to the recycling of these materials reducing the production-related fuel and electricity input requirements. Among C&D materials, recycling of non-ferrous metals and plastics showed the highest potential to reduce the total energy footprint.

Table 1. Energy savings of construction materials (TJ).

Material	Recycling	% Share	Landfilling	% Share	Incineration	% Share
Cardboard	0.271	3.6%	−0.004	5.6%	0.000	27.2%
Non-ferrous	5.720	75.5%	−0.002	2.8%	N/A	N/A
Ferrous Metal	0.156	2.1%	−0.002	2.8%	N/A	N/A
Concrete	0.241	3.2%	−0.024	37.0%	N/A	N/A
Plastic	0.978	12.9%	−0.002	2.7%	0.000	13.2%
Wood	0.093	1.2%	−0.008	12.3%	0.000	59.6%
Glass	0.010	0.1%	0.000	0.2%	N/A	N/A
Drywall	0.084	1.1%	−0.004	6.1%	N/A	N/A
Asphalt	0.027	0.4%	−0.020	30.3%	N/A	N/A

On the other hand, recycling of some C&D wastes, such as wood, drywall and cardboard, did not have a significant impact on minimizing the net energy consumption compared to other C&D materials. Although concrete and asphalt have the highest percentage contributions to the total waste amount, their recycling did not significantly reduce the net energy footprint due to high energy consumption during the recycling process. Additionally, landfilling and incineration did not have significant impacts on the overall energy footprint compared to the recycling of waste materials. Therefore, it is likely to conclude that a high recycling of metals and plastics, which accounts for almost 90% of the total energy savings of the nine construction materials, will be critical for reducing the net energy footprint of the C&D management systems.

4.1.2. GHG Emission Savings

In addition to energy consumption, the current study quantified the amount of GHG emissions and savings related to C&D waste recycling. We utilize the definition of carbon footprint as the total emissions of carbon dioxide or GHGs expressed in terms of CO_2 equivalents related to recycling, landfilling and incineration of per ton C&D waste. Based on the analysis results, recycling of ferrous and non-ferrous metals are found to have significant benefits for reducing total GHG emissions with a total percent share of 79.3% (see Table 2). This is because recycling of these metals reduced the amount of electricity and fuel inputs, which are highly utilized for the production of metal products. Additionally, on-site emissions are decreased when using recycled metals instead of virgin resources. Recycling of other C&D materials, such as paper, glass and cardboard, also contribute to reductions in the net GHG emissions.

Table 2. GHG emission savings of construction materials (Mt CO_2-eqv.).

Material	Recycling	% Share	Landfilling	% Share	Incineration	% Share
Cardboard	9.430	7%	−2.714	6%	-26.718	18%
Non-ferrous	38.142	30%	−1.370	3%	N/A	N/A
Ferrous Metal	61.732	49%	−1.370	3%	N/A	N/A
Concrete	−23.800	-	−17.918	37%	N/A	N/A
Plastic	15.733	12%	−1.318	3%	−32.250	22%
Wood	−7.910	-	−5.955	12%	−90.400	61%
Glass	0.436	0%	−0.108	0%	N/A	N/A
Drywall	0.562	0%	−2.963	6%	N/A	N/A
Asphalt	−8.930	-	−14.663	30%	N/A	N/A

On the contrary, recycling of C&D wastes, such as wood, drywall and asphalt, did not have a significant impact on GHG savings when compared to other C&D materials. Moreover, recycling of asphalt and concrete is not found to be an environmentally-friendly option due to increasing GHG emissions. This is because recycling of these materials requires high fuel consumption during crushing and emitted on-site GHG emissions in this process. When incineration was more closely analyzed, wood, plastic and cardboard resulted in additional GHG emissions. For this reason, combustion of these materials is not found to be an environmentally-friendly waste treatment option due to increased GHG emissions. Therefore, recycling has a positive impact on reduced energy consumption, and a high recycling of metal and plastic materials will have a key importance for decreasing the net carbon footprint of the LEED construction management strategies.

4.1.3. Water Savings

Water footprint analysis results are also presented for recycling, landfilling and incineration of per ton C&D waste. Recycling of non-ferrous metals and asphalt is found to be beneficial due to reductions in overall water consumption (see Table 3). This is because recycling of these materials reduced water consumption for the direct and indirect processes required for the production of these materials.

Table 3. Water savings of construction materials (kgal).

Material	Recycling	% Share	Landfilling	% Share	Incineration	% Share
Cardboard	0.201	0.2%	−1.102	5.6%	−0.048	27.3%
Non-ferrous	5.707	5.2%	−0.556	2.8%	N/A	N/A
Ferrous Metal	0.130	0.1%	−0.556	2.8%	N/A	N/A
Concrete	−0.211	-	−7.276	37.0%	N/A	N/A
Plastic	0.945	0.9%	−0.535	2.7%	−0.023	12.9%
Wood	−0.058	-	−2.418	12.3%	−0.106	59.8%
Glass	0.007	0.0%	−0.044	0.2%	N/A	N/A
Drywall	0.009	0.0%	−1.203	6.1%	N/A	N/A
Asphalt	101.988	93.6%	−5.954	30.3%	N/A	N/A

Among these C&D materials, asphalt showed the highest potential (94% of total share) for reducing the overall water footprint. This is due to recycling of asphalt producing a large amount of natural aggregate, which requires a large amount of water during its mining process. On the other hand, recycling of C&D wastes, such as wood, drywall, concrete and cardboard, did not show a significant contribution to water footprint savings. When compared to recycling, landfilling did not help the environmental sustainability, due to increasing water consumption.

It is important to note that landfilling of concrete and asphalt showed a higher water footprint value compared to other materials. This is because the waste composition of concrete and asphalt wastes was found to be the highest among C&D materials, and disposal of these materials through landfilling required a higher amount of water. In conclusion, a high recycling of metals, glass, plastic and paper should be encouraged by policy makers to diminish the net water footprint of the building-related construction wastes.

4.2. Optimization Results

Figure 3 presents the optimal recycling percentage of construction wastes with respect to each single objective. Mathematical optimization results show that 100% of cardboard, ferrous and non-ferrous metals, plastic and glass wastes should be recycled for maximizing economic and environmental benefits. For GHG gas reductions and water footprint savings, recycling of concrete waste, which accounts for 37% of total waste, is not found to be a feasible option. However, recycling a small portion of concrete makes a positive contribution to net economic savings and energy minimization. In addition, 100% recycling of asphalt waste is found to be a feasible policy when economic and GHG savings are under consideration. For economic savings, 100% recycling of ferrous and non-ferrous metals, asphalt, plastic and glass is suggested, whereas recycling of drywall is not found to be an economically feasible recycling strategy. On the other hand, approximately 100% of produced drywall should be recycled to maximize energy and GHG emissions savings, as well (see Figure 3).

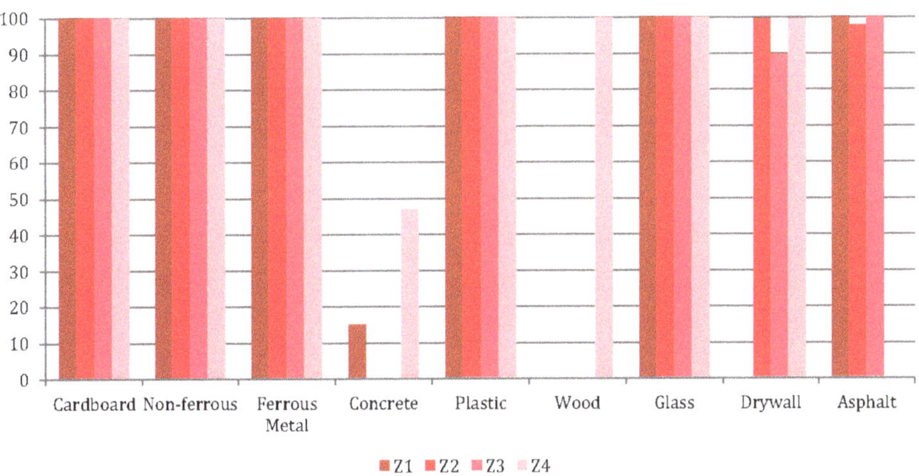

Figure 3. Optimal recycling percentages of C&D wastes for single objectives.

After generating the output of this single objective optimization, a compromise programming model is used to determine the optimal set of recycling amounts when considering the maximization of all objectives, simultaneously. For this model, three points of the compromise set, such as a = 1, 2 and ∞, are calculated for decision analysis, and the percentage recycling rates are presented in Figure 4 for each waste material. Multi-criteria optimization results revealed that recycling of wood and concrete is not found to be a feasible solution when all objectives are aimed to be maximized. However, 100% recycling ferrous and non-ferrous metals, plastic, glass and cardboard will be a sound policy from economic and environmental perspectives. The recycling of a small portion of wood is suggested for a compromise solution in which *a* is infinity. However, this recycling rate is found to be negligible when compared to the recycling rates of other construction materials. According to optimization results, recycling of over 90% of drywall is suggested for the three points in the set. Consequently, by using the results of the LCA model in conjunction with the multi-objective optimization model, the decision makers will have a better understanding of optimum recycling rates of each building waste material (see Figure 4).

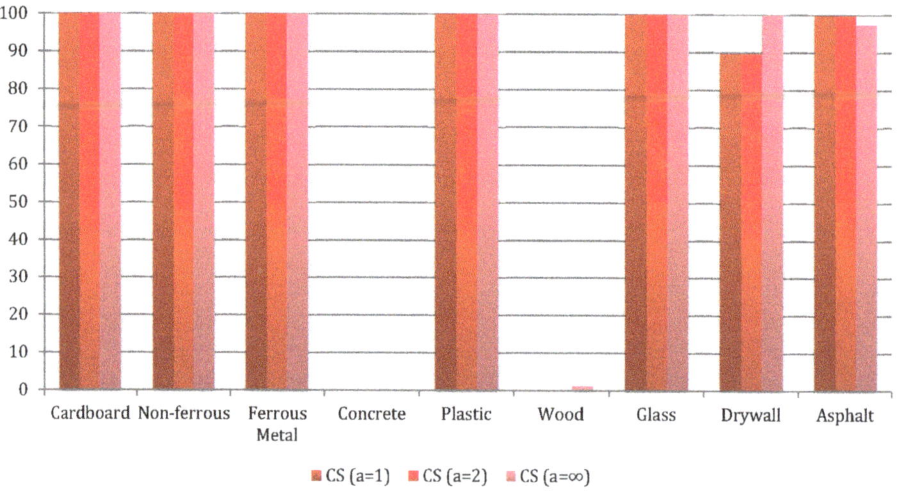

Figure 4. Optimal recycling percentages of C&D wastes for compromise solutions (CS).

5. Conclusions, Limitations and Future Work

The overarching goal of this research was to offer a decision making methodology for recycling of building materials, specifically for LEED-certified green buildings. This study is the first attempt to combine the EIO-based LCA model with multi-objective decisions analysis for construction waste management options. First, the EIO-based hybrid LCA model was used for quantifying the environmental impacts of building materials associated with different waste management scenarios. Second, a multi-criteria optimization model was developed to propose sustainable waste management strategies considering both environmental and economic impacts all together. Our analysis shows that recycling of asphalt and concrete increased the net carbon footprints due to high fuel consumption and emissions during the recycling process. Moreover, recycling of these materials has a minimum impact on the net energy footprint reductions. On the contrary, ferrous and non-ferrous metals are critical for reducing the net carbon footprint of waste management systems. In addition to that, 100% recycling of metals, plastic, glass and cardboard will be an economically- and environmentally-sound policy based on multi-objective optimization results. Even though the case study focuses on a LEED-certified university building, the proposed approach can be robustly integrated for other types of buildings as long as comprehensive C&D waste data are readily available.

It is critical to note that this work presents an integrated decision making framework combining optimization and LCA methods. Our results are based on recycling of 50% of total C&D debris for a selected LEED-certified university building in the United States. Based on assumptions made and the collected data, recycling of concrete is not found to be a feasible option when environmental and economic impacts are considered simultaneously. In other words, recycling of other C&D materials, such as ferrous and nonferrous metals, plastic and glass, is found to be a better strategy, and policy makers should give priority to these materials for recycling. On the other hand, our results still showed that recycling of concrete has a positive contribution to the economy and environment in terms of cost and energy savings. Keeping in mind that among different types of C&D wastes, the composition of concrete waste is found to be higher than 80% of the volume of C&D waste in many countries, such as Australia and Japan, to minimize the concrete waste generated from construction activities, recycling of concrete waste is one of the best methods, and many studies propose that recycling concrete as aggregate for new concrete production can provide a cost-effective method for the construction industry and help save the environment [33,34].

It is certain that the net environmental impacts of C&D waste management are not limited to the findings presented in this study. The application domain of the proposed approach can be extended to other types of buildings, which is a horizontal research extension. Additionally, the vertical depth of the methodology can be further extended by adding different weight scenarios for economic *vs.* environmental impact domains for policy making. Ecological impacts due to toxic releases, hazardous waste generation and ground water pollution can also be considered for analyzing the environmental impacts of different waste treatment options. Additionally, a multi-criteria decision making model is used to select the most appropriate building-related waste materials to maximize environmental and economic savings. However, the social impacts of C&D waste recycling are still critical and can be considered for a more comprehensive sustainability analysis. Therefore, we plan to develop a multi-criteria-based decision making model to consider all economic, social and environmental impacts of waste recycling strategies for the future. In this way, it is possible to have an optimized solution for C&D recycling by considering the triple-bottom-line sustainability impacts. In addition, as a future research direction, further analyzing alternative buildings from different universities worldwide could provide a more comprehensive framework that can be used for sensitivity analysis purposes. Consequently, this research provides an important decision making model, which offers vital guidance for policy makers when developing environmentally- and economically-sound waste management policies.

Author Contributions: Author Contributions: Murat Kucukvar carried out the analyses, including work related to life cycle inventory, data collection and processing. Gokhan Egilmez contributed to the multi-criteria optimization model development and literature review parts. Omer Tatari supervised the research and contributed to the framework of the LCA methodology. All of the authors contributed to preparing and approving the manuscript.

Conflicts of Interest: Conflicts of Interest: The authors declare no conflict of interest.

References

1. USEPA. *Estimating 2003 Building-Related Construction and Demolition Materials Amounts*; USEPA: Washington, DC, USA, 2009.
2. Townsend, T.; Wilson, C.; Beck, B. *The Benefits of Construction and Demolition Materials Recycling in the United States*; University of Florida: Gainesville, FL, USA, 2014.
3. USGBC. *LEED 2009 for New Construction and Major Renovations Rating System*; U.S. Green Building Council: Washington, DC, USA, 2009.
4. Tatari, O.; Kucukvar, M. Cost premium prediction of certified green buildings: A neural network approach. *Build. Environ.* **2010**, *46*, 1081–1085.
5. Lave, L.B.; Hendrickson, C.T.; Conway-Schempf, N.M.; McMichael, F.C. Municipal Solid Waste Recycling Issues. *J. Environ. Eng.* **1999**, *125*, 1–16.
6. Gustavsson, L.; Sathre, R. Variability in energy and carbon dioxide balances of wood and concrete building materials. *Build. Environ.* **2006**, *41*, 940–951.
7. Petersen, A.K.; Solberg, B. Environmental and economic impacts of substitution between wood products and alternative materials: A review of micro-level analyses from Norway and Sweden. *For. Policy Econ.* **2005**, *7*, 249–259.
8. Lasvaux, S.; Habert, G.; Peuportier, B.; Chevalier, J. Comparison of generic and product-specific Life Cycle Assessment databases: Application to construction materials used in building LCA studies. *Int. J. Life Cycle Assess.* **2015**, *20*, 1473–1490.
9. Dixit, M.K.; Culp, C.H.; Fernandez-Solis, J.L. Embodied energy of construction materials: Integrating human and capital energy into an IO-based hybrid model. *Environ. Sci. Technol.* **2015**, *49*, 1936–1945.
10. Pierucci, A. LCA evaluation methodology for multiple life cycles impact assessment of building materials and components. *Tema:Tempo Mater. Arch.* **2015**, *1*, 1–6.
11. Cabeza, L.F.; Rincón, L.; Vilariño, V.; Pérez, G.; Castell, A. Life cycle assessment (LCA) and life cycle energy analysis (LCEA) of buildings and the building sector: A review. *Renew. Sustain. Energy Rev.* **2014**, *29*, 394–416.
12. Dixit, M.K.; Fernández-Solís, J.L.; Lavy, S.; Culp, C.H. Need for an embodied energy measurement protocol for buildings: A review paper. *Renew. Sustain. Energy Rev.* **2012**, *16*, 3730–3743.

13. Van den Heede, P.; de Belie, N. Environmental impact and life cycle assessment (LCA) of traditional and "green" concretes: Literature review and theoretical calculations. *Cem. Concr. Compos.* **2012**, *34*, 431–442.
14. Ortiz, O.; Castells, F.; Sonnemann, G. Sustainability in the construction industry: A review of recent developments based on LCA. *Constr. Build. Mater.* **2009**, *23*, 28–39.
15. University of Central Florida Sustainability and Energy Management Department. LEED GOLD Physical Sciences Building: Material and Resources Credits. 2012. Available online: http://www.sustainable.ucf.edu/?q=node/108 (accessed on 5 March 2015).
16. Hendrickson, C.; Lave, L.; Matthews, H. *Environmental Life Cycle Assessment of Goods and Services: An Input-Output Approach*; Routledge: London, UK, 2006.
17. Suh, S.; Huppes, G. Methods for Life Cycle Inventory of a product. *J. Clean. Prod.* **2005**, *13*, 687–697.
18. Egilmez, G.; Kucukvar, M.; Tatari, O. Sustainability assessment of U.S. manufacturing sectors: An economic input output-based frontier approach. *J. Clean. Prod.* **2013**, *53*, 91–102.
19. Egilmez, G.; Kucukvar, M.; Tatari, O. Supply chain sustainability assessment of the U.S. food manufacturing sectors: A life cycle-based frontier approach. *Resour. Conserv. Recycl.* **2014**, *82*, 8–20.
20. Kucukvar, M.; Egilmez, G.; Tatari, O. Evaluating environmental impacts of alternative construction waste management approaches using supply-chain-linked life-cycle analysis. *Waste Manag. Res.* **2014**, *32*, 500–508.
21. Noori, M.; Kucukvar, M.; Tatari, O. A macro-level decision analysis of wind power as a solution for sustainable energy. *Int. J. Sustain. Energy* **2013**, *34*, 629–644.
22. Tatari, O.; Nazzal, M.; Kucukvar, M. Comparative sustainability assessment of warm-mix asphalts: A thermodynamic based hybrid life cycle analysis. *Resour. Conserv. Recycl.* **2012**, *58*, 18–24.
23. Kucukvar, M.; Noori, M.; Egilmez, G.; Tatari, O. Stochastic decision modeling for sustainable pavement designs. *Int. J. Life Cycle Assess.* **2014**, *19*, 1185–1199.
24. Onat, N.C.; Kucukvar, M.; Tatari, O.; Zheng, Q.P. Combined Application of Multi-Criteria Optimization and Life-Cycle Sustainability Assessment for Optimal Distribution of Alternative Passenger Cars in US. *J. Clean. Prod.* **2015**, *30*, 1–17.
25. Chang, N.-B. *Systems Analysis for Sustainable Engineering*; McGraw-Hill: New York, NY, USA, 2011.
26. Lindo Inc. LINDO Systems—Optimization Software: Integer Programming, Linear Programming, Nonlinear Programming, Stochastic Programming, Global Optimization. 2012. Available online: http://www.lindo.com/ (accessed on 10 May 2012).
27. Christensen, T.H.; Bhander, G.; Lindvall, H.; Larsen, A.W.; Fruergaard, T.; Damgaard, A.; Manfredi, S. Experience with the use of LCA-modelling (EASEWASTE) in waste management. *Waste Manag. Res.* **2007**, *25*, 257–262.
28. Denison, R.A. Environmental life-cycle comparisons of recycling, landfilling, and incineration: a review of recent studies. *Annu. Rev. Energy Environ.* **1996**, *21*, 191–237.

29. USEPA. Waste Reduction Model (WARM). 2010. Available online: http://www.epa.gov/climatechange/wycd/waste/calculators/Warm_Form.html (accessed on 3 March 2015).
30. Diaz, R.; Warith, M. Life-cycle assessment of municipal solid wastes: Development of the WASTED model. *Waste Manag.* **2006**, *26*, 886–901.
31. NREL. *U.S. Life Cycle Inventory Database: Transport, Train, Diesel Powered*; NREL: Washington, DC, USA, 2010.
32. Carnegie Mellon University Green Design Institute. Economic Input-Output Life Cycle Assessment (EIO-LCA), US 2002 Industry Benchmark Model. 2012. Available online: http://www.eiolca.net/ (accessed on 5 March 2015).
33. Tam, V.W. Economic comparison of concrete recycling: A case study approach. *Resour. Conserv. Recycl.* **2008**, *52*, 821–828.
34. Tam, V.W. Comparing the implementation of concrete recycling in the Australian and Japanese construction industries. *J. Clean. Prod.* **2009**, *17*, 688–702.

Evaluation of the Carbon Dioxide Uptake of Slag-Blended Concrete Structures, Considering the Effect of Carbonation

Han-Seung Lee and Xiao-Yong Wang

Abstract: During the production of concrete, cement, water, aggregate, and chemical and mineral admixtures will be used, and a large amount of carbon dioxide will be emitted. Conversely, during the decades of service life of reinforced concrete structures, carbon dioxide in the environment can ingress into concrete and chemically react with carbonatable constitutes of hardened concrete, such as calcium hydroxide and calcium silicate hydrate. This chemical reaction process is known as carbonation. Carbon dioxide will be absorbed into concrete due to carbonation. This article presents a numerical procedure to quantitatively evaluate carbon dioxide emissions and the absorption of ground granulated blast furnace slag (GGBFS) blended concrete structures. Based on building scales and drawings, the total volume and surface area of concrete are calculated. The carbon dioxide emission is calculated using the total volume of concrete and unit carbon dioxide emission of materials. Next, using a slag blended cement hydration model and a carbonation model, the carbonation depth is determined. The absorbed carbon dioxide is evaluated using the carbonation depth of concrete, the surface area of concrete structures, and the amount of carbonatable materials. The calculation results show that for the studied structure with slag blended concrete, for each unit of CO_2 produced, 4.61% of carbon dioxide will be absorbed during its 50 years of service life.

Reprinted from *Sustainability*. Cite as: Lee, H.-S.; Wang, X.-Y. Evaluation of the Carbon Dioxide Uptake of Slag-Blended Concrete Structures, Considering the Effect of Carbonation. *Sustainability* **2016**, *8*, 312.

1. Introduction

Portland cement is the principle hydraulic binder used in modern concrete. The production of one ton of ordinary Portland cement (OPC) generates 0.55 ton of chemical CO_2 and requires an additional 0.39 ton of CO_2 in fuel emissions, accounting for a total of 0.94 ton of CO_2 [1]. The word's yearly cement accounts for nearly 7% percent of global CO_2 emissions [1]. On the other hand, ground granulated blast furnace slag (GGBFS), which is a byproduct of the steel industry, has been increasingly used in the concrete industry as a mineral admixture to partially replace cement. Slag blended concrete has many advantages, such as higher resistance against sulfate and seawater attack, higher late age strength, lower materials cost, and lower CO_2 emissions [1].

Carbon dioxide in the environment can ingress into concrete and chemically react with carbonatable constitutes of hardened concrete, such as calcium hydroxide and calcium silicate hydrate. This chemical reaction process is called carbonation. Carbonation presents both advantages and disadvantages to reinforced concrete structures [2]. Carbonation can reduce the porosity of concrete and improve the compressive strength in the carbonated region of concrete. On the other hand, carbonation decreases the alkaline in concrete and induces the corrosion of steel rebar [1].

Many investigations have been conducted regarding the experimental and theoretical study of the carbonation of slag blended concrete and the life cycle of the carbon dioxide emission of structural concrete.

Regarding the carbonation of slag blended concrete, Sulapha [3], Elke [4], Sisomphon [5], Monkman [6], and Bernal [7] experimentally found that carbonation of slag-blended concrete correlates to water to binder ratios, slag replacement ratios, and concrete curing methods. By increasing the replacement of slag, the carbonation depth of concrete will increase. Higher carbonation depth does not necessarily mean higher absorption of CO_2. It might mean that there is faster diffusion of CO_2 due to there being less CSH to react with. Reducing the water to binder ratio and extending the initial curing period before carbonation tests can reduce the carbonation depth of slag-blended concrete. Papadakis [8,9] proposed chemical reaction equations for cement–mineral admixture blends and evaluated the contents of carbonatable materials and porosity of concrete. Furthermore, carbonation depth was calculated by considering both concrete chemical components and environmental conditions.

Regarding the life cycle assessment of CO_2 emission, Hasanbeigi [10], Gartner [11], Miller [12], Roh [13], Kim [14], and Tae [15,16] analyzed carbon dioxide emissions for buildings with different concrete mixing proportions, different building types, and different life cycle stages. On the other hand, some numerical algorithms have been proposed to make an optimum design regarding the reduction of carbon dioxide emission. Using the evolution algorithm, Kim [17] and Roh [18] selected the optimal concrete mix design method, which minimizes the CO_2 emission of an apartment house. Ji [19] proposed three methods (eco-efficiency, environmental priority strategy system, and certified emissions reduction price) to support the decision-making processes that simultaneously consider cost and CO_2 emissions. Yepes [20] proposed a hybrid glowworm swarm algorithm and optimized the cost and CO_2 emissions of concrete beam roads.

However, the carbon dioxide uptake from carbonation was not considered in References [3–20]. Carbonation is a lengthy chemical reaction process and will proceed continuously during the service life of buildings. Carbon dioxide will be absorbed into concrete due to carbonation. Compared with abundant research regarding carbonation and the carbon dioxide emission life cycle [3–20], the study of carbon dioxide uptake is relatively limited. In recent years, References [21–25] presented some analysis models of the carbon dioxide uptake from carbonation.

Using the carbonation reaction model, Lee [21] analyzed carbon dioxide uptake due to carbonation in the building use stage. García-Segura [22] made life cycle greenhouse gas emissions of blended cement concrete, considering carbonation and durability. The carbon dioxide uptake due to carbonation during the use stage and after demolition stage was considered.

However, in Lee's [21], García-Segura's [22], and other researchers' [3–20] studies, they assumed that cement is completely hydrated (*i.e.*, the hydration degree is 100%) regardless of the water to cement ratio. Lagerblad [23] and Yang [24] reported that concrete with lower water to cement ratios has a slower rate of hydration and a lower ultimate degree of hydration. Yang [24] proposed a Portland cement hydration model, evaluated the contents of carbonatable materials, and calculated carbonation depth and carbon dioxide uptake during the use stage and recycling of demolished concrete. Kashef-Haghighi [25] proposed a mathematical model, evaluated the hydration degree of Portland cement, and calculated carbon dioxide uptake in accelerated carbonation curing.

However, the hydration model proposed by Yang [24] and Kashef-Haghighi [25] is only valid for Portland cement. For slag-blended cement, due to the coexistence of cement hydration and slag reaction, Yang's [24] or Kashef-Haghighi's [25] hydration model is not valid.

In this study, to overcome the weak points in former studies [3–25], we propose a slag-blended cement hydration model, calculate carbonatable materials' content and porosity, and evaluate carbonation depth and carbon dioxide uptake. The flowchart of the proposed numerical procedure is shown in Figure 1. The input parameters of the numerical procedure are the shopping drawing of buildings and environmental conditions. By using shopping drawings, the total volume and surface area of concrete are calculated. By using the total volume of concrete and unit carbon dioxide emission of materials, carbon dioxide emissions are calculated. Then, by using a carbonation model considering material properties and environmental conditions, the carbonation depth is determined. By using the carbonation depth of concrete, the surface area of concrete structures, and the amount of carbonatable materials, the absorbed carbon dioxide is calculated. Finally, the ratio between mass of absorbed CO_2 and mass of emitted CO_2 is calculated.

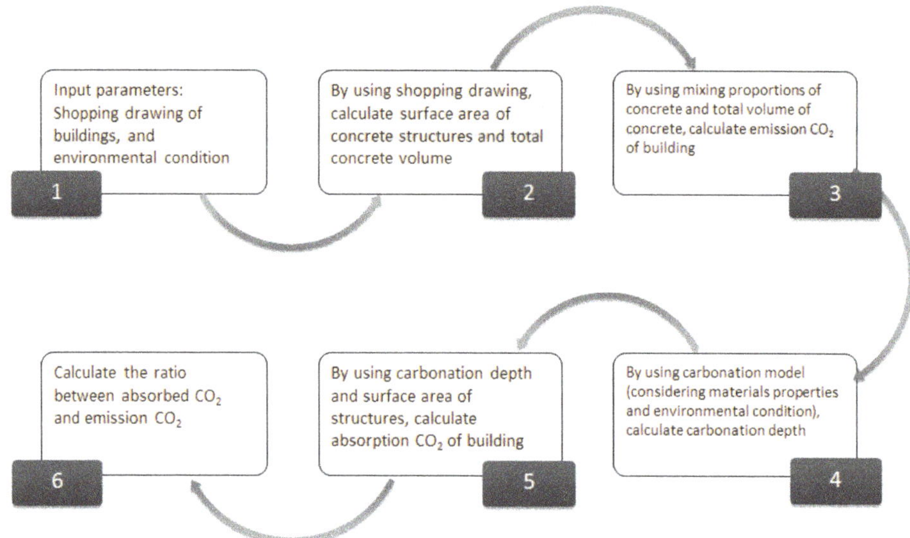

Figure 1. Flowchart of numerical process.

The contributions of this article are summarized as follows: first, propose a slag-blended cement hydration model and calculate reaction degrees of cement and slag. Second, evaluate the carbonatable materials content and porosity using the reaction degrees of binders. Third, calculate carbonation depth and carbon dioxide uptake of slag-blended concrete, considering material properties and environmental conditions.

2. Evaluation of CO_2 Emissions and CO_2 Uptake of Slag-Blended Concrete

2.1. CO_2 Emissions

Emissions of CO_2 from Portland cement production include direct emissions and indirect emissions. As shown in Table 1, direct CO_2 emissions mainly come from carbonate decomposition from raw material and the burning of cement kiln fuel. Indirect CO_2 emissions mainly come from electricity consumption. Considering both direct and indirect CO_2 emissions, to produce 1 ton of ordinary Portland cement, 0.93 ton of CO_2 will be emitted [1].

During the production of concrete, cement, water, aggregate, mineral admixtures, and superplasticizer will be used. Table 2 shows a summary of CO_2 emission factors for concrete production. The CO_2 emission content of 1 kg of GGBFS is 0.034 kg [1], which is much lower than that of ordinary Portland cement. The CO_2 emission from concrete production can be calculated as follows:

$$CO_2 - e = CO_2 - e_M + CO_2 - e_T + CO_2 - e_P, \quad (1)$$

where CO_2-e is the total CO_2 emission from concrete production, CO_2-e_M is the emission of CO_2 from concrete materials, CO_2-e_T is the emission of CO_2 from transport, and CO_2-e_P is the emission of CO_2 from the mixing of concrete. CO_2-e_M is the sum of CO_2 emissions from various components of concrete, such as CO_2 emissions from producing cement, water, fine aggregate, coarse aggregate, mineral admixtures, and superplasticizer. Table 3 shows an example of CO_2 emissions during the concrete production process. The input parameters are concrete mixing proportions, the CO_2 emissions factor, the transportation cycle, and distance. The output result is total CO_2 emissions from the concrete production process.

Table 1. Emission factor of CO_2 in Portland cement production process [1].

Emission Relationship	Emission Style	CO_2 Emission Factor (t/t)
Direct emission	Carbonate decomposition from raw material	0.527
	Kiln dust calcining	0.009
	Organic carbon burning of raw material	0.012
	Burning of cement kiln fuel	0.235
Indirect emission	Cement clinker electricity consumption Cement flour electricity consumption	0.15

Table 2. Summary of CO_2 emissions for concrete production [1].

OPC	GGBFS	Sand	Gravel	Water	Superplasticizer	Truck	Concrete Mixing
9.31×10^{-1} (kg/kg)	3.40×10^{-2} (kg/kg)	0.0037 (kg/kg)	0.0028 (kg/kg)	1.12×10^{-4} (kg/kg)	0.25 (kg/kg)	3×10^{-5} (kg/kg· km)	0.007 (kg/m^3)

Table 3. Example of CO_2 emissions during the concrete production process.

	Material			Transport		
Unit	A	B	C = A × B	D	E	F = A × D × E
Item	kg	CO_2 kg/kg	kg	km	CO_2 kg/kg· km	kg
OPC	300	0.931	279.3	25	3.00×10^{-5}	0.225
Sand	890	0.0037	3.293	15	3.00×10^{-5}	0.401
Gravel	970	0.0028	2.716	15	3.00×10^{-5}	0.436
Water	150	1.12×10^{-4}	0.0168	-	-	-
	sum		285.33		sum	1.062
Concrete mixing	1 m^3	0.71 kg/m^3	0.71			
Total			287.102 = 285.33 + 1.062 + 0.71			

2.2. CO_2 Uptake from Carbonation of Concrete

Carbon dioxide will be absorbed into concrete due to carbonation. Concrete carbonation is a complicated physicochemical process. The process consists of several steps, such as the diffusion of gaseous phase CO_2 from the air environment into concrete pores, CO_2 dissolution in the water film of concrete pores, the dissolution of solid calcium hydroxide (CH) in concrete pore water, the diffusion of dissolved CH in concrete pore water, CH reaction with dissolved CO_2, and the reaction of CO_2 with calcium silicate hydrate (CSH). The chemical reaction of carbonation is shown as follows:

$$Ca(OH)_2 + CO_2 \rightarrow CaCO_3 + H_2O \quad (2)$$

$$(3CaO \cdot 2SiO_2 \cdot 3H_2O) + 3CO_2 \rightarrow 3CaCO_3 \cdot 2SiO_2 \cdot 3H_2O. \quad (3)$$

As shown in Equations (2) and (3), carbonation closely relates to the compound compositions of concrete, such as the amount of carbonatable materials CH and CSH. Moreover, carbonation also relates to the concrete porosity because concrete pores are necessary paths for the diffusion of atmospheric CO_2 into the concrete. Hence, to evaluate concrete carbonation depth, an accurate evaluation of the concrete material properties is necessary, such as CH content, CSH content, and porosity.

Concrete material properties closely relate to mixing proportions and the hydration process. Our former research [26,27] originally proposed a blended hydration model for slag-blended concrete. The hydration model starts with concrete mixing proportions. The age-dependent material properties of concrete, such as the carbonatable materials content and porosity, can be quantitatively calculated. The hydration model has a wide application range for concrete, with different water to binder ratios, different slag replacement levels, and different curing methods [26,27].

In our proposed hydration model, the hydration degree of cement and reaction degree of slag are adopted as fundamental indicators to evaluate properties of hardening slag-blended concrete. The hydration degree of cement (α) is defined as the ratio of the mass of hydrated cement to the mass of cement in the mixing proportion. The value of the hydration degree of cement (α) ranges between 0 and 1. $\alpha = 0$ means cement hydration does not start and $\alpha = 1$ means all the cement has been hydrated. The hydration degree of cement can be determined using an integration method in the time domain ($\alpha = \int_0^t \left(\frac{d\alpha}{dt}\right) dt$, where t is time; $\frac{d\alpha}{dt}$ is the rate of cement hydration. The detailed equation for $\frac{d\alpha}{dt}$ is available in our former research [26,27]). Similarly, the reaction degree of slag (α_{SG}) is defined as the ratio of the mass of reacted slag to the mass of slag in the mixing proportion. The value of the reaction degree of slag (α_{SG}) ranges between 0 and 1. $\alpha_{SG} = 0$ means the slag reaction does not start and $\alpha_{SG} = 1$ means all the slag has reacted. The reaction degree of slag can also be determined using an integration method in the time domain ($\alpha_{SG} = \int_0^t \left(\frac{d\alpha_{SG}}{dt}\right)dt$, where $\frac{d\alpha_{SG}}{dt}$ is the rate of the slag reaction. The detailed equation for $\frac{d\alpha_{SG}}{dt}$ is available in our former research [26,27]).

For slag-blended concrete, calcium hydroxide and calcium silicate hydrate are carbonatable materials. Cement hydration produces calcium hydroxide, while the slag reaction consumes calcium hydroxide. Considering the production and consumption of calcium hydroxide, the amounts of calcium hydroxide in cement–slag can be determined as follows [26,27]:

$$CH(t) = RCH_{CE} \times C_0 \times \alpha - \nu_{SG} \times \alpha_{SG} \times P, \quad (4)$$

where RCH_{CE} denotes the mass of calcium hydroxide produced by 1-unit mass cement hydration; C_0 is the mass of cement; α denotes the degree of cement hydration; v_{SG} denotes the stoichiometric ratio of the mass of CH to slag (v_{SG} = 0.22 [28,29]); α_{SG} denotes the degree of reaction of slag; and P is the mass of mineral mixtures. $RCH_{CE} \times C_0 \times \alpha$ considers the production of calcium hydroxide from cement hydration, while $\nu_{SG} \times \alpha_{SG} \times P$ considers the consumption of calcium hydroxide from the slag reaction.

For slag-blended concrete, the calcium silicate hydrate (CSH) content, which is the most critical parameter to strength development, can be calculated as a function of the binder content, reaction degree of binders, the weight fraction of reactive silica in slag, and the weight fractions of the SiO_2 in the slag and cement. The amount of CSH in hardening slag-blended concrete can be calculated as follows [26,27]:

$$CSH(t) = 2.85(f_{S,C} \times C_0 \times \alpha + \gamma_s \times f_{S,P} \times P \times \alpha_{SG}), \quad (5)$$

where $f_{S,C}$ is the weight fraction of silica in cement and $f_{S,P}$ is the weight fraction of silica in slag; and γ_S is the weight fraction of the reactive SiO_2 in the slag. The coefficient 2.85 is the mass ratio between the molar weight of CSH and the weight of oxide SiO_2 in CSH. $f_{S,C} \times C_0 \times \alpha$ considers the CSH production from cement hydration, while $\gamma_S \times f_{S,P} \times P \times \alpha_{SG}$ considers the CSH production from the slag reaction.

The porosity of hydrating blends is reduced due to the Portland cement hydration, reaction of slag, and carbonation of concrete. The porosity, ε, can be estimated as follows [26,27]:

$$\varepsilon(t) = \frac{W}{\rho_W} - 0.25 \times C_0 \times \alpha - 0.3 \times \alpha_{SG} \times P - \Delta\varepsilon_C, \quad (6)$$

where ε is porosity, W is water content, and ρ_W is the density of water. $0.25 \times C_0 \times \alpha$ considers the porosity reduction from cement hydration, while $0.3 \times \alpha_{SG} \times P$ considers the porosity reduction from the slag reaction; $\Delta\varepsilon_C$ is the reduction of porosity due to the carbonation of concrete.

The calculation results from Equations (4)–(6) can used as input parameters for carbonation depth calculation models. For the usual range of parameters (especially for relative humidity higher than 55%, where CO_2 diffusion controls the carbonation process [8,9]), a carbonation front will take place that distinguishes concrete as one of two different parts: a fully carbonated part and one part in which concrete carbonation has not started at all. The distance between this carbonation front and the outer concrete surface is called the carbonation depth, and for the most common one-dimensional cases, its evolution with time is given by a simple analytical expression, in terms of the composition and the environmental conditions. The evolution of concrete carbonation depth x_c with time t is calculated as follows [8,9,26,27]:

$$x_c = \sqrt{\frac{2D_C[CO_2]_0 t}{0.33\text{CH} + 0.214\text{CSH}}} \tag{7}$$

$$D_C = A\left(\frac{\varepsilon}{\frac{C_0}{\rho_c} + \frac{P}{\rho_{SG}} + \frac{W_0}{\rho_w}}\right)^a \left(1 - \frac{RH}{100}\right)^{2.2}, \tag{8}$$

where $[CO_2]_0$ is the ambient concentration of CO_2 at the concrete surface and A and a are reaction parameters. 0.33CH + 0.214CSH in the denominator of Equation (7) denotes the CO_2 uptake ability of concrete. RH in Equation (8) is the ambient relative humidity (because carbonation generally occurs at the surface region of concrete, Papadakis [8,9] assumed that the relative humidity in the carbonated zone is equal to that in the ambient environment). The effect of relative humidity on the rate of hydration can be considered using a reduction factor $\beta_{RH} = \left[\frac{RH - 0.55}{0.45}\right]^4$ for RH > 0.55 and $\beta_{RH} = 0$ for RH < 0.55 [26,27,30,31]). In Equations (4)–(6), to consider the further hydration of binders during the concrete carbonation period, items α and α_{SG} should be multiplied by β_{RH}. The influence of temperature on carbonation depth can be considered using the activation energy [8,9,26,27,32–36].

The CO_2 uptake due to concrete carbonation can be determined as follows:

$$CO_2 - u = x_c \times S \times (0.33CH + 0.214CSH), \tag{9}$$

where $CO_2 - u$ is the CO_2 uptake due to carbonation and S is the surface area of the building. In Equation (9), the unit of x_c is m, the unit of S is m^2, and the unit of carbonatable materials CH and CSH is kg/m^3; hence, the unit of $CO_2 - u$ is kg (m × m^2 × kg/m^3 = kg).

The CO_2 uptake ratio χ due to carbonation can be calculated as follows:

$$\chi = \frac{CO_2 - u}{CO_2 - e}. \tag{10}$$

The calculation steps for determining the CO_2 uptake ratio are summarized as follows:

(1) Using building scales and shopping drawings, calculate the total volume and surface area of concrete. The carbon dioxide emissions are calculated using the total volume of concrete and unit carbon dioxide emission of materials (Equation (1)).
(2) Using the blended cement hydration model, calculate carbonatable materials content and the porosity of concrete (Equations (4)–(6)).
(3) Using the carbonation model, calculate the carbonation depth of concrete (Equations (7) and (8)).
(4) Using the CO_2 uptake model, calculate the content of CO_2 uptake due to carbonation (Equation (9)).
(5) Based on CO_2 emission content and CO_2 uptake content, calculate the CO_2 uptake ratio due to carbonation (Equation (10)).

3. Experimental Study of Accelerated Carbonation of Slag-Blended Concrete

Accelerated carbonation tests are widely used to evaluate the carbonation durability of concrete. The CO_2 concentration employed in accelerated carbonation tests is much higher than that in the natural environment. Papadakis [8,9] compared the carbonation of concrete under both a natural environment (0.03% CO_2 by volume) and accelerated carbonation tests (7% and 50% CO_2 by volume). Papadakis [8,9] found that Equation (7) is valid for concrete with different CO_2 concentrations. Because the effect of CO_2 concentration on carbonation is considered in Equation (7), the coefficients in Equation (7) do not vary with CO_2 concentration.

To verify the carbonation model, a laboratory experimental study was carried out on accelerated carbonation tests of slag-blended concrete with different water to binder ratios (0.3 and 0.5) and slag replacement levels (30% and 50%). The compound compositions of cement and slag are shown in Table 4. The mixing proportions of concrete are shown in Table 5. The size of prism specimens is 10 cm × 10 cm × 40 cm. After 28 days of sealed curing, the specimens were put into a carbonation chamber, in which the CO_2 concentration was much higher than that of the natural environment. During accelerated carbonation tests, five sides of prism specimens were sealed with epoxy and one side of each specimen was exposed to CO_2. Hence, one-dimensional

CO_2 ingression occurred. The temperature in the carbonation chamber was 20 °C, the relative humidity in the carbonation chamber was 60%, and the CO_2 concentration in the carbonation chamber was 10%. After 1, 4, 8, 13, and 26 weeks of exposure, the carbonation depth was measured via phenolphthalein spraying.

Table 4. Chemical compositions and physical properties of binder materials.

Used Materials		OPC	GGBFS
Chemical compositions (%)	SiO_2	19.29	35.1
	Al_2O_3	5.16	15.02
	Fe_2O_3	2.87	0.53
	CaO	61.68	43.0
	MgO	4.17	5.59
	SO_3	2.53	0.06
	K_2O	0.92	0.28
	Na_2O	0.205	0.24
Physical properties	Fineness (cm^2/g)	3200	4500
	Specific gravity	3.15	2.9
	Loss on ignition	1.49	0.02

Table 5. Mix proportions of concrete.

Specimens	Water to Binder Ratio	Slag Replacement Levels	Unit Weight (kg/m^3)					
			Water	OPC	GGBFS	Sand	Gravel	Superplasticizer (%)
0.5S0	0.5	0	174	344	0	811	941	0.68
0.5S30	0.5	30%	174	241	103	811	941	0.68
0.5S50	0.5	50%	174	172	172	811	941	0.68
0.3S0	0.3	0	154	514	0	752	845	4.11
0.3S30	0.3	30%	154	360	154	752	845	4.11
0.3S50	0.3	50%	154	257	257	752	845	4.11

Figure 2 shows the reaction degree of cement (α) in cement–slag blends. For slag-blended concrete, the addition of slag will improve the water to cement ratio. Consequently, the reaction degree of cement in cement–slag blends is higher than in plain cement specimens [26–31]. The more slag additions, the higher the cement hydration degree. With a lower water to binder ratio (Figure 2b), a higher replacement of slag will increase the hydration degree of the cement.

Figure 3 shows the reaction degree of slag (α_{SG}) in cement–slag blends. When the replacement ratio of slag increases from 30% to 50%, the alkali-activated effect on the slag reaction will be weakened and the reaction degree of slag will decrease. When the water to binder ratio decreases from 0.5 (Figure 3a) to 0.3 (Figure 3b), due to the reduction of the capillary water concentration and available deposit space of reaction products, the reaction degree of the slag will decrease [26–31].

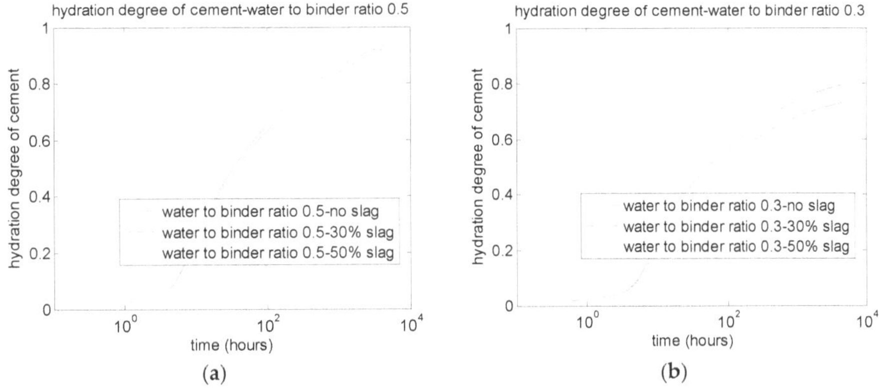

Figure 2. Reaction degree of cement. (**a**) Water to binder ratio of 0.5; (**b**) water to binder ratio of 0.3.

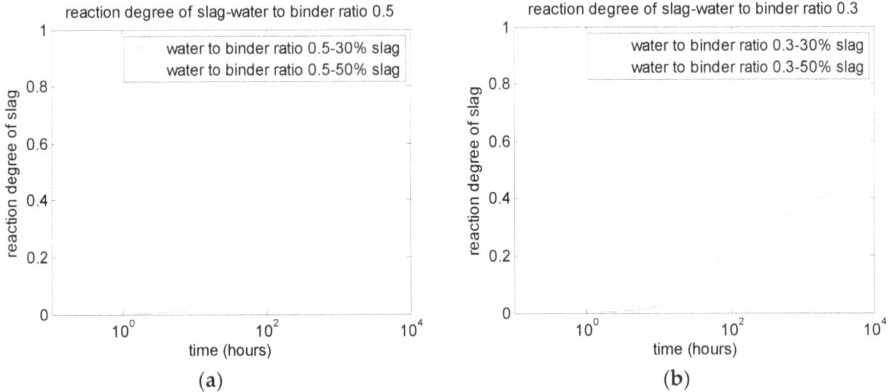

Figure 3. Reaction degree of slag. (**a**) Water to binder ratio of 0.5; (**b**) water to binder ratio of 0.3.

Using the concrete mixing proportions and reaction degrees of cement and slag, the calcium hydroxide (CH) contents can be calculated using Equation (4). As shown in Figure 4, for slag-blended concrete, due to the consumption of CH from the slag reaction, the CH content of slag-blended concrete is much lower than that of the control concrete. When slag replacement levels increase, CH content will decrease. When the water to binder ratio decreases from 0.5 (Figure 4a) to 0.3 (Figure 4b), the CH contents will increase [26–31].

Using the concrete mixing proportions and reaction degrees of cement and slag, calcium silicate hydrate (CSH) contents can be calculated by using Equation (5). As shown in Figure 5, for slag-blended concrete, in early stages, mainly because the reaction degree of slag is lower than that of cement, the CSH contents of slag-blended concrete are lower than those of the control concrete. At a late stage, mainly because the SiO_2 content in the slag is higher than that in OPC, the CSH contents of slag-blended concrete can surpass those of the control concrete. When slag replacement levels increase, the time corresponding to strength surpassing will delay. When the water to binder ratio decreases from 0.5 (Figure 5a) to 0.3 (Figure 5b), the CSH contents will increase [26–31].

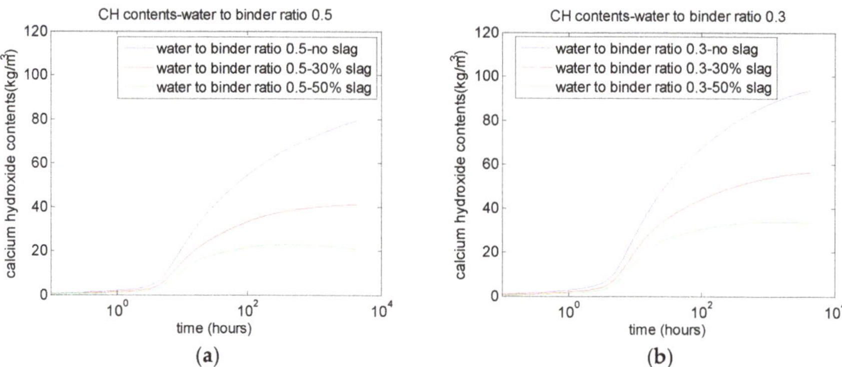

Figure 4. Calcium hydroxide (CH) contents. (**a**) Water to binder ratio of 0.5; (**b**) water to binder ratio of 0.3.

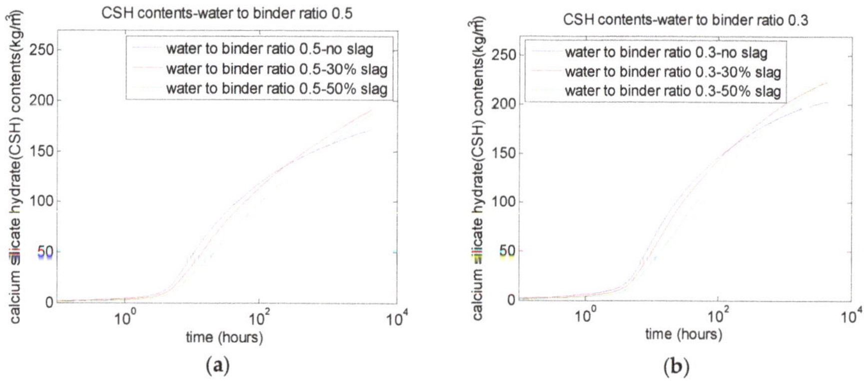

Figure 5. Calcium silicate hydrate (CSH) contents. (**a**) Water to binder ratio of 0.5; (**b**) water to binder ratio of 0.3.

Using the concrete mixing proportions and reaction degrees of OPC and GGBFS, porosity reduction in paste due to the hydration of binders can be calculated using Equation (6). As shown in Figure 6, for OPC–slag blends, because the reaction rate of slag is slower than that of OPC, the porosity of slag blended paste is higher than that of the control paste. When the water to binder ratio decreases from 0.5 (Figure 6a) to 0.3 (Figure 6b), the porosity will decrease [26–31].

After determination of the carbonatable materials content and porosity of concrete, the carbonation depth of concrete can be calculated using Equation (7) ($A = 6.5 \times 10^{-6}$ and $a = 3.6$). Figure 7 shows experimental *vs.* analytical results. The analysis results generally agree with the experimental results. When the slag replacement levels increase, the carbonation depth increases and more CO_2 is absorbed into the concrete. When the water to binder ratio decreases from 0.5 (Figure 7a) to 0.3 (Figure 7b), the carbonation depth of concrete decreases significantly [26–31].

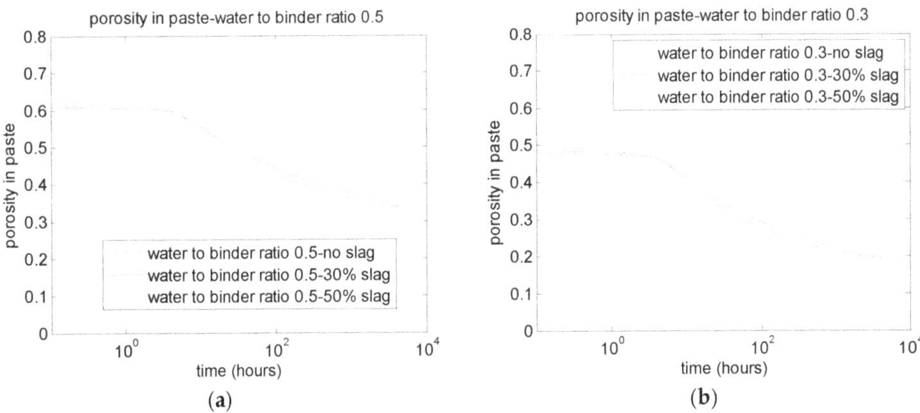

Figure 6. Porosity in paste. (**a**) Water to binder ratio of 0.5; (**b**) water to binder ratio of 0.3.

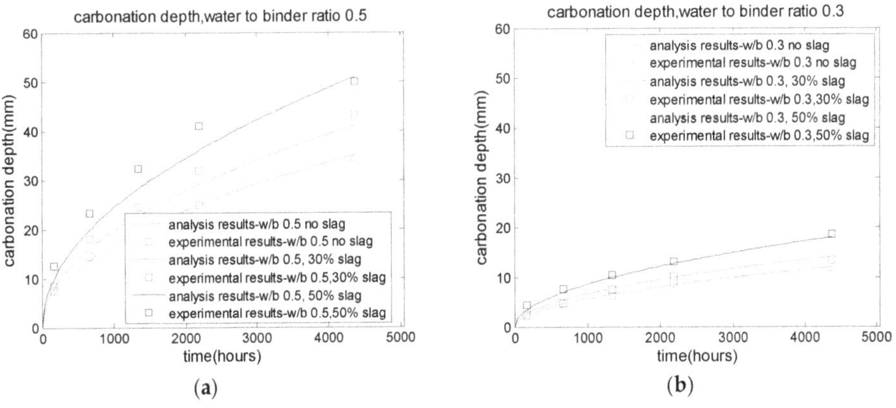

Figure 7. Carbonation depth of concrete. (**a**) Water to binder ratio of 0.5; (**b**) water to binder ratio of 0.3.

4. CO_2 Uptake in Real Buildings

4.1. *Effect of Finishing Materials and Cracks on Carbonation*

For calculating carbonation depth, Equation (7) is valid for sound concrete without cracks or finishing materials. For concrete in real buildings, due to loading and environmental effects, cracks frequently occur. The diffusivity of CO_2 in air is much higher than that in concrete and surface cracks will aggravate carbonation [32]. On the other hand, finishing materials, such as mortar and waterproof coatings, are widely used to improve the durability performance of concrete structures [33–36]. Considering the effects of cracks and finishing materials, the calculation equation for carbonation depth in concrete can be modified as follows [33–36]:

$$x_c' = \beta_{cr} \times \beta_{fm} \times x_c, \tag{11}$$

where x_c' is the carbonation depth considering the effects of cracks and finishing materials and β_{cr} considers the aggravation effect of cracks on carbonation. β_{cr} is higher than 1.0 and relates to crack characteristics such as crack width, crack depth, and crack spacing distance [32]. β_{fm} considers the suppression effect of finishing materials on carbonation. β_{fm} is lower than 1.0 and relates to characteristics of finishing materials, such as material type, depth of finishing materials, and environment influence [33–36]. Yoda [35] conducted field investigations of the carbonation of slag-blended concrete via 40 years of natural aging and the preventive effect of finishing materials. Yoda [35] found that the carbonation suppression coefficients β_{fm} for 20 mm mortar, 5 mm coating, and tile are 0.166, 0.4, and 0.1, respectively. On the other hand, note that Equation (11) is an empirical equation. Equation (11) does not accurately simulate the diffusion and carbonation reaction process. The effect of covering materials on carbonation should be further studied. First, CO_2 diffusivity in covering materials and the chemical reaction between covering materials and CO_2 should be measured. Second, the diffusion–reaction process in cover materials and substrate concrete should be modeled. Third, the continuity conditions on the interface between cover materials and substrate concrete should be established.

4.2. *CO_2 Uptake in Real Buildings during Use Stage*

In this study, a 30-story apartment complex was selected to evaluate the CO_2 absorption of concrete. The 1st floor to 3rd floor are a shopping mall and the 4th floor to 30th floor are for residual usage. The building floor plan (scale = 1/500) is shown in Figure 8 and the building elevation plan is shown in Figure 9. The structure type is a frame-shear wall structure.

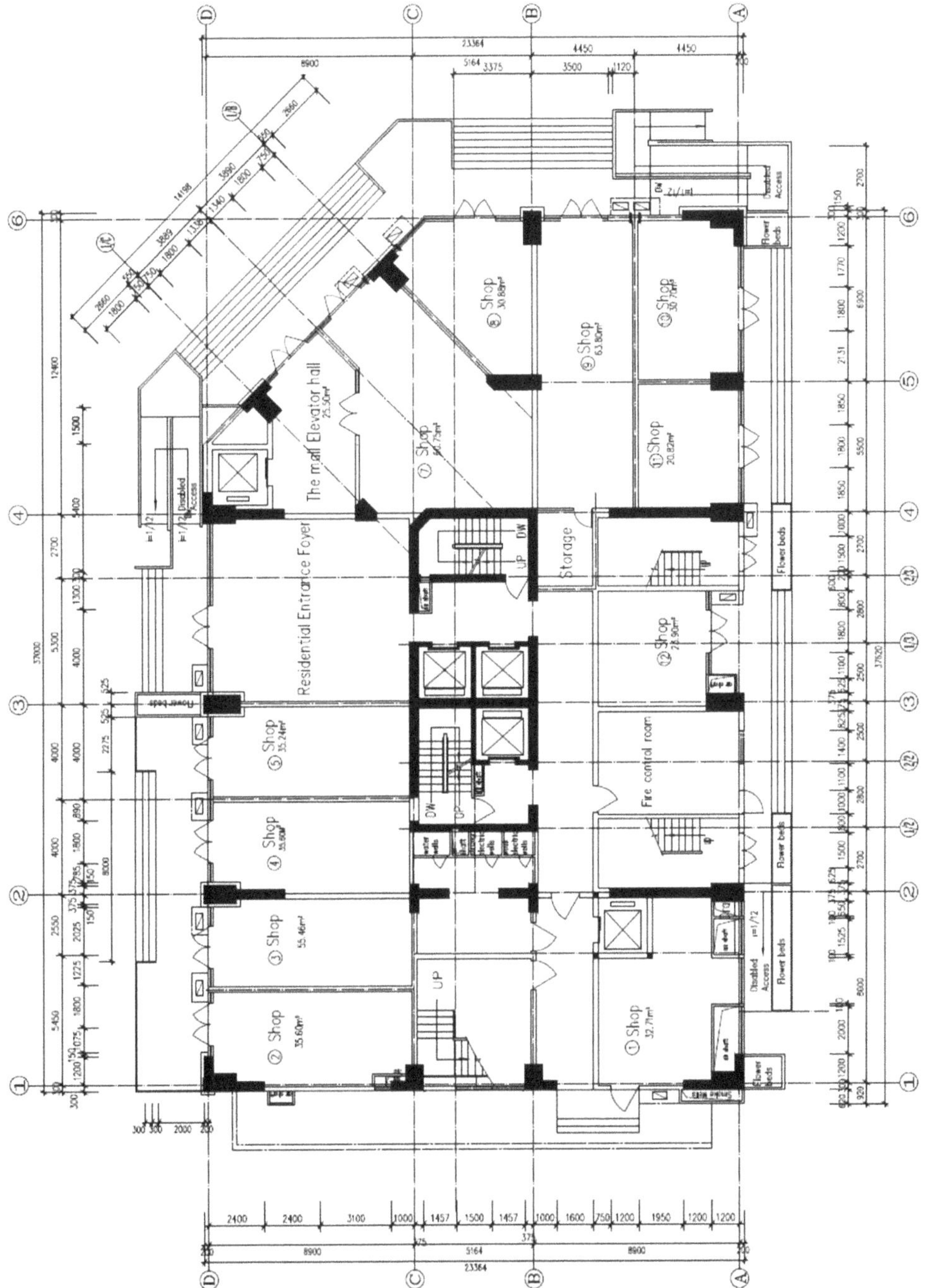

Figure 8. Plan view of first floor of the building.

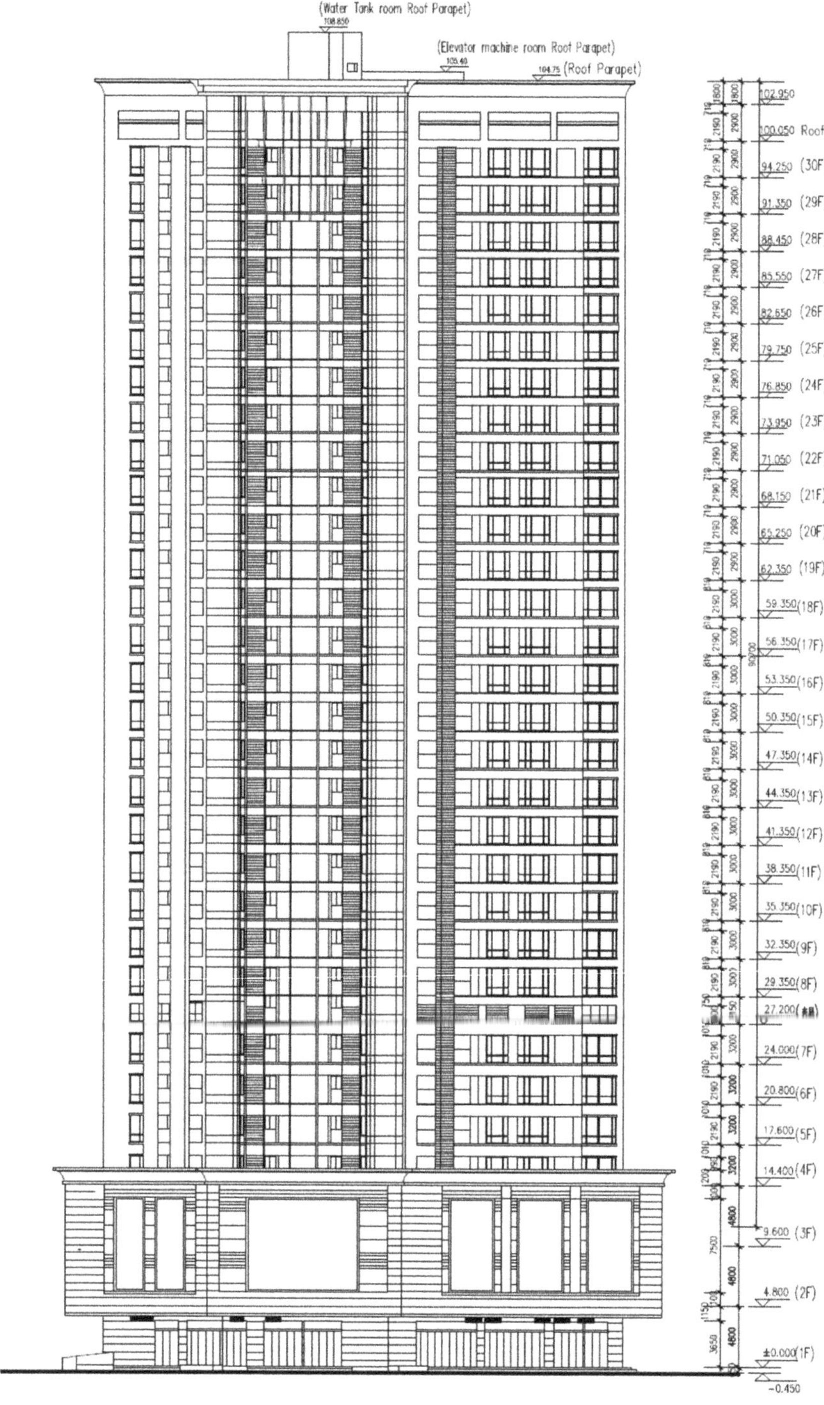

Figure 9. Elevation view of the building.

Environmental conditions: The environmental data were taken from the Korean meteorological administration [37]. The site of the building is in the southeast of Korea. The average CO_2 concentration of the indoor environment is 0.08% and of the outdoor environment is 0.04%. The average relative humidity of the indoor environment is 55% and of the outdoor environment is 65%. The average temperature of the indoor environment is 18.5 °C. For the local region of the building, the average outdoor temperature in spring (from March to May), in summer (from June to August), in autumn (from September to November), and in winter (from December to February) is 15.7 °C, 26.0 °C, 19.6 °C, and 7.1 °C ,respectively. The average outdoor temperature over one year is approximately 17.1 °C.

Material properties: The mixing proportions of concrete are shown in Table 6. The slag replacement level is 40%. From the 1st floor to 10th floor, the compressive strength of concrete is 64.8 MPa, and from the 11th floor to 30th floor, the compressive strength of concrete is 52.3 MPa. The transport distances for OPC, GGBFS, fine aggregate, coarse aggregate, and superplasticizer are 25 km, 25 km, 20 km, 20 km, and 10 km, respectively. Using building drawings, we can calculate the volume of concrete for different floors and different structural members, such as columns, slabs, beams, stairs, and shear walls. As shown in Table 7, the total volume of concrete is 10,769 m^3. Similarly, using building drawings, we can calculate the surface area of concrete for different floors and different structural members. As shown in Table 8, the total surface area of concrete is 79,228 m^2. The slab and shear wall have a flat shape and larger surface area than that of other structural elements. Four-millimeter polymer waterproof coating is used as the finishing material (β_{fm} = 0.48) [33–36]. The average crack depth of a structural member is 0.18 mm (β_{cr} = 1.5) [33–36].

Table 6. Mixing proportions of concrete in building.

Floor	Strength (MPa)	Water (kg/m^3)	OPC (kg/m^3)	GGBFS (kg/m^3)	Fine Aggregate (kg/m^3)	Coarse Aggregate (kg/m^3)	Superplasticizer
1F~10F	64.8	165	248	165	780	874	3.5%
11F~30F	52.3	174	209	140	811	941	1.0%

Table 7. Volume of concrete for different floors.

	Concrete Volume (m^3)					
Floor	Column	Slab	Shear Wall	Beam	Stairs	Total
1F–3F	77.32	160.65	150.88	34.12	6.38	
4F–30F	51.55	160.65	100.59	34.12	4.25	
Total	1623.72	4819.5	3168.48	1023.6	133.98	10,769.28

Table 8. Surface area of concrete for different floors.

	Area of Each Structure (m^2)					
Floor	**Column**	**Slab**	**Shear Wall**	**Beam**	**Stairs**	**Total**
1F–3F	364.50	1475.90	893.03	243.08	59.55	
4F–30F	243.03	1475.90	595.35	243.08	39.70	
Total	7654.5	44,277	18,753.63	7292.4	1250.55	79,228.08

Figure 10 shows the carbonation depth of different floors. The compressive strength of concrete in floors 1–10 (64.8 MPa) is higher than that in floors 11–30 (52.3 MPa). After 50 years of exposure, the calculated carbonation depth for floors 1–10 is approximately 14 mm, while the calculated carbonation depth for floors 11–30 is approximately 22 mm. When the compressive strength of the concrete increases, the carbonation depth decreases.

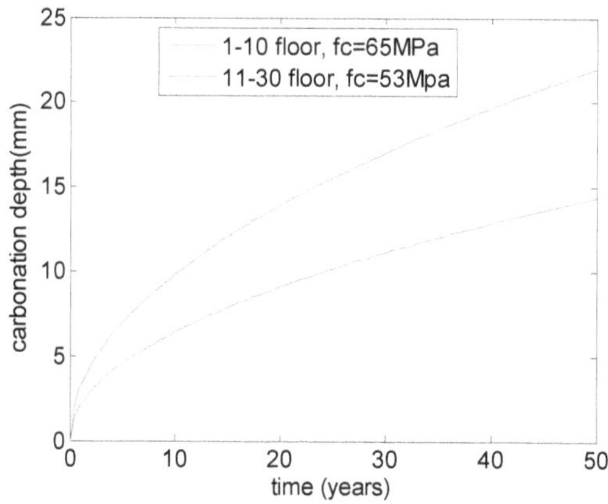

Figure 10. Carbonation depth of different floors.

Figure 11 shows the CO_2 uptake of different floors. After 50 years of exposure, by using Equation (9), the sum of CO_2 uptake for floors 1–10 is approximately 3.17×10^4 kg, while the sum of CO_2 uptake for floors 11–30 is approximately 8.13×10^4 kg. Hence, the CO_2 uptake for the total building is determined to be approximately 11.3×10^4 kg.

Figure 12 shows the CO_2 uptake ratio of all buildings. Using Equation (1), we can calculate the sum of CO_2 emission for floors 1–10 (approximately 9.43×10^5 kg) and for floors 11–30 (approximately 15.07×10^5 kg). The CO_2 emission for the total building is approximately 24.5×10^5 kg. Furthermore, using Equation (10), we can calculate the CO_2 uptake ratio of all buildings. After 50 years, the ratio between the absorbed CO_2 and the emitted CO_2 is approximately

4.61% ($11.3 \times 10^4/24.5 \times 10^5 = 4.61\%$). As shown in Table 1, using slag in the concrete industry can reduce CO_2 emissions. As shown in Figure 7, using slag will increase the carbonation depth of concrete and absorb more CO_2 from the surrounding environment. On the other hand, note that after the carbonation depth of concrete exceeds the protective layer of steel rebar, corrosion of steel rebar will be initiated. More attention with respect to carbonation-induced corrosion should be paid to slag-blended concrete.

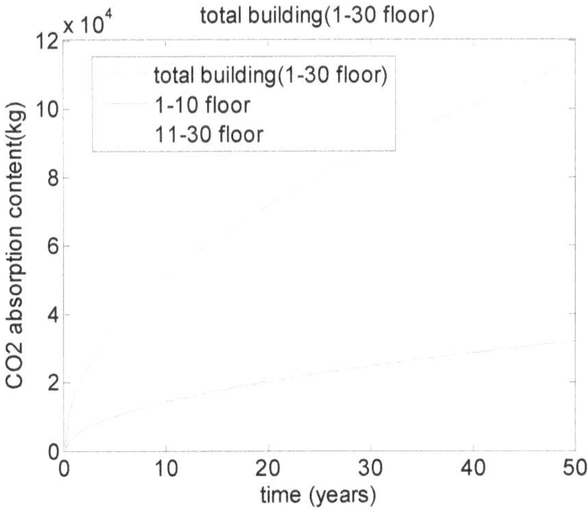

Figure 11. CO_2 uptake of different floors.

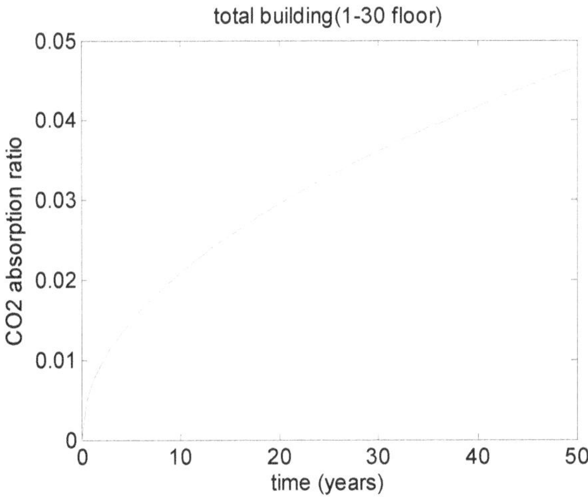

Figure 12. CO_2 uptake ratio of building.

4.3. Discussion about Life Cycle Assessment, Service Life, Carbonation, and CO_2 Uptake after Demolitions

The life cycle of the construction system consists of four stages: production stage, construction stage, use stage, and demolition stage. The production stage involves obtaining raw materials and processing them, transport to the concrete plant, concrete mixing, steel bar production, and transport to the building site. The construction stage is the structure building process. The use stage is the longest-lasting stage. Finally, in the demolition stage, the structure is pulled down and the demolished concrete is crushed and recycled. During the use stage, carbon dioxide will be absorbed due to carbonation. During the demolition stage, after concrete is demolished and crushed, the surface area of concrete will significantly increase, fresh uncarbonated concrete will be exposed to the environmental air, a new cycle of carbonation will begin, and more carbon dioxide will be absorbed [22].

On the other hand, durability, such as carbonation, must be considered in the use stage. The service life of RC structures consists of two distinct phases. The first phase is the initiation of corrosion; CO_2 penetrates the concrete cover and leads to the loss of reinforcement passivity. The second phase is the propagation of corrosion. Once a limiting state is reached, beyond which the consequences of corrosion cannot be tolerated, the service life ends. We can extend the technical lifetime of an RC structure via maintenance or repair, but this will involve high costs. It is important to consider service life in the life cycle assessment of buildings [22]. As shown in Figure 10, after 50 years of exposure, the carbonation depth of concrete (22 mm) is less than the concrete cover depth (30 mm). This building is in the corrosion initiation phase.

In this article, the crushed concrete is assumed to be used as back filler after demolition. The average diameter of crushed concrete in Korea is 25 mm [24], the average CO_2 concentration below the ground is 0.05% [24], and the average relative humidity below the ground is 75% [24]. Using carbonation rate relations, the crushed concrete takes 59.51 years to fully carbonate. Considering carbonation both during the use stage (50 years) and after the demolition stage (59.51 years), the total carbon dioxide uptake ratio is 19.21%. On the other hand, if we only consider carbonation during the use stage (50 years), the carbon dioxide uptake ratio is 4.61%. Hence, a major CO_2 uptake will take place when the concrete structures are demolished.

5. Conclusions

(1) This paper presents a numerical procedure for quantitatively evaluating the amount of carbon dioxide emissions and absorption for slag-blended concrete structures. The analysis presents theoretical innovations: first, using a slag-blended cement hydration model, we calculate the reaction degrees of cement and slag and evaluate concrete material properties, such as carbonatable materials content and

porosity; second, using a micro-structure-based carbonation model, we calculate the carbonation depth of slag-blended concrete; third, using building drawings and carbonation depth, we calculate the carbon dioxide uptake of concrete. The proposed numerical procedure can be applied for evaluating the CO_2 uptake of buildings with different building types, different concrete mixing proportions, and different environment conditions.

(2) Using slag in the concrete industry can reduce CO_2 emissions, increase the carbonation depth of concrete, and uptake more CO_2 from the surrounding environment. A real building case study that considered CO_2 uptake performance was carried out. The calculation results show that for concrete containing 40% slag as binders, 4.61% (113 tons) carbon dioxide will be absorbed during 50 years of service life. CO_2 uptake ability closely relates to the surface area of structural elements. Slabs and shear walls have a flat shape and larger surface area than other structural elements and make significant contributions to CO_2 uptake. On the other hand, a major CO_2 uptake will take place when the concrete structures are demolished and crushed because of the increase in the exposed surface area of uncarbonated concrete to the air.

(3) In the life cycle assessment of concrete buildings, different phases of service life, such as the initiation of corrosion and the propagation of corrosion, should be taken into account. Because slag-blended concrete shows higher carbonation depth than control concrete, more attention with respect to carbonation-induced corrosion should be paid to slag-blended concrete.

(4) The proposed numerical procedure is not perfect and has some limitations: first, the diffusion–reaction process in finishing materials and substrate concrete needs further study. Second, concrete durability includes many aspects, such as freezing and thawing, chloride penetration, carbonation, sulfate attack, and corrosion. The interactions between carbonation and other durability aspects require further study. Third, reinforced concrete structures have many structural styles, such as buildings, bridges, and dams. This paper focuses on the carbon dioxide uptake of RC buildings. For other structural styles, the carbon dioxide uptake amount may differ from that presented in this study.

Acknowledgments: This research was supported by the Basic Science Research Program through the National Research Foundation of Korea (NRF), funded by the Ministry of Science, ICT and Future Planning (No. 2015R1A5A1037548).

Author Contributions: Han-Seung Lee performed the accelerated carbonation experiment; Xiao-Yong Wang analyzed the experimental results; Han-Seung Lee and Xiao-Yong Wang wrote the paper.

Conflicts of Interest: The authors declare no conflict of interest.

References

1. Jung, Y.B.; Yang, K.H. Mixture-Proportioning Model for Low-CO_2 Concrete Considering the Type and Addition Level of Supplementary Cementitious Materials. *J. Korea Concr. Inst.* **2015**, *27*, 427–434.
2. Pacheco-Torgal, F.; Miraldo, S.; Labrincha, J.A.L.; de Brito, J. An overview on concrete carbonation in the context of eco-efficient construction. *Constr. Build. Mater.* **2012**, *36*, 141–150.
3. Sulapha, P.; Wong, S.F.; Wee, T.H.; Swaddiwudhipong, S. Carbonation of concrete containing mineral admixtures. *ASCE J. Mater. Civ. Eng.* **2003**, *15*, 134–143.
4. Elke, G.; van den Philip, H.; de Nele, B. Carbonation of slag concrete: Effect of the cement replacement level and curing on the carbonation coefficient—Effect of carbonation on the pore structure. *Cem. Concr. Compos.* **2013**, *35*, 39–48.
5. Sisomphon, K.; Franke, L. Carbonation rates of concretes containing high volume of pozzolanic materials. *Cem. Concr. Res.* **2007**, *37*, 1647–1653.
6. Monkman, S.; Shao, Y. Carbonation Curing of Slag-Cement Concrete for Binding CO_2 and Improving Performance. *J. Mater. Civ. Eng.* **2010**, *22*, 296–304.
7. Bernal, S.A.; Provis, J.L.; Gutierrez, R.M.; Deventer, J.S.J. Accelerated carbonation testing of alkali-activated slag/metakaolin blended concretes: Effect of exposure conditions. *Mater. Struct.* **2015**, *48*, 653–669.
8. Papadakis, V.G.; Tsimas, S. Effect of supplementary cementing materials on concrete resistance against carbonation and chloride ingress. *Cem. Concr. Res.* **2000**, *30*, 291–299.
9. Demis, S.; Papadakis, V.G. A software-assisted comparative assessment of the effect of cement type on concrete carbonation and chloride ingress. *Comput. Concr.* **2012**, *10*, 391–407.
10. Hasanbeigi, A.; Price, L.; Lin, E. Emerging energy-efficiency and CO_2 emission-reduction technologies for cement and concrete production: A technical review. *Renew. Sustain. Energy Rev.* **2012**, *16*, 6220–6238.
11. Gartner, E.M. Industrially interesting approaches to "low-CO_2" cements. *Cem. Concr. Res.* **2004**, *34*, 1489–1498.
12. Miller, S.A.; Horvath, A.; Monteiro, P.J.M.; Ostertag, C.P. Greenhouse gas emissions from concrete can be reduced by using mix proportions, geometric aspects, and age as design factors. *Environ. Res. Lett.* **2015**, *10*, 114017.
13. Roh, S.; Tae, S.; Shin, S. Development of building materials embodied greenhouse gases assessment criteria and system (BEGAS) in the newly revised Korea Green Building Certification System (G-SEED). *Renew. Sustain. Energy Rev.* **2014**, *35*, 410–421.
14. Kim, R.; Tae, S.; Yang, K.; Kim, T.; Roh, S. Analysis of lifecycle CO_2 reduction performance for long-life apartment house. *Environ. Prog. Sustain. Energy* **2015**, *34*, 555–566.
15. Tae, S.; Baek, C.; Shin, S. Life cycle CO_2 evaluation on reinforced concrete structures with high-strength concrete. *Environ. Impact Assess. Rev.* **2011**, *31*, 253–260.
16. Tae, S.; Shin, S.; Woo, J.; Roh, S. The development of apartment house life cycle CO_2 simple assessment system using standard apartment houses of South Korea. *Renew. Sustain. Energy Rev.* **2011**, *15*, 1454–1467.

17. Kim, T.; Tae, S.; Roh, S. Assessment of the CO_2 emission and cost reduction performance of a low-carbon-emission concrete mix design using an optimal mix design system. *Renew. Sustain. Energy Rev.* **2013**, *25*, 729–741.
18. Roh, S.; Tae, S.; Shin, S.; Woo, J. Development of an optimum design program (SUSB-OPTIMUM) for the life cycle CO_2 assessment of an apartment house in Korea. *Build.Environ.* **2014**, *73*, 40–54.
19. Ji, C.; Hong, T.; Park, H.S. Comparative analysis of decision-making methods for integrating cost and CO_2 emission—Focus on building structural design. *Energy Build.* **2014**, *72*, 186–194.
20. Yepes, V.; Marti, J.V.; Garcia-Segura, T. Cost and CO_2 emission optimization of precast-prestressed concrete U-beam road bridges by a hybrid glowworm swarm algorithm. *Autom. Constr.* **2015**, *49*, 123–134.
21. Lee, S.; Park, W.; Lee, H. Life cycle CO_2 assessment method for concrete using CO_2 balance and suggestion to decrease $LCCO_2$ of concrete in South-Korean apartment. *Energy Build.* **2013**, *58*, 93–102.
22. Garcia Segura, T.; Yepes, V.; Alcala, J. Life cycle greenhouse gas emissions of blended cement concrete including carbonation and durability. *Int. J. Life Cycle Assess.* **2014**, *19*, 3–12.
23. Lagerblad, B. *Carbon Dioxide Uptake during Concrete Life Cycle—State of the Art*; Background Report; Nordic Innovation Center: Palo Alto, CA, USA, 2006.
24. Yang, K.; Tae, S. Carbonation and CO_2 uptake of concrete. *Environ. Impact Assess. Rev.* **2014**, *46*, 43–52.
25. Kashef-Haghighi, S.; Shao, Y.; Ghoshal, S. Mathematical modeling of CO_2 uptake by concrete during accelerated carbonation curing. *Cem. Concr. Res.* **2015**, *67*, 1–10.
26. Wang, X.Y.; Lee, H.S. Modeling the hydration of concrete incorporating fly ash or slag. *Cem. Concr. Res.* **2010**, *40*, 984–996.
27. Lee, H.S.; Wang, X.Y.; Zhang, K.N.; Koh, K.T. Analysis of the Optimum Usage of Slag for the Compressive Strength of Concrete. *Materials* **2015**, *8*, 1213–1229.
28. Maekawa, K.; Ishida, T.; Kishi, T. *Multi-Scale Modeling of Structural Concrete*; Taylor & Francis: London, UK; New York, NY, USA, 2009.
29. Ishida, T.; Luan, Y.; Sagawa, T.; Nawa, T. Modeling of early age behavior of blast furnace concrete based on micro-physical properties. *Cem. Concr. Res.* **2011**, *41*, 1357–1367.
30. Van Breugel, K. Numerical simulation of hydration and microstructural Development in hardening cement-based materials (I) theory. *Cem. Concr. Res.* **1995**, *25*, 319–331.
31. Van Breugel, K. Numerical simulation of hydration and microstructural development in hardening cement-based materials: (II) applications. *Cem. Concr. Res.* **1995**, *25*, 522–530.
32. Kwon, S.J.; Na, U.J. Prediction of Durability for RC Columns with Crack and Joint under Carbonation Based on Probabilistic Approach. *Int. J. Concr. Struct. Mater.* **2011**, *5*, 11–18.
33. Lee, H.S.; Wang, X.Y. Prediction of the Carbonation Depth of Concrete with a Mortar Finish. *Key Eng. Mater.* **2008**, *385–387*, 633–636.

34. Lee, S.H.; Lee, H.S.; Park, K.B. Study on an FEM Analysis to Evaluate Restrain-Performance of Surface-Finishes for Carbonation. *Key Eng. Mater.* **2007**, *348–349*, 477–480.
35. Yoda, A. Carbonation of Portland blast furnace slag cement concrete by 40 year natural aging and preventive effect of finishing materials. *Cem. Sci. Concr. Technol.* **2002**, *56*, 449–454.
36. Lee, W.J. A Study about the Deterioration Behavior of Surface Finishing Materials and the Depression Effect for Carbonation in the Reinforced Concrete Building. Ph.D. Thesis, The University of Tokyo, Tokyo, Japan, 2009.
37. Korea Meteorological Administration. Available online: http://www.kma.go.kr/ (accessed on 28 March 2016).

Sensitivity Analysis on the Impact Factors of the GSHP System Considering Energy Generation and Environmental Impact Using LCA

Taehoon Hong, Jimin Kim, Myeongsoo Chae, Joonho Park, Jaemin Jeong and Minhyun Lee

Abstract: The world is facing a crisis due to energy depletion and environmental pollution. The ground source heat pump (GSHP) system, the most efficient new/renewable energy (NRE) system that can reduce the load of heating/cooling equipment in a building, can be used to address this crisis. Designers and contractors have implemented such systems depending on their experience, although there are many factors that affect the performance of the GSHP system. Therefore, this study aimed to conduct a sensitivity analysis on the impact factors in terms of energy generation and environmental impact. This study was conducted as follows: (i) collecting the impact factors that affect the GSHP system's performance; (ii) establishing the GSHP system's scenarios with the impact factors; (iii) determining the methodology and calculation tool to be used for conducting sensitivity analysis; and (iv) conducting sensitivity analysis on the impact factors of the GSHP system in terms of energy generation and environmental impact using life cycle assessment. The results of this study can be used: (i) to establish the optimal design strategy for different application fields and different seasons; and (ii) to conduct a feasibility study on energy generation and environmental impact at the level of the life cycle.

Reprinted from *Sustainability*. Cite as: Hong, T.; Kim, J.; Chae, M.; Park, J.; Jeong, J.; Lee, M. Sensitivity Analysis on the Impact Factors of the GSHP System Considering Energy Generation and Environmental Impact Using LCA. *Sustainability* **2016**, *8*, 376.

1. Introduction

Today, the world is facing a crisis caused by energy depletion and environmental pollution. The world's energy consumption is expected to be three-fold higher than its present level by 2035, with 60% of the energy to be consumed by developing countries, like China, India and Middle East countries, with the current rapid increase in global energy consumption. Alongside this rapid increase in global energy consumption is a dearth of fossil fuels. The reserves to production ratios are 52.5 years for oil, 54.1 years for natural gas and 110 years for coal, as calculated by British Petroleum in 2015 [1]. This suggests that the world's fossil fuel reserves can be depleted [2,3]. In response to this, the world's major developed countries organized the United Nations Framework Convention on Climate Change based on the recognition of the need for greenhouse gas (GHG) reduction and energy savings. In the Conference of Parties 21 held on 12 December 2015, the "Parties' Agreement" was adopted to replace the existing Kyoto Protocol, which is to expire in 2020. Consequently, various countries around the world established a national carbon emission reduction target (CERT) (*i.e.*, 20% by 2020 compared to the 1990 level in the EU, 34% by 2020 compared to the 1990 level in the U.K., 17% by 2020 compared to the 2005 level in the U.S., and 15% by 2020 compared to the 2005 level in Japan) and have established policies to achieve their respective targets. To keep pace with such a global trend, the South Korean government established a plan to reduce the country's GHG emissions by 37% (850.6 million tons CO_2-eq) by 2030 compared to the current estimates [4–12].

Along with the efforts to overcome the crisis due to the depletion of energy resources and to achieve the goal of reducing the GHG emissions, there has been a surge of interest globally in new/renewable energy (NRE) [13–18]. NRE accounted for 18% of the global electricity generation based on the level in 2009, and the energy-generating facilities constituted 25% (1230 GW of 4800 GW). According to the "NRE Medium-Term Market Report" released by International Energy Agency (IEA), the power generation from NRE is expected to increase by 40% compared to the 2011 level. This is a high growth rate, exceeding the growth rate from 2006 to 2011, and two-thirds of the new power plants are expected to be owned and operated by non-OECD countries. As the NRE-related technologies are entering a virtuous cycle of cost reduction owing to global competition, it is now possible to transfer these technologies to developing countries [19–23].

Meanwhile, the U.S. and EU have put more emphasis on NRE-related power stations than on fossil energy-related power stations since 2008. According to a report released by the United Nations Environment Program, NRE has already accounted for more than 50% of the power generation in the U.S. and Europe [24–28]. In other words, IEA forecasted that the NRE production will be double the present amount in 2035. That is, the proportion of electricity production through NRE is expected to increase by 33%, similar to the power generation from fossil fuels, by 2035 [29,30].

The South Korean government has proceeded with a number of projects for the revitalization and institutional stabilization of NRE dissemination in South Korea in line with the global trend. The basic direction of the projects is to expand the NRE market and to thus induce private investment. It also established the New and Renewable Energy Centre under the control of the Korea Energy Management Corporation (KEMCO) and implemented a system of investment incentives, such as the provision of financial support for NTE businesses, the "1 Million Green Homes Project" and feed-in-tariff. In addition, it promoted a mandatory system of public institutions and a renewable portfolio agreement to enhance the leading role of the public sector in NRE promotion. Furthermore, it built a foundation for the support of research and industrialization, such as the establishment of Core Technology Development Center to improve technologies and enhance product qualities and designated specialized colleges by area to foster skilled manpower systematically [31–36].

The ground source heat pump (GSHP) system can reduce the load of heating/cooling equipment in a building most efficiently and is regarded as a system that ensures reasonable operation and maintenance costs of the NRE systems [37,38]. Despite these advantages, however, the GSHP system also has a disadvantage: it involves excessive initial investment costs. Therefore, it is very important to accurately predict the energy load reduction of the target facility through the introduction of the GSHP system prior to the introduction of the said system and, thus, to evaluate the return on investment.

However, the reduction of the load of heating/cooling equipment effects varies depending on: (i) the regional factors (*i.e.*, ground heat capacity, ground temperature, ground thermal conductivity); (ii) the system factor; and (iii) the design factors (*i.e.*, borehole length, number of boreholes) in the case of load reduction by the GSHP system. There is no accurate analysis of the design variables and prediction performance, and in most cases, the analysis depends on the experience of the GSHP system design and construction companies in the existing GSHP system introduction process. It is considered that this trend exists until today, as the introduction rate of the system is relatively low, but as the rate of mandatory NRE introduction increases, an accurate analysis of the alternatives by key impact factor will be more important.

In the previous relevant studies, the factors affecting the GSHP system performance were analyzed in terms of: (i) energy generation; (ii) economic effect; and (iii) environmental impact [39–46]. Fujii *et al.* (2012) conducted a study for the optimization model of the slinky-coil horizontal ground heat exchanger (GHE) to be applied in the U.S. and Canada. They measured the average supply temperature by fraction to be applied to the burial depth of the GHE and the GSHP system and conducted a sensitivity analysis in terms of energy generation by measuring the seasonal heat exchange rate according to the soil type and supply direction of the fluid within the GHE [39]. Casasso *et al.* (2014) selected the impact factors that affect the GHE performance and conducted sensitivity analysis. The GHE length selection was found to be the most important factor in terms of the economic effect, and the U-pipe spacing and grout materials turned out to have an effect on the entire system's performance [40]. Kim *et al.* (2015) performed an economic and environmental assessment for the optimization design of GHE. They conducted an analysis in terms of life cycle cost and LCA by creating a total of five scenarios according to the entering water temperature (EWT). In terms of the environmental impact, the best result was obtained at 25 °C, and in terms of the economic effect, it was achieved at 30 °C [41]. Cosentino *et al.* (2015) analyzed the variables of the thermal energy storage system of boreholes and presented the optimal heat charging system considering a variety of design conditions. In the study, the analysis period was assumed to be 10 and 20 years, and in the case of 10 years, the case in which the borehole length was set to 125 m and the interval to 7 m showed a 76% better result in terms of energy efficiency than the case where the interval was set to 3 m. In the case of 20 years, when the borehole length was set to 125 m and the interval to 7 m, the energy efficiency was further enhanced by 81.4% compared to the case where the interval was set to 3 m [42]. Hepbasli (2002) evaluated the operating performance of the heating/cooling system on the target facility, in which the GSHP system was installed, to analyze the economic effect of the actual system. Compared to the conventional heating/cooling system, the GSHP system was disadvantageous in terms of the initial investment cost, but it was found to be more economical than the conventional system in terms of energy consumption [43]. Boyaghchi *et al.* (2015) constructed a new combined cooling heating and power (CCHP) system and analyzed its heating/cooling performance using the GSHP and solar photovoltaic systems through optimization. It was found through the analysis that the performance of the system varies greatly depending on the type of refrigerant used inside the CCHP system [44]. Essen and Inalli (2009) predicted the energy performance through artificial neural networks by utilizing the experiment data on the heating/cooling performance of the GSHP system [45]. Alavy *et al.* (2013) analyzed the heating/cooling energy savings according to the applicable percentage of the GSHP system with ten target facilities. The analysis of the initial investment cost, payback period and operating costs revealed that the GSHP system is highly economical, as it assumes more than 80% of the entire load of heating/cooling equipment on average [46].

As mentioned earlier, various studies have been conducted on several variables that affect the GSHP system performance, but no research has been done that comprehensively considers energy generation and environmental impact in the sensitivity analysis. In addition, the study on the environmental impact analysis was limited to evaluating the CO_2 emission reduction due to the change in performance that occurs when the GSHP system is installed. Sensitivity analysis is the study of how the uncertainty in the output of a system can be apportioned to different sources of uncertainty in its inputs [47–49]. Therefore, this study aimed to conduct sensitivity analysis on the impact factors of the GSHP system in terms of energy generation and environmental impact. This study was conducted as follows: (i) collecting the impact factors affecting the GSHP system's performance; (ii) establishing the GSHP system's scenarios with the impact factors; (iii) determining the methodology and calculation tool to be used for conducting sensitivity analysis; and (iv) conducting sensitivity analysis on the impact factors in terms of energy generation and environmental impact using LCA (refer to Figure 1). The results of this study can be used in future research (*i.e.*, development of an analysis model for the GSHP system) as the impact factors to be intensively considered for the efficient design and analysis of the GSHP system.

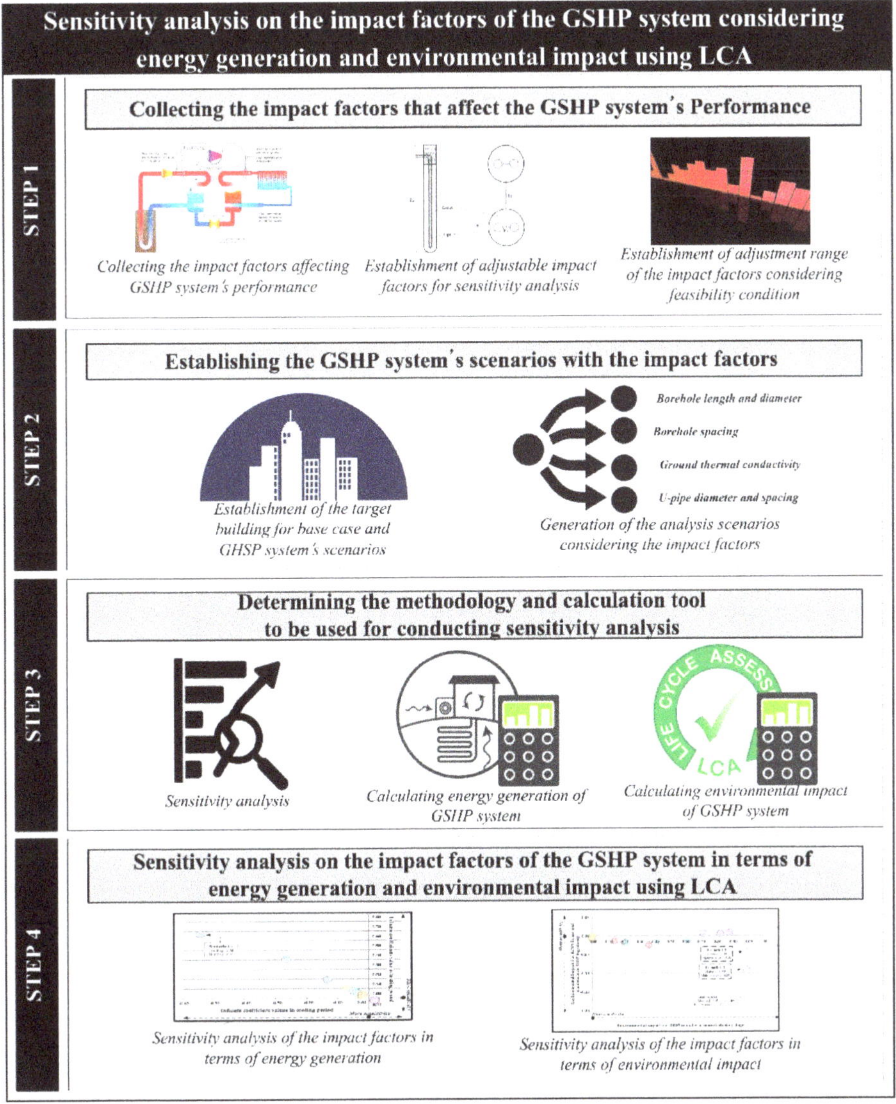

Figure 1. Framework of the sensitivity analysis on the impact factors of the GSHP system considering energy generation and environmental impact.

2. Materials and Methods

2.1. Collecting the Impact Factors Affecting the GSHP System's Performance

The GSHP system is a highly efficient NRE system that uses the ground heat energy through the GHE connected to the ground to reduce the load of heating/cooling equipment in a building and provides heat through a heat pump. In this study, the adjustable impact factors to be considered in the design of the GSHP system were divided as follows: (i) regional factors; (ii) system factors; and (iii) design factors (refer to Table 1) [41]. In addition to the above factors, operation factors and other factors (e.g., control type, schedule of operation and setting of EWT) exist, and those could critically affect the energy generation and environmental impact of the GSHP system [50]. In this study, however, the impact factors affecting the GSHP system independently are selected. Of the aforementioned factors, control type and schedule of operation are influenced by the combination of the other factors and an operating environment according to the type of buildings. Because those are impact factors of a different level, other methods are required to analyze them. Therefore, the study of aforementioned factors should be conducted in future research.

Table 1. Overview of the impact factors.

Classification	Impact Factor (Unit)
Regional factors	Ground temperature (°C), soil type, ground thermal conductivity (W/m·K), ground heat capacity (kJ/K·m^3) [51–54]
System factors	Capacity (kW), power input (kW), heat of rejection (kW), heat of extraction (kW), coefficient of performance, energy efficiency rating [55–57]
Design factors	Borehole length (m), borehole spacing (m), borehole diameter (mm), U-pipe position (mm), number of boreholes, arrangement, grout conductivity (W/m·K), borehole thermal resistance (K/(W/m)), U pipe type, U pipe diameter, fluid type, flow rate (L/s), EWT (°C) [58–67]

First, the regional factors were classified as non-adjustable impact factors because they are external factors that affect the performance of the GSHP system in a given environment (*i.e.*, ground temperature, soil type, ground thermal conductivity and ground heat capacity). Second, the system factors were classified as non-adjustable impact factors because the capacity is determined by the heading/cooling load of a building, and the manufacturer's technical capability determines the cost and coefficient of performance. Third, the design factors were classified as impact factors that can adjust the detailed parameters in the design. Among the design factors, however, borehole thermal resistance was classified as a non-adjustable factor, as it appears in the form of a combination of the borehole diameter, borehole thermal conductivity, U-pipe diameter and U-pipe position. The EWT was also classified as a non-adjustable factor, as it is a result value obtained by the combination of all of the design factors. Meanwhile, the fluid type was excluded as it has an insignificant effect on the energy generation and environmental impact of GHE [41]. The borehole arrangement was also excluded because it is limited to the installation site area and site type, and the flow rate was excluded, as it is determined by the total installation capacity of the GSHP system. Accordingly, the adjustable maximum/minimum range of the detailed parameters was calculated with the rest of the impact factors, except for the above-mentioned factors (refer to Table 2).

Table 2. Adjustable range of impact factors.

Impact Factors	Unit	Range			
Borehole Length	m	50	100	150	200
Grout Thermal Conductivity	W/m·K	0.875	0.99	1.2116	1.6
Borehole Spacing	m	4	5	6	7
Borehole Diameter	mm	125	150	175	200
U-pipe Diameter	mm	25	32	40	50
U-pipe Position	-	A	AS	B	C

- Category 1 (borehole length): The borehole length represents the installed capacity of GHE. In this study, the installation range was set to 50 to 200 m. The GSHP system uses geothermal energy and shows a tendency for an up to a 50 m change in the temperature of the ground surface due to the solar radiation to occur, but an increase in temperature due to the ground heat occurs linearly from less than 50 m. The maximum design length of the GHE of the vertical closed loop is limited to 200 m, and the GHE length design programs used globally, such as "Ground loop design (GLD)" and "professional ground loop heat exchanger design (GLHEPro)", show that the design of less than 200 m represents a reliable result [68].
- Category 2 (grout thermal conductivity): This represents the degree of heat transfer of GHE. As the higher the thermal conductivity is, the lower the borehole resistance, the efficiency of GHE increases. The degree of grout thermal conductivity varies depending on the mixing ratio of bentonite, silica sand and water. Table A1 shows the grout thermal conductivity according to the silica sand ratio at 20% bentonite.
- Category 3 (borehole spacing): As the heat capacity varies depending on the soil type, the optimal distance where the performance of GHE does not decrease due to the crossing of the range of the ground heat that each borehole releases and absorbs is required. In the existing study results and construction field,

it was designed to be more than 4 m, and if it is less than 4 m, the severe heat interference that occurs between the boreholes (the ground temperature rises due to the heat generated from GHE in the ground) will have an adverse effect on the performance of the GSHP system. In addition, it is impossible to perform construction in a very wide range due to the restrictions on the area. Therefore, in this study, the maximum borehole spacing was limited to 7 m after an interview with experts and based on the existing records on the installation [69].

- Category 4 (borehole diameter): It is designed considering the volume of the grout that fills the borehole and U-pipe through which the fluid flows. As the higher the thickness is, the higher the borehole thermal resistance and the lower the performance and, if it is too thick, the borehole or internal components that protect the grout can be damaged, it should be designed to have an appropriate thickness. The borehole diameter can be adjusted according to the technical skills applied in the construction, and as the wider the borehole diameter is, the higher the material costs, the higher the resistance of the borehole, the lower the thermal conductivity and, therefore, the lower the efficiency. The minimum borehole diameter size is normally determined by the diameter of the U-pipe, and the maximum borehole diameter is calculated to ensure that the U-pipe is safely embedded and that proper heat exchange is done. In this study, as the maximum size of the U-pipe diameter is 50 mm, the minimum size of the borehole diameter was calculated to be 125 mm in order to exceed the sum of the two sides of the U-tube diameter. The maximum range was limited to 200 mm based on the standards of the GHSP association in the U.K., the International Ground Source Heat Pump Association (IGSHPA) [70,71].
- Category 5 (U-pipe diameter): There is a need for a combination that satisfies the heat load considering the low velocity of the fluid that flows inside the pipe and the subsequent heat transfer. Strength and durability exceeding a certain level are required for the part with which the fluid has direct contact. The U-pipe diameter is determined based on the standard predetermined at the time of construction. The currently-used U-pipe diameters are 25 mm, 32 mm, 40 mm and 50 mm, and 32 mm is mainly used.
- Category 6 (U-pipe position): This refers to the interval at which the U-pipe is installed, and the proper interval is needed to achieve sufficient heat exchange. Figure 2 shows the U-pipe position that can be introduced, and it is normally designed as the B type.

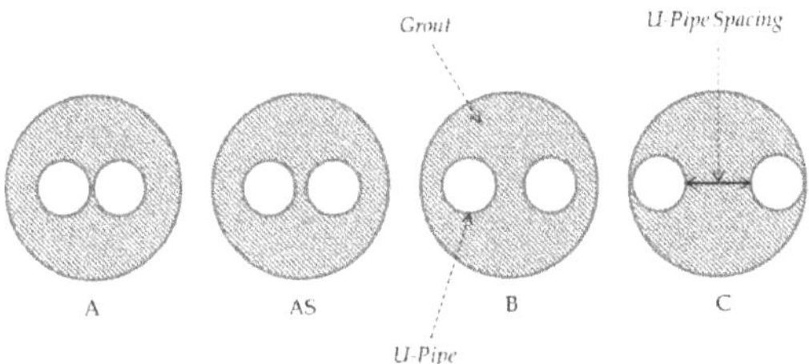

Figure 2. U-pipe installation position.

2.2. *Establishing the GSHP System's Scenario Considering the Range of Impact Factors*

2.2.1. Establishment of the Target Building for the Base Case and the Analysis Scenarios

In this study, the target building was selected to fix the regional and system factors, in addition to the adjustable impact factors. The target building was selected based on the following criteria:

- Among the buildings that require the introduction of the NRE system, a building that facilitates the construction of the GSHP system was selected. According to the 2013 statistics of the Korea Energy Agency, energy-intensive buildings, which use energy of more than 2000 TOE (tons of oil equivalent) per building annually, consumed a total of 2,307,000 TOE and emitted a total of 10,083,000 t CO_2 in 2012. Among various types of buildings, universities consumed a total of 336,000 TOE and emitted a total of 1,397,000 t CO_2, which accounts for about 15% of the energy consumed in the entire building sector [72]. In addition, in most cases, the newly-constructed buildings within a university have adopted the NRE system, and especially, the GSHP system has actively been introduced [73].
- In this study, the GSHP system was analyzed, with a focus on the vertical closed-loop type among the several existing GSHP system types. This is because the vertical closed-loop type can yield more accurate simulation results compared to the other types, and the vertical closed-loop type accounts for more than 60% of all of the types used, especially in South Korea [74–77]. Therefore, a building in which the vertical-closed-loop-type GSHP system is installed was selected as the target building in this study.
- For sensitivity analysis, a building where the daily use of the GSHP system is nearly constant, *i.e.*, a residential building in which the daily electricity consumption does not change significantly, was selected as the target building [78].

Finally, a dormitory building in a university in Seoul where the GSHP system had been installed was selected for the sensitivity analysis. The regional factors, like the underground environment and the building heating/cooling load, were defined, and the specifications of the heat pump system were also defined. Meanwhile, the specifications of the existing GSHP system were set as a base case for the sensitivity analysis with respect to the target building in which the GSHP system had been installed. The lot area of the target building is 26,298 m^2, and the gross area is 6612 m^2. The heating and cooling loads are 283.1 and 349.2 kW, respectively, and the installation capacities of the existing GSHP system are 352.206 kW (heating) and 364.42 kW (cooling), respectively (refer to Tables 3 and 4). The GSHP system's scenarios were established with GSHP system's base case and the range of impact factors (refer to Table 5).

Table 3. Overview of the target building.

Category	University Facility
Year established	2014
Location	Seoul
Building type	Residential facility
Electricity system	On-grid
Heating system	Individual heating
Progressive tax	No
Gross floor area	6612 m^2
Major energy service	GSHP system
Installation of capacity	Heating: 352.2 kW/cooling: 364.4 kW
Borehole	Length: 150 m/ Number of borehole: 40

Table 4. Overview of the GSHP system's base case.

Classification	Borehole Length (m)	Grout Thermal Conductivity (W/m· K)	Borehole Spacing (m)	Borehole Diameter (mm)	U-Pipe Diameter (mm)	U-Pipe Spacing (mm)
Base case	150	0.99	5	150	32	B

2.2.2. Validation of the Designed Model

This was analyzed via CV(RMSE) to verify the validity of the model [79]. The CV(RMES) was calculated by Equation (1):

$$CV(RMSE) = \frac{\sqrt{\sum_{i=1}^{n} (MEC_i - SEC_i)^2 \times \frac{1}{n}}}{\sum_{i=1}^{n} MEC_i \times \frac{1}{n}} \times 100 \quad (1)$$

where *MEC* is the measured energy consumption (kWh); *SEC* is the simulation-based energy consumption (kWh) during 40 years; and n is the number of compared data (months).

Table 5. The GSHP system's scenarios with the impact factors.

Classification		Borehole Length (m)	Grout Thermal Conductivity (W/m·K)	Borehole Spacing (m)	Borehole Diameter (mm)	U-Pipe Diameter (mm)	U-Pipe Spacing (mm)
Category 1 Borehole Length (m)	Scenario 1-1	50	0.99	5	150	32	B = 28.67
	Scenario 1-2	100	0.99	5	150	32	B = 28.67
	Scenario 1-3	150	0.99	5	150	32	B = 28.67
	Scenario 1-4	200	0.99	5	150	32	B = 28.67
Category 2 Grout Thermal Conductivity (W/m·K)	Scenario 2-1	150	0.875	5	150	32	B = 28.67
	Scenario 2-2	150	0.99	5	150	32	B = 28.67
	Scenario 2-3	150	1.2116	5	150	32	B = 28.67
	Scenario 2-4	150	1.6	5	150	32	B = 28.67
Category 3 Borehole spacing (m))	Scenario 3-1	150	0.99	4	150	32	B = 28.67
	Scenario 3-2	150	0.99	5	150	32	B = 28.67
	Scenario 3-3	150	0.99	6	150	32	B = 28.67
	Scenario 3-4	150	0.99	7	150	32	B = 28.67
Category 4 Borehole diameter (mm)	Scenario 4-1	150	0.99	5	125	32	B = 28.67
	Scenario 4-2	150	0.99	5	150	32	B = 28.67
	Scenario 4-3	150	0.99	5	175	32	B = 28.67
	Scenario 4-4	150	0.99	5	200	32	B = 28.67
Category 5 U-Pipe Diameter (mm)	Scenario 5-1	150	0.99	5	150	25	B = 28.67
	Scenario 5-2	150	0.99	5	150	32	B = 28.67
	Scenario 5-3	150	0.99	5	150	40	B = 28.67
	Scenario 5-4	150	0.99	5	150	50	B = 28.67
Category 6 U-Pipe spacing (mm)	Scenario 6-1	150	0.99	5	150	32	A = 0
	Scenario 6-2	150	0.99	5	150	32	AS = 3.17
	Scenario 6-3	150	0.99	5	150	32	B = 28.67
	Scenario 6-4	150	0.99	5	150	32	C = 86

Scenarios 1-3, 2-2, 3-2, 5-2, 6-3 stand for the base case; scenarios were established by considering the range of impact factors.

The actual energy consumption data of the building do not exist, because the service life of the building is less than one year. Therefore, in this study, the actual heating/cooling load of the building and the COP data of GSHP that was installed in the building were used to predict the energy consumption data of the actual building. The validation was conducted as the following steps: (i) estimating the monthly actual load of the facility (refer to Table A2); (ii) using the coefficient of performance measured through the experiment with GSHP installed in the building; during the cooling period, the COP were measured as 6.46, 5.09 and 4.16, respectively, when the EWT was at 15, 25 and 32 °C; during the heating period, the COP were measured as 3.83, 4.2 and 4.63, respectively, when the EWT was at 5, 10 and 15 °C (refer to Figures A1 and A2); (iii) estimating monthly electricity consumption for 40 years through simulation (refer to Table A3); (iv) predicting the electricity consumption with the result of Steps (i) and (ii); and (v) comparing the value of predicting and comparing the predicted value with the simulation result (refer to Figure 3). The CV(RESE) value was measured as 8.39%; therefore, it was proven that the model for the GSHP system's scenario was feasible.

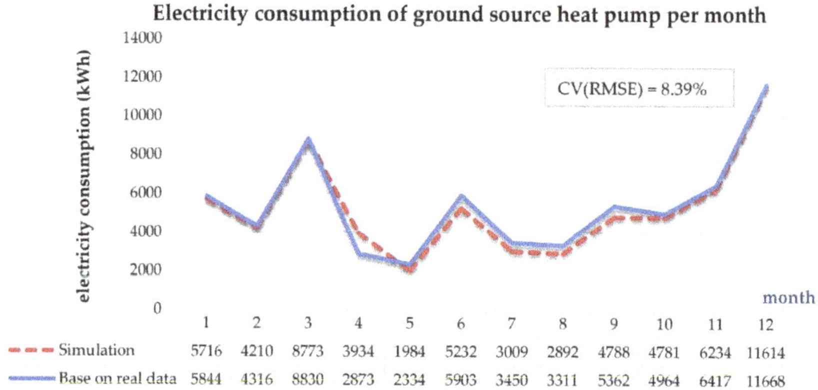

Figure 3. Comparing the electricity consumption of the ground source heat pump for CV(RMSE).

2.3. Sensitivity Analysis on the Energy Generation and Environmental Impact of the GSHP System

2.3.1. Sensitivity Analysis

The previous relevant studies defined sensitivity analysis as consisting of a quantitative comparison of the changes in the outputs to the changes in the inputs [80–83]. In this study, the regional and system factors of the GSHP system were fixed, and the design factors (e.g., borehole length, grout thermal conductivity, borehole spacing, borehole diameter, U-pipe diameter and U-pipe position) of the base case were configured as inputs (e.g., scenarios by category) according to the range to establish the GSHP system's scenarios (refer to Table 5). The sensitivity analysis on the environmental impact and energy generation of the GSHP system was performed by changing only one design factor. The sensitivity influence coefficient (IC) has been used in several studies as one of the most adequate forms of sensitivity coefficients in the assessment of sensitivity (refer to Equation (2)) [80–83].

$$IC = (\Delta OP/OP_{BC})/(\Delta IP/IP_{BC}) \quad (2)$$

where IC is the influence coefficient, $\triangle OP$ is the change in output resulting from a $\triangle IP$ change in input, OP_{BC} is the base case output value and IP_{BC} is the base case output value.

This approach resulted in a unit-less form of sensitivity coefficient, which was important for comparing parameters with different units (e.g., m for the borehole length, W/m·K for the grout thermal conductivity, m for the borehole spacing and mm for the borehole diameter). For example, an IC value of +0.2 for the borehole length means that a 1% increase in borehole length will lead to a 0.2% increase in the performance of the GSHP system. Meanwhile, a negative IC value of −0.2 for the borehole diameter means that a 1% increase in the borehole diameter will lead to a 0.2% decrease in the performance of the GSHP system.

2.3.2. Calculating the Energy Generation of the GSHP System

The process of calculating the energy generation of the GSHP system is presented herein for the sensitivity analysis on the impact factors of the GSHP system in terms of energy generation. The EWT generated from GHE and supplied to the heat pump was measured to calculate the heat efficiency of the GSHP system. EWT is an indicator of the final performance resulting from the combination of impact factors and serves as the heat pump inlet temperature from GHE that satisfies the energy demand of a facility. EWT exhibits higher efficiency when provided as a higher temperature during the heating season and as a lower temperature during the cooling season. Therefore, the load of the heat pump can be reduced, and thus, it is possible to reduce the electricity energy consumption of a building. In this study, "GLHEPro", the GHE simulation and design program, was used to calculate the EWT and analyze the scenarios [84,85]. "GLHEPro", which is used internationally, uses the combination of a variety of formulas and a database of the various materials of the GSHP system to design the borehole depth. In addition, through "GLHEPro", the energy generation of GHE and the annual/monthly energy consumption of the heat pump of a building can be analyzed. In this study, the energy consumption pattern, the result of the simulation, was established to come up with the GSHP system's analysis scenarios.

2.3.3. Calculating the Environmental Impact of the GSHP System

The LCA methodology is still under discussion and continues in the ISO meeting; the basic configuration is as Figure 4. In this study, LCA was used to analyze the environmental impact of the GSHP system, and the method of assessment consists of the following steps that the ISO 14040 defined [86–88].

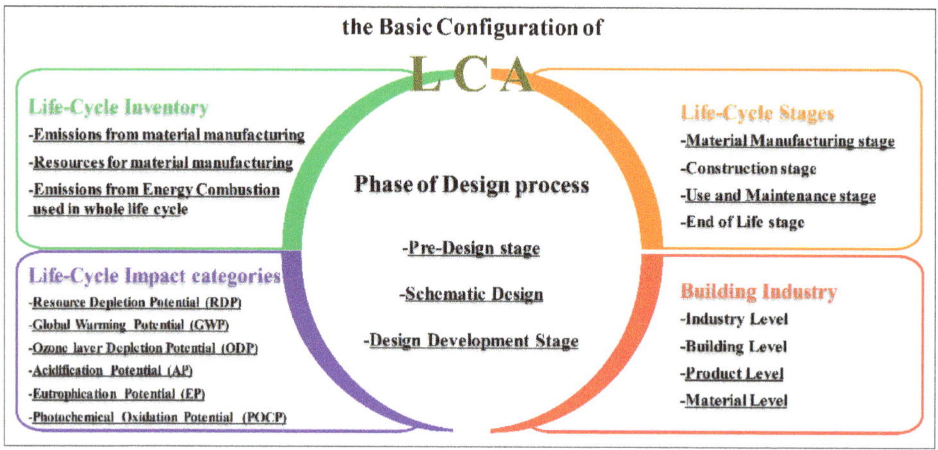

Figure 4. The basic configuration of LCA.

- Step 1. Practical unit and scope: The target and scope to conduct LCA should be clear. In this study, the GSHP system supplied from the material manufacturing and the use and maintenance phase for the whole life cycle is defined as the practical unit, and the relevant data for the whole life cycle is defined as the scope to conduct LCA.
- Step 2. Life cycle inventory (LCI) analysis: The environmental impact substances can be calculated by following LCI steps. First, the energy source amount used to manufacture the components of the GSHP system over the life cycle was calculated using input-output LCA. Second, the environmental impact substances produced in the contaminant and energy production process were measured using the process-based LCA method studied in former research with the domestic LCI database established in South Korea [87,88].
- Steps 3 and 4. Life cycle impact assessment (LCIA) and results: LCIA defines the environmental impacts using following phases: (i) classification; (ii) characterization; (iii) normalization; and (iv) weighting [86,87]. The characterization factor of each category is required for calculating the characterized impact. In this study, the scenarios of the GSHP system were analyzed in terms of environmental impact with the following categories (e.g., resource depletion potential (RDP), global warming potential (GWP), ozone

layer depletion potential (ODP), acidification potential (AP), eutrophication potential (EP) and photochemical oxidation potential (POCP)). The characterized impacts are calculated using Equation (3).

$$ICC_n = \sum_{n} E_s \times CF_{s,n} \quad (3)$$

where ICC_n is the impact category's (n) characterized impact, E_s is the emission of substance (s) and $CF_{s,n}$ is the substance's characterization factor (s) to impact category (n).

3. Results and Discussion

The impact factors of the GSHP system were analyzed through sensitivity analysis, considering the energy generation and environmental impact. Sensitivity analyses were performed on one base case and 24 scenarios in six categories. Based on the simulation results, the average maximum and minimum EWT and the monthly electricity consumption were analyzed to compare the performance of the GSHP system in the average heating period with that in the average cooling period to calculate in terms of energy generation. Furthermore, the environmental impact was calculated using LCA for the material manufacturing and use and maintenance stages.

3.1. Sensitivity Analysis on Impact Factors of the GSHP System in Terms of Energy Generation

The results of the sensitivity analysis on the impact factors of the GSHP system in terms of energy generation are summarized in Figure 5 and Table 6, which show the average minimum/maximum EWT and IC values calculated for the different categories of impact factors. These IC values showed the changing influence of the impact factors on energy generation (*i.e.*, EWT) of the GSHP system, where the EWT exhibits higher efficiency if provided as a high temperature in the heating period and as a low temperature in the cooling period. Accordingly, as the IC values of the cooling period become greater as negative values, the influence level is large, and as the IC values of the heating period become greater as positive values, the influence level is large. For the results, Category 1 (borehole length) showed the most influential impact factors for the GSHP system. Specifically, the IC of Scenario 1-1 in the cooling and heating periods was −0.54 and 0.67, respectively. In the cooling period, when the borehole length was reduced by as much as 100 m (66.7%), the EWT increased by as much as 6.82°C (36%) compared to that of the base case. In the heating period, when the borehole length was reduced by as much as 100 m (66.7%), the EWT decreased by as much as 5.63 °C (44.5%) compared to that of the base case.

Meanwhile, Category 6 (U-pipe spacing) showed the least influential impact factors for the GSHP system. Specifically, the IC of Scenario 6-1 in the cooling and heating periods was −0.02 and 0.03, respectively. In the cooling period, when U-pipe spacing was reduced by as much as 25.5 mm (88.8%), the EWT was increased by as much as 0.39 °C (2.1%) compared to that of the base case. In the heating period, the U-pipe spacing was reduced by as much as 25.5 mm (88.8%), and the EWT was decreased by as much as 0.39°C (3.1%) compared to that of the base case.

The results of other impact factors are as follows. In the case of Category 2 (grout thermal conductivity), the IC of Scenario 2-1 in the cooling and heating periods was −0.07 and 0.10, respectively. When grout thermal conductivity was reduced by as much as 0.115 W/m·K (11.6%) in the cooling period, the EWT was increased by as much as 0.15 °C (0.8%) compared to that of the base case. When the grout thermal conductivity was reduced by as much as 0.115 W/m·K (11.6%) in the heating period, the EWT was decreased by as much as 0.15 °C (1.1%) compared to that of the base case. In the case of Category 3 (borehole spacing), the IC of Scenario 3-1 in the cooling and heating periods were −0.07 and 0.10, respectively. When the borehole spacing was reduced by as much as 1 m (20%) in the cooling period, the EWT was increased by as much as 0.26 °C (1.4%) compared to that of the base case. When the bore spacing was reduced by as much as 1 m (20%) in the heating period, the EWT was decreased by as much as 0.26 °C (2.1%) compared to that of the base case. In the case of Category 5 (U-pipe diameter), the IC of Scenario 5-1 in the cooling and heating periods were −0.05 and 0.07, respectively. When the U-pipe diameter was reduced by as much as 7 mm (21.9%) in the cooling period, the EWT was increased by as much as 0.2 °C (1.1%) compared to that of the base case. When the U-pipe diameter was reduced by as much as 7 mm (21.9%) in the heating period, the EWT was decreased by as much as 0.2 °C (1.6%) compared to that of the base case. In the case of Category 4 (borehole diameter), the IC of Scenario 4-1 in the cooling and heating periods was 0.02 and −0.03, respectively. When the bore diameter was reduced by as much as 25 mm (17%) in the cooling period, the EWT was decreased by as much as 0.07 °C (0.4%) compared to that of the base case. When the bore diameter was reduced by as much as 25 mm (16.7%) in the heating period, the EWT was increased by as much as 0.07 °C (0.6%) compared to that of the base case.

To sum up, the borehole length was determined to be the most influential impact factor, and the borehole diameter and U-pipe spacing were determined to be the least influential impact factors in terms of the energy generation of the GSHP system (refer to Figure 5).

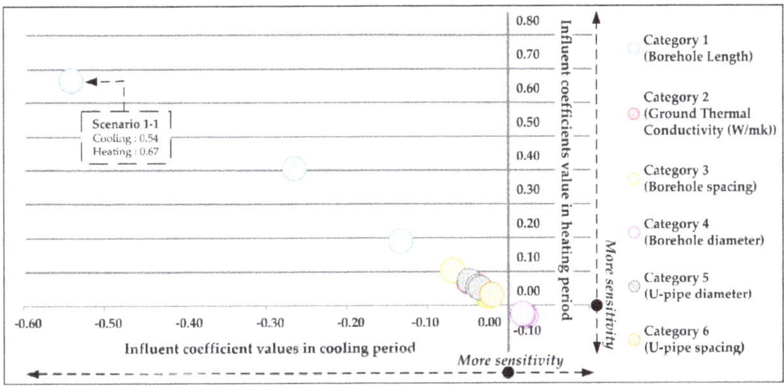

Figure 5. Influence coefficient values of different impact factors in terms of energy generation.

Table 6. Influence coefficient values of different impact factors in terms of energy generation.

Classification of Impact Factors		Avg Min. EWT (°C)	Avg. Max. EWT (°C)	Influence Coefficient Values	
		Cooling Period	Heating Period	Cooling Period	Heating Period
Category 1 borehole length (m)	Scenario 1-1	25.75	7.02	−0.54	0.67
	Scenario 1-2	20.6	10.94	−0.26	0.41
	Scenario 1-3	18.93	12.65	-	-
	Scenario 1-4	18.09	13.45	−0.13	0.19
Category 2 grout thermal conductivity (W/m·K)	Scenario 2-1	19.08	12.5	−0.07	0.10
	Scenario 2-2	18.93	12.65	-	-
	Scenario 2-3	18.72	12.85	−0.05	0.07
	Scenario 2-4	18.5	13.08	−0.04	0.06
Category 3 borehole spacing (m)	Scenario 3-1	19.19	12.39	−0.07	0.10
	Scenario 3-2	18.93	12.65	-	-
	Scenario 3-3	18.83	12.72	−0.03	0.03
	Scenario 3-4	18.75	12.78	−0.02	0.03
Category 4 borehole diameter (mm)	Scenario 4-1	18.86	12.72	0.02	−0.03
	Scenario 4-2	18.93	12.65	-	-
	Scenario 4-3	18.99	12.59	0.02	−0.03
	Scenario 4-4	19.04	12.54	0.02	−0.03
Category 5 U-pipe diameter (mm)	Scenario 5-1	19.13	12.45	−0.05	0.07
	Scenario 5-2	18.93	12.65	-	-
	Scenario 5-3	18.75	12.82	−0.04	0.05
	Scenario 5-4	18.56	13.01	-0.03	0.05
Category 6 U-pipe spacing (mm)	Scenario 6-1	19.32	12.26	−0.02	0.03
	Scenario 6-2	19.27	12.31	−0.02	0.03
	Scenario 6-3	18.93	12.65	-	-
	Scenario 6-4	18.32	13.21	−0.02	0.02

Scenarios 1-3, 2-2, 3-2, 5-2 and 6-3 stand for the base case; each scenario is described in Table 5.

3.2. Sensitivity Analysis on the Impact Factors of the GSHP System in Terms of the Environmental Impact

The results of the sensitivity analysis on the impact factors of the GSHP system in terms of the environmental impact (*i.e.*, six impact categories) are summarized in

Figure 6, Figure A3 and Tables 7, A4–A7. In Section 2.3.3, the practical unit of LCA is defined as "the GSHP system supplied from the material manufacturing and use and maintenance stages for the whole service life." Therefore, the environmental impact was assessed in two stages: (i) the material manufacturing stage; and (ii) the use and maintenance stage. Table 7 shows the results of the sensitivity analysis in terms of RDP. Tables A4–A7 show the results of the sensitivity analysis on the other environmental impact categories.

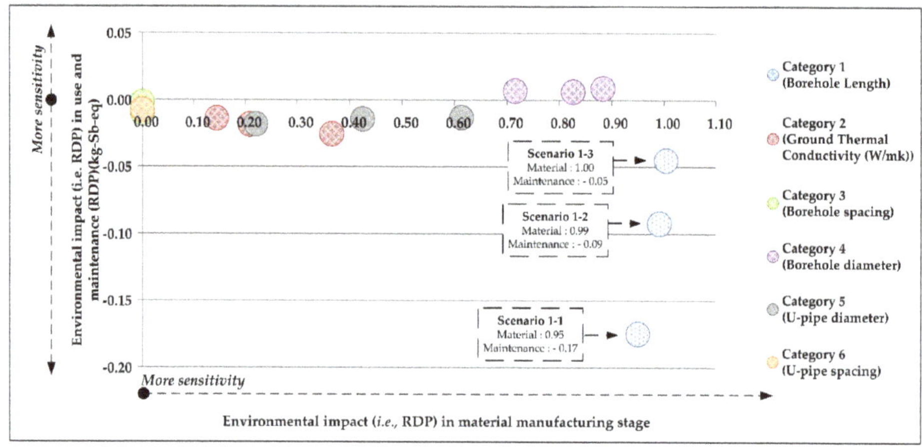

Figure 6. Influence coefficient values of different impact factors in terms of environmental impact (*i.e.*, resource depletion potential (RDP)).

As shown in Table 7, Category 1 (borehole length) showed the most influential impact factors for the GSHP system in the material manufacturing and use and maintenance stages. Specifically, the IC of Scenario 1-4 in the material manufacturing stage and the IC of Scenario 1-1 in the use and maintenance stage were 1.00 and −0.17, respectively. In the material manufacturing stage, when the borehole length increased by as much as 50 m (33.3%) compared to that of the base case (150 m), the RDP increased by as much as 170 kg-Sb-eq (33.7%) compared to that of the base case (505 kg-Sb-eq). In the use and maintenance stage, when the borehole length was reduced by as much as 100 m (66.7%) compared to that of the base case (150 m), the RDP decreased by as much as 320 kg-Sb-eq (11.6%) compared to that of the base case (2168 kg-Sb-eq). Meanwhile, Category 3 (borehole spacing) showed the least influential impact factors for the GSHP system. Specifically, the ICs of Scenario 3-4 in the material manufacturing and use and maintenance stages were 0.00 and −0.002, respectively. In the material manufacturing stage, even when the borehole spacing increased by as much as 2 m (40%) compared to that of the base case (5 m), the RDP showed no change compared to that of the base case (505 kg-Sb-eq). In the use and maintenance stage, when the borehole spacing was reduced by as much as 2 m (40%) compared to that of the base case (5 m), the RDP decreased by as much as 2 kg-Sb-eq (0.2%) compared to that of the base case (2168 kg-Sb-eq).

The results of other impact factors are as follows. In the case of Category 2 (grout thermal conductivity), the ICs of Scenario 2-1 in the material manufacturing and use and maintenance stages were 0.37 and −0.03, respectively. In the material manufacturing stage, even when the grout thermal conductivity was decreased by as much as 0.115 W/m·K (11.6%) compared to that of the base case, the RDP was decreased by as much as 21 kg-Sb-eq (4.2%) compared to that of the base case (505 kg-Sb-eq). In the use and maintenance stage, when the grout thermal conductivity was reduced by as much as 0.115 W/m·K (11.6%) compared to that of the base case, the RDP was increased by as much as 6 kg-Sb-eq (0.3%) compared to that of the base case (2168 kg-Sb-eq). In the case of Category 4 (borehole diameter), the ICs of Scenario 4-1 in the material manufacturing and use and maintenance stages were 0.88 and 0.01, respectively. In the material manufacturing stage, even when the borehole diameter was decreased by as much as 25 mm (16.7%) compared to that of the base case, the RDP was decreased by as much as 74 kg-Sb-eq (14.7%) compared to that of the base case (505 kg-Sb-eq). In the use and maintenance stage, when the borehole diameter was reduced by as much as 25 mm (16.7%) compared to that of the base case, the RDP was decreased by as much as 3 kg-Sb-eq (0.1%) compared to that of the base case (2168 kg-Sb-eq). In the case of Category 5 (U-pipe diameter), the IC of Scenario 5-4 in the material manufacturing stage and the IC of Scenario 5-1 in the use and maintenance stage were 0.61 and −0.02, respectively. In the material manufacturing stage, when the U-pipe diameter was increased by as much as 18 mm (56.3%) compared to that of the base case, the RDP was increased by as much as 175 kg-Sb-eq (34.7%) compared to that of the base case (505 kg-Sb-eq). In the use and maintenance stage, when the U-pipe diameter was reduced by as much as 7 mm (21.9%) compared to that of the base case, the RDP was increased by as much as 8 kg-Sb-eq (0.4%) compared to that of the base case (2168 kg-Sb-eq). In the case of Category 6 (U-pipe spacing), the ICs of Scenario 6-2 in the material manufacturing and use and maintenance stages were 0.00 and −0.01, respectively. In the material manufacturing stage, even when the U-pipe spacing was decreased by as much as 25.2 mm (88.8%) compared to that of the base case, the RDP showed no change compared to that of the base case (505 kg-Sb-eq). In the use and maintenance stage, when the U-pipe spacing was reduced by as much as 25.2 mm (88.8%) compared to that of the base case, the RDP was increased by as much as 14 kg-Sb-eq (0.6%) compared to that of the base case (2168 kg-Sb-eq). As shown in Tables A4–A7 other environmental impact categories showed similar trends.

Table 7. Influence coefficient values of different impact factors in terms of environmental impact (*i.e.*, resource depletion potential (RDP)).

Classification		Values	Environmental Impact (RDP) (kg-Sb-eq)			Influent Coefficient (IC) of Environmental Impact (RDP)		
			Material Manufacturing	Use and Maintenance	Sum	Material Manufacturing	Use and Maintenance	Sum
Category 1 borehole length (m)	Scenario 1-1	50	185	2421	2606	0.95	−0.17	0.04
	Scenario 1-2	100	338	2235	2573	0.99	−0.09	0.11
	Scenario 1-3	150	505	2168	2673	-	-	-
	Scenario 1-4	200	675	2135	2810	1.00	−0.05	0.15
Category 2 grout thermal conductivity (W/m·K)	Scenario 2-1	0.875	484	2174	2658	0.37	−0.03	0.05
	Scenario 2-2	0.99	505	2168	2673	-	-	-
	Scenario 2-3	1.2116	529	2159	2688	0.21	−0.02	0.02
	Scenario 2-4	1.6	550	2149	2700	0.14	−0.01	0.02
Category 3 borehole spacing (m)	Scenario 3-1	4	505	2172	2678	0	−0.01	−0.01
	Scenario 3-2	5	505	2168	2673	-	-	-
	Scenario 3-3	6	505	2167	2672	0	−0.002	−0.002
	Scenario 3-4	7	505	2166	2672	0	−0.002	−0.002
Category 4 borehole diameter (mm)	Scenario 4-1	125	431	2165	2596	0.88	0.01	0.17
	Scenario 4-2	150	505	2168	2673	-	-	-
	Scenario 4-3	175	566	2170	2736	0.72	0.01	0.14
	Scenario 4-4	200	645	2172	2817	0.83	0.01	0.16
Category 5 U-pipe diameter (mm)	Scenario 5-1	25	481	2176	2657	0.22	−0.02	0.03
	Scenario 5-2	32	505	2168	2673	-	-	-
	Scenario 5-3	40	559	2160	2719	0.42	−0.01	0.07
	Scenario 5-4	50	680	2152	2832	0.61	−0.01	0.11
Category 6 U-pipe spacing (mm)	Scenario 6-1	0	505	2185	2690	0	−0.01	−0.01
	Scenario 6-2	3.17	505	2182	2688	0	−0.01	−0.01
	Scenario 6-3	28.37	505	2168	2673	-	-	-
	Scenario 6-4	86	505	2142	2647	0	−0.01	0

Scenarios 1-3, 2-2, 3-2, 5-2 and 6-3 stand for the base case; Sb stands for Antimony (stomic number 51); each scenario is described in Table 5.

In summary, the borehole length was determined to be the most influential impact factor, and the borehole spacing and U-pipe spacing were determined to be the least influential impact factors in terms of the environmental impact of the GSHP system (refer to Figure 6).

4. Conclusions

This study aimed to conduct sensitivity analysis on the impact factors of the GSHP system in terms of energy generation and environmental impact. This study was conducted as follows: (i) collecting the impact factors affecting the GSHP system's performance; (ii) establishing the GSHP systems' scenarios with the impact factors; (iii) determining the methodology and calculation tool to be used for conducting sensitivity analysis; and (iv) sensitivity analysis on the impact factors in terms of energy generation and environmental impact using LCA. A dormitory building in a university in Seoul where the GSHP system had been installed was selected for the sensitivity analysis. The average maximum and minimum EWT and the monthly electricity consumption were analyzed, which could enable the comparison of the performances of the GSHP system during the heating and cooling periods based on the simulation results. Furthermore, the environmental impact was calculated using LCA for the material manufacturing and use and maintenance stages.

In terms of the energy generation of the GSHP system, the borehole length was determined to be the most influential impact factor, showing influence coefficient values of −0.54 and 0.67, respectively, during cooling and heating period. These values are relatively high compared to those of other impact factors, such as the grout thermal conductivity (−0.07 and 0.10, respectively, during the cooling and heating period), borehole spacing (−0.07 and 0.10, respectively, during the cooling and heating period), borehole diameter (0.02 and −0.03, respectively, during the cooling and heating period) and U-pipe diameter (−0.05 and 0.07, respectively, during the cooling and heating period), which indicate that the borehole length influences the energy generation of the GSHP the most. On the other hand, the U-pipe spacing was determined to be the least influential impact factor, showing influence coefficient values of −0.02 and 0.03, respectively, during the cooling and heating period. In terms of the environmental impact of the GSHP system, the borehole length was again determined to be the most influential impact factor, showing influence coefficient values of 1.00 and −0.17, respectively, in the material manufacturing and use and maintenance. These values are relatively high compared to those of other impact factors, such as the grout thermal conductivity (0.37 and −0.03, respectively, in the material manufacturing and use and maintenance), borehole spacing (−0.07 and 0.10, respectively, in the material manufacturing and use and maintenance), borehole diameter (0.88 and 0.01, respectively, in the material manufacturing and use and maintenance) and U-pipe diameter (0.61 and −0.02, respectively, in the material manufacturing and use and maintenance), which indicate that the borehole length influences the environmental impact of the GSHP the most. On the other hand, the borehole spacing and U-pipe spacing were determined to be the two least influential impact factors, showing influence coefficient values of zero and −0.01, respectively, in the material manufacturing and use and maintenance for both impact factors. To sum up, the borehole length was determined to be the most influential impact factor in terms of both energy generation and environmental impact. Meanwhile, U-pipe spacing was determined to be the least influential impact factor in terms of both energy generation and environmental impact

The results of this study can be used: (i) to establish the optimal design strategy for different application fields and different seasons; and (ii) to conduct a feasibility study on energy generation and environmental impact at the level of the life cycle.

Research on the following is recommended for future studies: (i) economic and environmental assessment for selecting the optimal implementation fraction of the GSHP system using the above-analyzed factors; and (ii) a multi-objective optimization system for the ultimate decision maker to analyze the uncountable scenarios in terms of several impact factors.

Acknowledgments: This work was supported by the National Research Foundation of Korea (NRF) grant funded by the Korea government (MSIP; Ministry of Science, ICT and Future Planning) (No. NRF-2015R1A2A1A05001657).

Author Contributions: All authors read and approved the manuscript. All authors contributed to this work, discussed the results and implications and commented on the manuscript at all stages. Taehoon Hong gave valuable advice on the establishment of the framework, as well as the design process. Jimin Kim made the model with the co-authors and helped to understand the optimization model more thoroughly. Myeongsoo Chae discussed the main idea behind the work and reviewed and revised the manuscript. Joonho Park led the development of the paper. Jaemin Jeong enhanced the quality of the manuscript at all stages. Minhyun Lee developed the revised manuscript.

Conflicts of Interest: The authors declare no conflict of interest.

Abbreviations

IEA	International Energy Agency
OECD	Organization for Economic Cooperation and Development
GHG	Greenhouse gas
CERT	Carbon emission reduction target
NRE	New/renewable energy
KEMCO	Korea Energy Management Corporation
GSHP	Ground source heat pump
GHE	Ground heat exchanger

IGSHPA	International Ground Source Heat Pump Association
CCHP	Combined cooling heating and power
LCA	Life cycle assessment
EWT	Entering water temperature
CV(RMSE)	Coefficient of variation of the root-mean-square error
COP	Coefficient of performance
IC	Influence coefficient
ISO	International Organization for Standardization
LCI	Life cycle inventory
LCIA	Life cycle impact assessment
RDP	Resource depletion potential
GWP	Global warming potential
ODP	Ozone layer depletion potential
AP	Acidification potential
EP	Eutrophication potential
POCP	Photochemical oxidation potential

Appendix

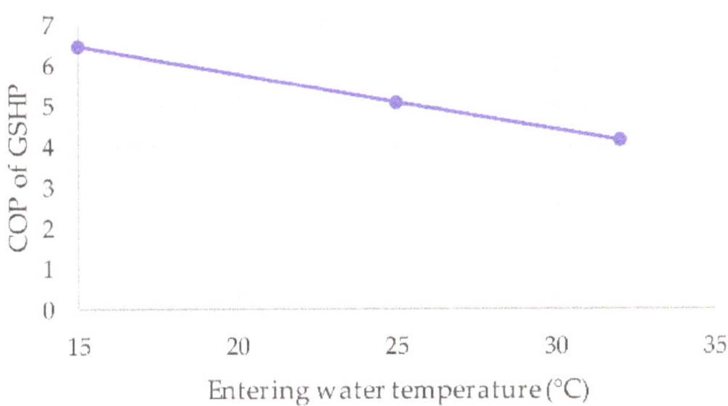

Figure A1. The coefficient of performance measured through experiment during the cooling period.

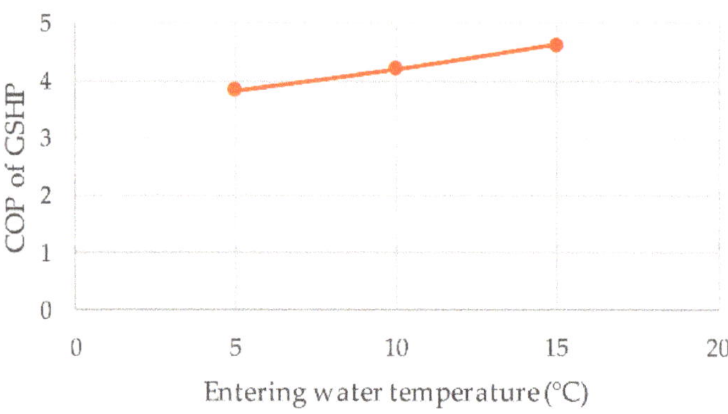

Figure A2. The coefficient of performance measured through experiment during the heating period.

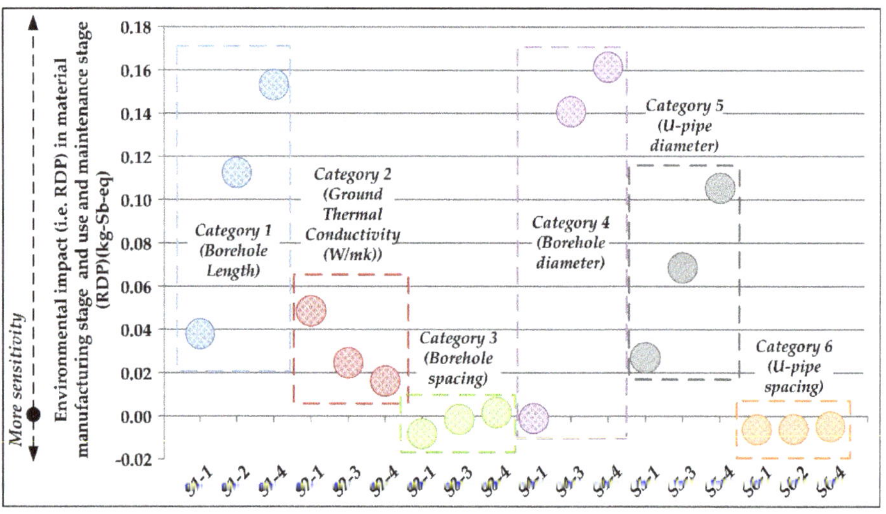

Figure A3. Influence coefficient values of different impact factors for environmental impact (*i.e.*, resource depletion potential (RDP)), sum of material manufacturing and use and maintenance stage.

Table A1. Ground thermal conductivity according to silica sand mass fraction.

20% Bentonite	Silica Sand Mass Fraction (%)										
	0	5	10	15	20	25	30	35	40	45	50
DY-100	0.7746	0.8619	0.9055	0.9408	0.9738	1.0567	1.1373	1.233	1.3438	1.4967	1.6107
DY-100S	0.7937	0.9072	0.9279	0.964	1.0157	1.0828	1.1653	1.2634	1.3769	1.5059	1.6504
Montigel F	0.7879	0.8831	0.9211	0.957	1.0082	1.0748	1.1508	1.2542	1.3668	1.4949	1.6383
EZ-SEAL	0.8067	0.9221	0.9431	0.9798	1.0323	1.1005	1.1844	1.32841	1.3995	1.5306	1.6774
Thermal Grout	0.8374	0.9571	0.979	1.0348	1.0716	1.1598	1.2295	1.3504	1.4527	1.5888	1.7531
Volcay Grout	0.7615	0.8746	0.9159	0.9554	0.9897	1.0884	1.1286	1.2116	1.2997	1.4839	1.6052

Table A2. The actual monthly load of the facility.

Month	Total Load (kWh)	Peak Load (kW)
January	26,946	236
February	19,913	193
March	40,572	142
April	18,521	67
May	14,800	85
June	35,807	213
July	21,312	306
August	20,424	293
September	32,226	191
October	22,965	80
November	29,633	107
December	53,586	188

Table A3. Monthly electricity consumption for 40 years through simulation.

Month	Electricity Consumption (kWh)	EWT (°C)
January	5716	13.9
February	4210	14.1
March	8773	13.0
April	3934	15.1
May	1984	15.9
June	5232	17.9
July	3009	17.1
August	2892	17.1
September	4788	18.3
October	4781	14.8
November	6234	14.3
December	11,614	12.8

Table A4. Influence coefficient values of different impact factors for environmental impact (*i.e.*, global warming potential (GWP)).

Classification		Values	Environmental Impact (GWP) ($kg\text{-}CO_2\text{-}eq$)			Influent Coefficient (IC) of Environmental Impact (GWP)		
			Material Manufacturing	Use and Maintenance	Sum	Material Manufacturing	Use and Maintenance	Sum
Category 1 borehole length (m)	Scenario 1-1	50	54,479	1,375,833	1,430,312	0.96	−0.17	−0.05
	Scenario 1-2	100	101,103	1,270,132	1,371,235	1.00	−0.09	0.03
	Scenario 1-3	150	151,374	1,232,166	1,383,539	-	-	-
	Scenario 1-4	200	201,926	1,213,583	1,415,509	1.00	−0.05	0.07
Category 2 grout thermal conductivity (W/m·K)	Scenario 2-1	0.875	148,573	1,235,809	1,384,382	0.16	−0.03	−0.01
	Scenario 2-2	0.99	151,374	1,232,166	1,383,539	-	-	-
	Scenario 2-3	1.2116	154,455	1,227,077	1,381,532	0.09	−0.02	−0.01
	Scenario 2-4	1.6	157,255	1,221,589	1,378,844	0.06	−0.01	−0.01
Category 3 borehole spacing (m)	Scenario 3-1	4	151,374	1,234,689	1,386,063	0.00	−0.01	−0.01
	Scenario 3-2	5	151,374	1,232,166	1,383,539	-	-	-
	Scenario 3-3	6	151,374	1,231,652	1,383,026	0.00	0.00	0.00
	Scenario 3-4	7	151,374	1,231,219	1,382,593	0.00	0.00	0.00
Category 4 borehole diameter (mm)	Scenario 4-1	125	123,985	1,230,380	1,354,365	1.09	0.01	0.13
	Scenario 4-2	150	151,374	1,232,166	1,383,539	-	-	-
	Scenario 4-3	175	169,503	1,233,546	1,403,050	0.72	0.01	0.08
	Scenario 4-4	200	194,961	1,234,689	1,429,649	0.86	0.01	0.10
Category 5 U-pipe diameter (mm)	Scenario 5-1	25	148,904	1,237,056	1,385,960	0.07	−0.02	−0.01
	Scenario 5-2	32	151,374	1,232,166	1,383,539	-	-	-
	Scenario 5-3	40	156,807	1,227,781	1,384,588	0.14	−0.01	0.00
	Scenario 5-4	50	169,058	1,223,115	1,392,172	0.21	−0.01	0.01
Category 6 U-pipe spacing (mm)	Scenario 6-1	0	151,374	1,241,712	1,393,086	0.00	−0.01	−0.01
	Scenario 6-2	3.17	151,374	1,240,452	1,391,825	0.00	−0.01	−0.01
	Scenario 6-3	28.37	151,374	1,232,166	1,383,539	-	-	-
	Scenario 6-4	86	151,374	1,217,255	1,368,629	0.00	−0.01	−0.01

Table A5. Influence coefficient values of different impact factors for environmental impact (*i.e.*, acidification potential (AP)).

Classification		Values	Environmental Impact (AP) (kg-SO_2-eq)			Influent Coefficient (IC) of Environmental Impact (AP)		
			Material Manufacturing	Use and Maintenance	Sum	Material Manufacturing	Use and Maintenance	Sum
Category 1 borehole length (m)	Scenario 1-1	50	296	2362	2658	0.98	−0.17	0.16
	Scenario 1-2	100	569	2181	2749	0.99	−0.09	0.22
	Scenario 1-3	150	851	2115	2966	-	-	-
	Scenario 1-4	200	1135	2083	3219	1.00	−0.05	0.26
Category 2 grout thermal conductivity (W/m·K)	Scenario 2-1	0.875	829	2122	2950	0.22	−0.03	0.05
	Scenario 2-2	0.99	851	2115	2966	-	-	-
	Scenario 2-3	1.2116	875	2107	2982	0.13	−0.02	0.02
	Scenario 2-4	1.6	897	2097	2994	0.09	−0.01	0.02
Category 3 borehole spacing (m)	Scenario 3-1	4	851	2120	2970	0.00	−0.01	−0.01
	Scenario 3-2	5	851	2115	2966	-	-	-
	Scenario 3-3	6	851	2114	2965	0.00	0.00	0.00
	Scenario 3-4	7	851	2114	2965	0.00	0.00	0.00
Category 4 borehole diameter (mm)	Scenario 4-1	125	709	2112	2822	1.00	0.01	0.29
	Scenario 4-2	150	851	2115	2966	-	-	-
	Scenario 4-3	175	943	2118	3061	0.65	0.01	0.19
	Scenario 4-4	200	1074	2120	3193	0.79	0.01	0.23
Category 5 U-pipe diameter (mm)	Scenario 5-1	25	829	2124	2952	0.12	−0.02	0.02
	Scenario 5-2	32	851	2115	2966	-	-	-
	Scenario 5-3	40	900	2108	3007	0.23	−0.01	0.06
	Scenario 5-4	50	1010	2100	3109	0.33	−0.01	0.09
Category 6 U-pipe spacing (mm)	Scenario 6-1	0	851	2132	2983	0.00	−0.01	−0.01
	Scenario 6-2	3.17	851	2130	2980	0.00	−0.01	−0.01
	Scenario 6-3	28.37	851	2115	2966	-	-	-
	Scenario 6-4	86	851	2090	2941	0.00	−0.01	0.00

Table A6. Influence coefficient values of different impact factors for environmental impact (*i.e.*, eutrophication potential (EP)).

Classification		Values	Environmental Impact (EP) (kg-PO_4^{3}-eq)			Influent Coefficient (IC) of Environmental Impact (EP)		
			Material Manufacturing	Use and Maintenance	Sum	Material Manufacturing	Use and Maintenance	Sum
Category 1 borehole length (m)	Scenario 1-1	50	30	440	469	0.97	−0.17	0.03
	Scenario 1-2	100	57	406	462	0.99	−0.09	0.10
	Scenario 1-3	150	84	394	478	-	-	-
	Scenario 1-4	200	113	388	500	1.00	−0.05	0.14
Category 2 grout thermal conductivity (W/m·K)	Scenario 2-1	0.875	81	395	476	0.32	−0.03	0.04
	Scenario 2-2	0.99	84	394	478	-	-	-
	Scenario 2-3	1.2116	88	392	480	0.18	−0.02	0.02
	Scenario 2-4	1.6	91	390	481	0.13	−0.01	0.01
Category 3 borehole spacing (m)	Scenario 3-1	4	84	394	479	0.00	−0.01	−0.01
	Scenario 3-2	5	84	394	478	-	-	-
	Scenario 3-3	6	84	393	478	0.00	0.00	0.00
	Scenario 3-4	7	84	393	478	0.00	0.00	0.00
Category 4 borehole diameter (mm)	Scenario 4-1	125	72	393	465	0.92	0.82	0.85
	Scenario 4-2	150	84	394	478	-	-	-
	Scenario 4-3	175	94	394	488	0.89	0.82	0.84
	Scenario 4-4	200	106	394	501	0.88	0.82	0.84
Category 5 U-pipe diameter (mm)	Scenario 5-1	25	81	395	476	0.19	−0.02	0.02
	Scenario 5-2	32	84	394	478	-	-	-
	Scenario 5-3	40	92	392	484	0.36	−0.01	0.05
	Scenario 5-4	50	109	391	500	0.52	−0.01	0.08
Category 6 U-pipe spacing (mm)	Scenario 6-1	0	84	397	481	0.00	−0.01	−0.01
	Scenario 6-2	3.17	84	396	481	0.00	−0.01	−0.01
	Scenario 6-3	28.37	84	394	478	-	-	-
	Scenario 6-4	86	84	389	473	0.00	−0.01	0.00

Table A7. Influence coefficient values of different impact factors for environmental impact (*i.e.*, photochemical oxidation potential (POCP)).

Classification		Values	Environmental Impact (POCP) (kg-C_2H_4-eq)			Influent Coefficient (IC) of Environmental Impact (POCP)		
			Material Manufacturing	Use and Maintenance	Sum	Material Manufacturing	Use and Maintenance	Sum
Category 1 borehole length (m)	Scenario 1-1	50	195	3.97	199	0.97	−0.17	0.96
	Scenario 1-2	100	370	3.66	374	0.99	−0.09	0.98
	Scenario 1-3	150	552	3.55	555	-	-	-
	Scenario 1-4	200	737	3.50	740	1.01	−0.05	1.00
Category 2 grout thermal conductivity (W/m·K)	Scenario 2-1	0.875	522	3.57	526	0.46	−0.03	0.46
	Scenario 2-2	0.99	552	3.55	555	-	-	-
	Scenario 2-3	1.2116	584	3.54	588	0.26	−0.02	0.26
	Scenario 2-4	1.6	552	3.56	555	0.00	0.00	0.00
Category 3 borehole spacing (m)	Scenario 3-1	4	552	3.56	555	0.00	−0.01	0.00
	Scenario 3-2	5	552	3.55	555	-	-	-
	Scenario 3-3	6	552	3.55	555	0.00	0.00	0.00
	Scenario 3-4	7	552	3.55	555	0.00	0.00	0.00
Category 4 borehole diameter (mm)	Scenario 4-1	125	482	3.55	485	0.76	0.01	0.76
	Scenario 4-2	150	552	3.55	555	-	-	-
	Scenario 4-3	175	611	3.56	615	0.64	0.01	0.64
	Scenario 4-4	200	688	3.56	691	0.74	0.01	0.73
Category 5 U-pipe diameter (mm)	Scenario 5-1	25	517	3.57	521	0.29	−0.02	0.28
	Scenario 5-2	32	552	3.55	555	-	-	-
	Scenario 5-3	40	628	3.54	631	0.55	−0.01	0.55
	Scenario 5-4	50	799	3.53	803	0.80	−0.01	0.79
Category 6 U-pipe spacing (mm)	Scenario 6-1	0	552	3.58	555	0.00	−0.01	0.00
	Scenario 6-2	3.17	552	3.58	555	0.00	−0.01	0.00
	Scenario 6-3	28.37	552	3.55	555	-	-	-
	Scenario 6-4	86	552	3.51	555	0.00	−0.01	0.00

References

1. British Petroleum (BP). *Statistical Review of World Energy 2015*; British Petroleum: London, UK, 2015.
2. Hong, T.H.; Koo, C.W.; Kim, H.J. A decision support model for improving a multi-family housing complex based on CO_2 emission from electricity consumption. *J. Environ. Manag.* **2012**, *112*, 67–78.
3. Koo, C.W.; Lee, M.H.; Hong, T.H.; Park, H.S. Development for a new energy efficiency rating system for existing residential buildings. *Energ. Policy* **2014**, *68*, 218–231.
4. Ji, C.W.; Hong, T.H.; Park, H.S. Comparative analysis of decision-making methods for integrating cost and CO_2 emission. *Energy Build.* **2014**, *72*, 186–194.
5. Breidenich, C.; Magraw, D.; Rowley, A.; Rubin, J.W. Kyoto Protocol to the United Nations Framework Convention on Climate Change; United Nations. *Am. J. Int. Law.* **1998**, *92*, 315–331.
6. Intergovernmental Panel on Climate Change (IPCC). *Climate Change 2014: Synthesis Report*; Adopted; I.P.C.C.: Copenhagen, Denmark, 2014; Available online: http://www.ipcc.ch/ (accessed on 15 April 2016).
7. Hong, T.H.; Kim, H.J.; Kwak, T.H. Energy-Saving Techniques for Reducing CO_2 Emissions in Elementary Schools. *J. Manag. Eng.* **2012**, *28*, 39–50.
8. Hong, T.H.; Kim, J.M.; Koo, C.W. LCC and $LCCO_2$ analysis of green roofs in elementary schools with energy saving measures. *Energy Build.* **2012**, *45*, 229–239.
9. Federal Energy Management Program. Ground-Source Heat Pumps Applied to Federal Facilities—Second Editon. 2001. Available online: http://smartenergy.illinois.edu/pdf/Archive/ GroundSourceHeatPumpApplication.pdf (accessed on 14 April 2016).
10. European Environment Agency (EEA). Trend and Projections in Europe 2014: Tracking Progress towards Europe's Climate and Energy Targets for 2020. Denmark, 2014. Available online: http://www. actu-environnement.com/media/pdf/news-23105-etude-eea-europe-climat-energie.pdf (accessed on 14 April 2016).
11. Department of State (DOS). *Fifth National Communication of the United States of America Under the United Nations Framework Convention on Climate Change: U.S. Climate Action Report 2010*; U.S. Department of State (DOS): Washington, DC, USA, 2010.
12. Jones, R.S.; Yoo, B.S. Korea's Green Growth Strategy: Mitigating Climate Change and Developing New Growth Engines. *OECD Economics Department Working Papers*, 2010, Volume 54. Available online: http://dx.doi.org/10.1787/5kmbhk4gh1ns-en (accessed on 15 April 2016).
13. Park, J.H.; Hong, T.H. Maintenance management process for reducing CO_2 emission in shopping mall complexes. *Energy Build.* **2011**, *43*, 894–904.
14. Renewable Energy Policy Network for the 21st Century (REN21). Global Status Report: Renewables 2015. 2015. Available online: http://www.ren21.net/wp-content/uploads/2015/07/REN12- GSR2015_Onlinebook_low1.pdf (accessed on 14 April 2016).

15. Renewable Energy Policy Network for the 21st Century (REN21). REN21 10 Year Report. 2014. Available online: http://www.ren21.net/Portals/0/documents/activities/Topical%20Reports/REN21_10yr.pdf (accessed on 15 April 2016).
16. Pereira, A.O., Jr.; Costa, R.C.; Vale Costa, C.; Marreco, J.M.; Rovere, E.L.L. Perspectives for the expansion of new renewable energy sources in Brazil. *Renew. Sust. Energ. Rev.* **2013**, *23*, 49–59.
17. Schiling, M.A.; Esmundo, M. Technology S-curves in renewable energy alternatives: Analysis and implications for industry and government. *Energ. Policy* **2009**, *37*, 1767–1781.
18. Greenpeace International. *Energy Revolution: A Sustainable World Energy Outlook 2015*; Greenpeace International: Brussels, Belgium, 2015.
19. Koo, C.W.; Hong, T.H.; Park, H.S.; Yun, G.C. Framework for the analysis of the potential of the rooftop photovoltaic system to achieve the net-zero energy solar buildings. *Prog. Photovolt.* **2014**, *22*, 462–478.
20. Park, H.S.; Hong, T.H. Analysis of South korea's economic growth, carbon dioxide emission, and energy consumption using the Markov switching. *Renew. Sust. Energ. Rev.* **2013**, *18*, 543–551.
21. International Energy Agency (IEA). *Medium-Term Renewable Energy Market Report 2015*; OECD Publishing: Paris, France, 2015.
22. International Energy Agency (IEA). *Medium-Term Renewable Energy Market Report 2014*; OECD Publishing: Paris, France, 2014.
23. International Energy Agency (IEA). *Medium-Term Renewable Energy Market Report 2012*; OECD Publishing: Paris, France, 2012.
24. Committee on Climate Change. *The Fourth Carbon Budget: Reducing Emissions Through the 2020s*; Committee on Climate Change: London, England, 2010; Available online: https://www.theccc.org.uk/ archive/aws2/4th%20Budget/CCC_4th-Budget_interactive.pdf (accessed on 15 April 2016).
25. Energy Information Administration (EIA). *Annual Energy Outlook 2015 with projections to 2040*; Energy Information Administration: Washington, DC, USA, 2015. Available online: http://www.eia.gov/forecasts/aeo/pdf/0383(2015).pdf (accessed on 15 April 2016).
26. Department of Energy and Climate Change (DECC). 2014 UK Greenhouse Gas Emissions: Provisional Figures. UK, 2015. Available online: https://www.gov.uk/government/uploads/system/uploads/ attachment_data/file/416810/2014_stats_release.pdf (accessed on 14 April 2016).
27. International Energy Agency (IEA). Energy Technology Perspectives 2015: Mobilising Innovation to Accelerate Climate Action. 2015. Available online: https://www.iea.org/publications/ freepublications/publication/EnergyTechnologyPerspectives2015ExecutiveSummaryEnglishversion.pdf (accessed on 14 April 2016).
28. Koo, C.W.; Hong, T.H.; Lee, M.H. Estimation of the monthly average daily solar radiation using geographic information system and advanced case-based. *Environ. Sci. Technol.* **2013**, *47*, 4829–4839.

29. International Energy Agency (IEA). *World Energy Outlook 2015*; International Energy Agency: Paris, France, 2015; Available online: http://www.iea.org/Textbase/npsum/WEO2015SUM.pdf (accessed on 15 April 2016).
30. Hong, T.H.; Koo, C.W.; Kim, H.J.; Park, H.S. Decision support model for establishing the optimal energy retrofit strategy for existing multi-family housing complexes. *Energ. Policy* **2014**, *66*, 157–169.
31. New Renewable Energy Center in the Korea Energy Management Corporation (KEMCO). Available online: http://www.energy.or.kr (accessed on 16 September 2015).
32. Ramli, M.A.M.; Twaha, S. Analysis of renewable energy feed-in tariffs in selected regions of the globe: Lessons for Saudi Arabia. *Renew. Sust. Energy. Rev.* **2015**, *45*, 649–661.
33. Cory, K.; Couture, T.; Kreycik, C. *Feed-in Tariff Policy: Design, Implementation, and RPS Policy Interactions*; National Renewable Energy Laboratory (NREL): Washington, DC, USA, 2009.
34. Hong, T.H.; Koo, C.W.; Park, J.H.; Park, H.S. A GIS(geographic information system)-based optimization model for estimating the electricity generation of the rooftop PV. *Energy* **2014**, *65*, 190–199.
35. Hong, T.H.; Koo, C.W.; Lee, S.U. Benchmarks as a tool for free allocation through comparison with similar projects: Focused on multi-family housing complex. *Appl. Energy* **2014**, *114*, 663–675.
36. Han, S.W.; Hong, T.H.; Lee, S.Y. Production prediction of conventional and global positioning system-based earthmoving systems using simulation and multiple. *Can. J. Civil. Eng.* **2008**, *35*, 574–587.
37. Bertani, R. Geothermal power generation in the world 2010–2014 updated report. *Geothermics* **2016**, *60*, 31–43.
38. Kalz, D.E.; Vellei, M.; Winiger, S. Energy and Efficiency Analysis of Heat Pump Systems in Nonresidential Buildings by Means of Long-Term Measurements. In Proceedings of the 11th REHVA World Congress and 8th International Conference on IAQVEC, Prague, Prague Czech Republic, 16–19 June 2013.
39. Fujii, H.; Yamasaki, S.; Maehara, T.; Ishikami, T.; Chou, N. Numerical simulation and sensitivity study of double-layer Slinky-coil horizontal ground heat exchangers. *Geothermics* **2013**, *47*, 61–68.
40. Casasso, A.; Sethi, R. Efficiency of closed loop geothermal heat pumps: A sensitivity analysis. *Renew. Energ.* **2014**, *62*, 737–746.
41. Kim, J.; Hong, T.; Chae, M.; Koo, C.; Jeong, J. An Environmental and Economic Assessment for Selecting the Optimal Ground Heat Exchanger by Considering the Entering Water Temperature. *Energies* **2015**, *8*, 7752–7776.
42. Cosentino, S.; Sciacovelli, A.; Verda, V.; Noce, G. Energy and exergy analysis of ground thermal energy storage: optimal charging time in different operating conditions. In Proceedings of the ECOS 2015: The 28th International Conference on Efficiency, Cost, Optimization, Simulation and Environmental Impact of Energy System, Pau, France, 30 June 2015.

43. Hepbasli, A. Performance evaluation of a vertical ground-source heat pump system in Izmir, Turkey. *Int. J. Energy. Res.* **2002**, *26*, 1121–1139.
44. Boyaghchi, F.A.; Chavoshi, M.; Sabeti, V. Optimization of a novel combined cooling, heating and power cycle driven by gothermal and sola energies using the water/CuO (copper oxide) nanofluid. *Energy* **2015**, *91*, 685–699.
45. Esen, H.; Inalli, M. Modelling of a vertical ground coupled heat pump system by using artificial neural networks. *Expert. Syst. Appl.* **2009**, *36*, 10229–10238.
46. Alavy, M.; Nguyen, H.V.; Leong, W.H.; Dworkin, S.B. A methodology and computerized approach for optimizing hybrid ground source heat pump system design. *Renew. Energy* **2013**, *57*, 404–412.
47. Saltelli, A. Sensitivity Analysis for Importance Assessment. *Risk. Anal.* **2002**, *22*, 1–12.
48. Saltelli, A.; Ratto, M.; Andres, T.; Campolongo, F.; Cariboni, J.; Gatelli, D.; Saisana, M.; Tarantola, S. *Global Sensitivity Analysis: the Primer*; John Wiley and Sons: Chichester, West sussex, Englnad, 2008.
49. Koo, C.W.; Hong, T.H.; Hyun, C.T.; Park, S.H.; Seo, J.O. A study on the development of a cost model based on the owner's decision making at the early stages of a construction project. *Int. J. Strateg. Prop. Manag.* **2010**, *14*, 121–137.
50. Kalz, D.E.; Pfafferott, J.; Herkel, S.; Wagner, A. Energy and efficiency analysis of environmental heat sources and sinks: In-use performance. *Renew. Energy* **2011**, *36*, 916–929.
51. Gao, J.; Zhang, X.; Liu, J.; Li, K.S.; Yang, J. Thermal performance and ground temperature of vertical pile-foundation heat exchangers: A case study. *Appl. Therm. Eng.* **2008**, *28*, 2295–2304.
52. Hepbasli, A.; Akdemir, O.; Hancioglu, E. Experimental study of a closed loop vertical ground source heat pump system. *Energy Convers. Manag.* **2003**, *44*, 527–548.
53. Zhang, Q.; Murphy, W.E. Measurement of thermal conductivity for three borehole fill materials used for GSHP. *ASHRAE Trans.* **2000**, *106*, 434.
54. Kharseh, M.; Altorkmany, L.; Nordell, B. Global warming's impact on the performance of GSHP. *Renew. Energy* **2011**, *36*, 1485–1491.
55. Jeon, J.; Lee, S.; Hong, D.; Kim, Y. Performance evaluation and modeling of a hybrid cooling system combining a screw water chiller with a ground source heat pump in a building. *Energy* **2010**, *35*, 2006–2012.
56. Madani, H.; Claesson, J. Retrofitting a variable capacity heat pump to a ventilation heat recovery system: Modeling and performance analysis. In Proceedings of the International Conference on Applied Energy, Singapore, 21–23 April 2010; pp. 649–658.
57. Hepbasli, A.; Akdemir, O. Energy and exergy analysis of a ground source (geothermal) heat pump system. *Energy Convers. Manag.* **2004**, *45*, 737–753.
58. Lamarche, L.; Kajl, S.; Beauchamp, B. A review of methods to evaluate borehole thermal resistances in geothermal heat-pump systems. *Geothermics* **2010**, *39*, 187–200.
59. Butler, D.K.; Curro, J.R., Jr. Crosshole seismic testing—Procedures and pitfalls. *Geophysics* **1981**, *46*, 23–29.

60. Fan, R.; Gao, Y.; Hua, L.; Deng, X.; Shi, J. Thermal performance and operation strategy optimization for a practical hybrid ground-source heat-pump system. *Energy Build.* **2014**, *78*, 238–247.
61. Zeng, H.; Diao, N.; Fang, Z. Efficiency of vertical geothermal heat exchangers in the ground source heat pump system. *J. Therm. Sci.* **2003**, *12*, 77–81.
62. Liu, X.L.; Wang, D.L.; Fan, Z.H. Modeling on heat transfer of a vertical bore in geothermal heat exchangers. *Build. Energy Environ.* **2001**, *2*, 1–3.
63. Beier, R.A.; Smith, M.D.; Spitler, J.D. Reference data sets for vertical borehole ground heat exchanger models and thermal response test analysis. *Geothermics* **2011**, *40*, 79–85.
64. Inalli, M.; Esen, H. Experimental thermal performance evaluation of a horizontal ground-source heat pump system. *Appl. Therm. Eng.* **2004**, *24*, 2219–2232.
65. Khan, M.A.; Wang, J.X. Development of a graph method for preliminary design of boreholeground-coupled heat exchanger in North Louisiana. *Energy Build.* **2015**, *92*, 389–397.
66. Jeon, J.; Lee, S.; Hong, D.; Kim, Y. Performance evaluation and modeling of a hybrid cooling system combining a screw water chiller with a ground source heat pump in a building. *Energy* **2010**, *35*, 2006–2012.
67. Sivasakthivel, T.; Murugesan, K.; Sahoo, P.K. Potential reduction in CO_2 emission and saving in electricity by ground source heat pump system for space heating applications-a study on northern part of India. *Procedia Eng.* **2012**, *38*, 970–979.
68. Le Feuvre, P. An Investigation into Ground Source Heat Pump Technology, Its UK Market and Best Practice in System Design. Ph.D. Thesis, Strathclyde University, UK, September 2007.
69. Lee, C.K. Effects of multiple ground layers on thermal response test analysis and ground-source heat pump simulation. *Appl. Energ.* **2011**, *88*, 4405–4410.
70. International Ground Source Heat Pump Association. Closed-Loop/Geothermal Heat Pump Systems Design and Installation Standards 2010 Edition. 2010. Available online: http://www.igshpa.okstate.edu/ pdf_files/publications/standards2010s.pdf (accessed on 14 April 2016).
71. Ground Source Heat Pump Association. *Closed-Loop Vertical Borehole Design, Installation and Materials Standards*; Ground Source Heat Pump Association: England, UK, 2011.
72. Korea Energy Management Corporation (KEMCO). *Annual End-Use Energy Statistics*; Korea Energy Management Corporation: Seoul, Korea, 2013; Available online: http://www.kemco.or.kr/ (accessed on 28 June 2015).
73. Korean Statistical Information Service(KOSIS). *New Renewable Energy Investigation of Supply Statistics Report*; Korean Statistical Information Service: Seoul, Korea, 2015; Available online: http://kosis.kr/ (accessed on 15 April 2016).
74. Yang, H.; Cui, P.; Fang, Z. Vertical-borehole ground-coupled heat pumps: A review of models and systems. *Appl. Energy* **2010**, *87*, 16–27.
75. Lee, J.Y. Current status of ground source heat pumps in Korea. *Renew. Sust. Energ. Rev.* **2009**, *13*, 1560–1568.

76. Hong, T.H.; Koo, C.W.; Kwak, T.H. Framework for the implementation of a new renewable energy system in an educational facility. *Appl. Energy* **2013**, *103*, 539–551.
77. Hong, T.H.; Koo, C.W.; Kwak, T.H.; Park, H.S. An economic and environmental assessment for selecting the optimum new renewable energy system for educational facility. *Renew. Sust. Energy Rev.* **2014**, *29*, 286–300.
78. Swan, L.G.; Ugursal, V.I. Modeling of end-use energy consumption in the residential sector: A review of modeling techniques. *Renew. Sust. Energy Rev.* **2009**, *13*, 1819–1835.
79. American Society of Heating, Refrigerating and Air-conditioning Engineers (ASHRAE). *ASHRAE Guideline 14–2002: Measurement of Energy and Demand Savings: American Society of Heating*; Refrigerating and Air-conditioning Engineers: Atlanta, GA, USA, 2002.
80. Lam, J.C.; Hui, S.C. Sensitivity analysis of energy performance of office buildings. *Build. Environ.* **1996**, *31*, 27–39.
81. Lam, J.C.; Wan, K.K.W.; Yang, L. Sensitivity analysis and energy conservation measures implications. *Energ. Convers. Manag.* **2008**, *49*, 3170–3177.
82. Heiselberg, P.; Brohus, H.; Hesselholt, A.; Rasmussen, H.; Seinre, E.; Thomas, S. Application of sensitivity analysis in design of sustainable buildings. *Renew. Energy* **2009**, *34*, 2030–2036.
83. Azar, E.; Menassa, C.C.A. Comprehensive analysis of the impact of occupancy parameters in energy simulation of office buildings. *Energy Build.* **2012**, *55*, 841–853.
84. Yavuzturk, C. Modeling of Vertical Ground Loop Heat Exchangers for Ground Source Heat Pump Systems. Ph.D. Thesis, Oklahoma State University, Stillwater, OK, USA, December 1999.
85. Spitler, J.D. GLHEPRO-A Design Tool for Commercial Building Ground Loop Heat Exchangers. In Proceedings of the Fourth International Heat Pumps in Cold Climates Conference, Quebec, QC, Canada, 17–18 August 2000.
86. The American Institute of Architects (AIA). *A Guide to Life Cycle Assessment of Buildings*; AIA: New York, NY, USA, 2010.
87. Kim, C.J.; Kim, J.M.; Hong, T.H.; Koo, C.W.; Jeong, K.B.; Park, H.S. A program-level management system for the life cycle environmental and economic assessment of complex building projects. *Environ. Impact Assess.* **2015**, *54*, 9–21.
88. Jeong, K.B.; Ji, C.W.; Koo, C.W.; Hong, T.H.; Park, H.S. A model for predicting the environmental impacts of educational facilities in the project planning phase. *J. Clean. Prod.* **2014**, *107*, 538–549.

Analysis of Environmental Impact for Concrete Using LCA by Varying the Recycling Components, the Compressive Strength and the Admixture Material Mixing

Taehyoung Kim, Sungho Tae and Chang U. Chae

Abstract: Concrete is a type of construction material in which cement, aggregate, and admixture materials are mixed. When cement is produced, large amounts of substances that impact the environment are emitted during limestone extraction and clinker manufacturing. Additionally, the extraction of natural aggregate causes soil erosion and ecosystem destruction. Furthermore, in the process of transporting raw materials such as cement and aggregate to a concrete production company, and producing concrete in a batch plant, substances with an environmental impact are emitted into the air and water system due to energy use. Considering the fact that the process of producing concrete causes various environmental impacts, an assessment of various environmental impact categories is needed. This study used a life cycle assessment (LCA) to evaluate the environmental impacts of concrete in terms of its global warming potential, acidification potential, eutrophication potential, ozone depletion potential, photochemical ozone creation potential, and abiotic depletion potential (GWP, AP, EP, ODP, POCP, ADP). The tendency was that the higher the strength of concrete, the higher the GWP, POCP, and ADP indices became, whereas the AP and EP indices became slightly lower. As the admixture mixing ratio of concrete increased, the GWP, AP, ODP, ADP, and POCP decreased, but EP index showed a tendency to increase slightly. Moreover, as the recycled aggregate mixing ratio of concrete increased, the AP, EP, ODP, and ADP decreased, while GWP and POCP increased. The GWP and POCP per unit compressed strength (1 MPa) of high strength concrete were found to be about 13% lower than that for its normal strength concrete counterpart. Furthermore, in the case of AP, EP, ODP, and ADP per unit compressed strength (1 MPa), high-strength concrete was found to be about 10%~25% lower than its normal strength counterpart. Among all the environmental impact categories, ordinary cement was found to have the greatest impact on GWP, POCP, and ADP, while aggregate had the most impact on AP, EP, and ODP.

Reprinted from *Sustainability*. Cite as: Kim, T.; Tae, S.; Chae, C.U. Analysis of Environmental Impact for Concrete Using LCA by Varying the Recycling Components, the Compressive Strength and the Admixture Material Mixing. *Sustainability* **2016**, *8*, 389.

1. Introduction

Concrete is a construction material manufactured by the mixing of cement, aggregate, mixed water, and admixture materials. In the process of producing cement, which is the main composition material for concrete, not only do natural resources such as limestone and clay become depleted, but environmental impact substances are also emitted during clinker manufacturing through pyro process due to large amounts of energy use [1]. Additionally, the extraction of natural aggregate can lead to soil erosion or ecosystem destruction, while the waste sludge and wastewater emitted from a concrete batch plant have harmful effects on the water ecosystem [2].

As concrete has several impacts on the environment, the selection of various environmental impact categories is needed. When only a single environmental impact is evaluated, the limited assessment could lead to false interpretations of concrete's eco-friendliness. Therefore, efforts and investment to develop a design standard from the perspective of the life cycle of concrete to minimize its environmental impact are being conducted. In environmentally advanced nations such as the U.K. and Sweden, the Royal Institute of British Architects (RIBA) [3] and the Swedish Environmental Management Council [4] are developing Product Category Rules (PCRs) and conducting certification focused on construction products from their production to their disuse. The development of the draft version [5] of ISO 13315-2 (environmental management for concrete and concrete structures), an international standard regarding concrete, is ongoing. However, in Korea, quantitative studies on the environmental impact in the concrete industry are at their starting points, and data for policymaking and technological support to turn the concrete industry into a sustainable industry are also lacking [6]. Most research has a strong tendency to be focused on global warming wherein greenhouse gas emissions are evaluated. In addition to those concerning global warming, studies on product category rules and standards development on acidification, ozone layer destruction, and eutrophication are also in their primary stages [7,8]. The object of this study is to evaluate the effects of the increase in compressive strength, admixture material mixing, recycled aggregate mixing, and the amount of binder on environmental impact in a quantitative manner. The environmental impact assessment on concrete was based on the life cycle assessment process suggested in the ISO 14040 series [9,10], and environmental impact assessment index was based on "Korean Eco-indicator methodology" suggested by the Ministry of Environment in Korea [11]. Environmental impacts (global warming, acidification, eutrophication, ozone depletion, photochemical ozone creation, and abiotic depletion) were evaluated in a quantitative manner using 1000 concrete designs of a mix proportion database, and the major causes of environmental impact were analyzed.

2. Method of Environmental Impact Assessment for Concrete

2.1. Goal and Scope Definition

The product selected for environmental impact assessment was ordinary concrete, and based on various functions of concrete, concrete structures and the formation of concrete products were selected as major functions. Concrete size of 1 m^3 was selected as the functional unit.

The product stage of concrete (Cradle to Gate) was selected as the system boundary for the life cycle assessment of concrete, as can be seen in Figure 1. Furthermore, the production stages of concrete were divided into raw material, transportation, and manufacturing stages, and the environmental impact of factors in each stage on air and water systems was evaluated [12].

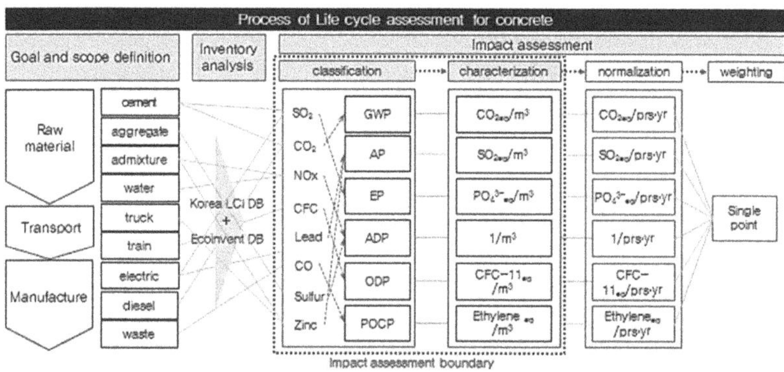

Figure 1. Process of environmental impact assessment.

2.2. Inventory Analysis

Based on the life cycle assessment ranges (system boundary) of concrete, input factors and output factors such as energy, raw material, product, and waste were analyzed. To this end, as can be seen from Table 1, the LCI DB (Life Cycle Index Database) on each of the input materials and energy sources in concrete production was investigated.

The LCI DB on the input materials and energy sources used in this life cycle assessment utilized the existing data of Korea's Ministry of Land, Infrastructure, and Transport [13] and Ministry of Environment [14]. As the LCI DB is different for each country, the DB offered in one's own country should be used. However, the LCI DB on ground granulated blast-furnace slag, fly ash, and admixture in Korea's LCI DB has not been established yet. Therefore, the DB [15] of ecoinvent, an overseas LCI DB, was used.

Table 1. Life Cycle Index (LCI) Database.

Division		Reference	Nation
Raw material	Cement	National LCI	Korea
	Coarse aggregate	National LCI	Korea
	Fine aggregate	National LCI	Korea
	Blast furnace slag	Ecoinvent	Swiss
	Fly ash	Ecoinvent	Swiss
	Water	National LCI	Korea
	Chemical admixture	Ecoinvent	Swiss
Energy	Electric	National LCI	Korea
	Diesel	National LCI	Korea
Transportation	Truck	National LCI	Korea

2.3. Environmental Impact Assessment

In general, an impact assessment is divided into the following stages: classification, in which list items elicited from the list analysis are gathered into corresponding impact categories; characterization, in which the items are categorized into impact categories, and the impact for each category is quantified; normalization, where the environmental impact of each impact category is divided by the total environmental impact of a specific area or period and, lastly; weighting, where the relative advantage among the impact categories is identified. Here, the classification and characterization stages are essential components as per the standards of ISO 14044, while the normalization and weighting stages can be applied as optional components. Currently, as normalization and weighting factors customized for concrete are not yet developed, this study evaluated up to the characterization stage. Environmental problems arising from this are global warming, ozone depletion, photochemical ozone creation, abiotic depletion, eutrophication, and acidification. Therefore, as seen in Table 2, based on the reference material and impact index of the six environmental impact categories such as Global Warming Potential (GWP), Abiotic Depletion Potential (ADP), Acidification Potential (AP), Eutrophication Potential (EP), Ozone Depletion Potential (ODP), and Photochemical Ozone Creation Potential (POCP), characterization values for each environmental impact category of concrete were calculated. As for the reference material and impact index for each environmental impact category, they were based on the database adopted by Korea's Ministry of Environment's Environmental Declaration [16], and classification and characterization were conducted using the LCI DB selected beforehand.

Classification consists of classifying and gathering impact materials according to environmental impact categories. Generally, when impact materials taken from LCI DB are classified by their environmental impact category, and when they are grouped according to the categories of environmental impact, the impact pattern of each material on the environment can be clearly identified. From the classification details of concrete LCI DB of this study, based on the reference material and impact index on each environmental impact category, Table 3 shows examples of ordinary cement and coarse aggregate. The IPCC (Intergovernmental Panel on Climate Change) guideline [17] defines 23 types in total, including carbon dioxide (CO_2), methane (CH_4), and nitrous oxide (N_2O), as substances that have an impact on global warming, while the reference material is carbon dioxide (CO_2). Resource depletion is based on the standard suggested by Guinee (1995) [18], and considers 89 types of resource items in total including crude oil, natural gas, and uranium (U). As for the acidification impact index, while it differs by regional characteristics and atmospheric environment, the impact index suggested by Heijung *et al.* and Hauschild and Wenzel [19] was applied as it is applicable to any region. Twenty-three impact materials in total, including sulfur dioxide (SO_2), hydrogen sulfide (H_2S), and hydrogen fluoride (HF), all appear as sulfur dioxide (SO_2), the reference material.

Table 2. Characterization value of composition material for concrete.

Composition Material	Unit	Environmental Impact Categories			
		GWP ($kg\text{-}CO_{2eq}$/unit)	AP ($kg\text{-}SO_{2eq}$/unit)	EP ($kg\text{-}PO_4^{3-}{}_{eq}$/unit)	POCP ($kg\text{-}Ethylene_{eq}$/unit)
Cement	kg	9.48×10^{-1}	1.28×10^{-3}	1.34×10^{-4}	2.43×10^{-3}
Fine aggregate	kg	1.49×10^{-3}	1.10×10^{-2}	1.92×10^{-3}	1.07×10^{-4}
Fly ash	kg	1.50×10^{-2}	1.16×10^{-4}	6.94×10^{-5}	6.57×10^{-5}
Water	kg	1.14×10^{-1}	1.94×10^{-4}	6.57×10^{-5}	4.86×10^{-7}

Table 3. Classification value of composition material for concrete. The six environmental impact categories are as follow: Global Warming Potential (GWP); Abiotic Depletion Potential (ADP); Acidification Potential (AP); Eutrophication Potential (EP); Ozone Depletion Potential (ODP); and Photochemical Ozone Creation Potential (POCP).

Inventory List	Environmental Impact Categories						Composition Material	
	GWP	ADP	AP	EP	ODP	POCP	Cement	Aggregate
Ammonia (NH_3)			■	■				6.95×10^{-7}
Carbon dioxide (CO_2)	■						9.31×10^{-1}	3.40×10^{-1}
CFC-11					■		2.05×10^{-9}	4.02×10^{-13}
Methane (CH_4)	■					■	1.71×10^{-2}	5.57×10^{-4}
Sulfur dioxide (SO_2)			■			■	1.27×10^{-2}	4.42×10^{-4}
Phosphate ($PO4^{3-}$)				■				4.22×10^{-8}

As for the eutrophication impact index, Heijung *et al.* and Hauschild and Wenzel's impact index was applied, just as it had been applied in the acidification impact index. Among a total of 11 types of impact materials such as phosphate (PO_4^{3-}), ammonia (NH_3), and nitrogen oxides (NO_x), the reference material was phosphate (PO_4^{3-}). In the case of ozone depletion, the impact index suggested by World Metrological Organization (WMO) [20] was selected, which considers a total of 23 types of impact materials such as CFC (chlorofluorocarbon)-11, Halon-1301, and CFC-114, and takes CFC-11 as the reference material. Among a total of 128 types of impact materials, including ethylene, NMVOC (Non-methane volatile organic compounds), and ethanol, photochemical ozone creation takes ethylene as reference material, and the impact index suggested by Derwent *et al.* (1998) [21] and Jenkin and Hayman (1999) [22] was applied. Characterization is the process of quantifying the environmental traits of classified impact materials according to each environmental impact category. In the classification stage, the corresponding impact materials in each environmental impact category were identified and linked to, but as the impact index of each impact material is different, it was hard to identify the extent of impact in a quantitative manner. Therefore, the characterization value of concrete was calculated in a quantitative manner by multiplying the environmental load of impact materials with the impact index for each environmental impact category, and adding all of them. This process is shown in Equation (1), and here, CI_i equals the characterization value, $Load_j$ equals the impact material j's environmental load, and $eqv_{i,j}$ equals the environmental impact index of the impact material j that is within the environmental impact category of i. Table 3 shows an example of the environmental impact characterization value of raw material used in the production of concrete in this study.

$$CI_i = \sum CI_{i,j} = \sum (Load_j \cdot eqv_{i,j}) \tag{1}$$

Here, CI_i is the size of impact that all the list items (j) included in the impact category i have on the impact category in which they are included. $CI_{i,j}$ is the size of impact that the list item j has on impact category i, $Load_j$ is the environmental load of the jth list item, and $eqv_{i,j}$ is the characterization coefficient value of jth list item within impact category I [23].

(1) Global Warming Potential (GWP)

Global warming is a phenomenon that refers to the rising average surface temperature of the Earth, primarily due to the increasing level of GHG (Greenhouse Gases) emissions. The standard substance for GWP is CO_2. Global warming causes changes in the terrestrial and aquatic ecosystems and in coastlines due to rising sea levels. The category indicator of GWP is expressed by Equation (2):

$$GWP = \sum Load(i) \times GWP(i) \tag{2}$$

where Load(i) is the experimental load of the global warming inventory item (i) and GWP(i) is the characterization factor of global warming inventory item (i).

(2) Ozone Depletion Potential (ODP)

Ozone depletion refers to the phenomenon of decreasing ozone density through the thinning of the stratospheric ozone layer (15–30 km altitude) as a result of anthropogenic pollutants. This leads to increased UV (Ultraviolet Ray) exposure of human skin, which implies a potential rise in incidence of melanoma. The standard substance for ODP is CFCs, and the category indicator of ODP is expressed by Equation (3):

$$\mathrm{ODP} = \sum \mathrm{Load(i)} \times \mathrm{ODP(i)} \tag{3}$$

where Load(i) is the experimental load of the ozone depletion inventory item (i) and ODP(i) is the characterization factor of inventory item (i) of the ozone depletion category.

(3) Acidification Potential (AP)

Acidification is an environmental problem caused by acidified rivers/streams and soil due to anthropogenic air pollutants such as SO_2, NH_3, and NO_x. Acidification increases mobilization and leaching behavior of heavy metals in soil and exerts adverse impacts on aquatic and terrestrial animals and plants by disturbing the food web. The standard substance for assessing AP is SO_2. The category indicator of AP is expressed by Equation (4):

$$\mathrm{AP} = \sum \mathrm{Load(i)} \times \mathrm{AP(i)} \tag{4}$$

where Load(i) is the experimental load of the acidification inventory item (i) and AP(i) is the characterization factor of inventory item (i) of the acidification category.

(4) Abiotic Depletion Potential (ADP)

Input materials (natural resources) required for concrete production are classified into renewable resources, such as groundwater and wood, and nonrenewable resources, such as minerals and fossil fuels. Abiotic depletion refers to the exhaustion of nonrenewable resources and the ensuing environmental impacts. The category indicator of ADP is expressed by Equation (5):

$$\mathrm{ADP} = \mathrm{Load(i)} \times \mathrm{ADP(i)} \tag{5}$$

where Load(i) is the environmental load of the ADP inventory item (i) and ADP(i) is the characterization factor for the ADP inventory item (i).

(5) Photochemical Oxidant Creation Potential (POCP)

Photochemical oxidant creation refers to the reaction of airborne anthropogenic pollutants with sunlight that produces chemical products such as ozone (O_3), leading to an increase in ground level ozone concentration; this causes smog that contains chemical compounds adversely affecting ecosystems and hazardous to human health and crop growth. Ethylene is used as the standard substance for POCP. The category indicator of POCP is expressed by Equation (6):

$$\text{POCP} = \sum \text{Load(i)} \times \text{POCP(i)} \tag{6}$$

where Load(i) is the environmental load of the POCP inventory item (i) and POCP(i) is the characterization factor for the POCP inventory item (i).

(6) Eutrophication Potential (EP)

Eutrophication is a phenomenon in which inland waters are heavily loaded with excess nutrients due to chemical fertilizers or discharged wastewater, triggering rapid algal grow and red tides. The standard substance for EP is PO_4^{3-}. The category indicator of EP is expressed by Equation (7):

$$\text{EP} = \sum \text{Load(i)} \times \text{EP(i)} \tag{7}$$

where Load(i) is the environmental load of the EP inventory item (i) and EP(i) is the characterization factor for the EP inventory item (i).

3. Environmental Impact Analysis of Concrete

3.1. Mix Design Database

In order to evaluate the life cycle environmental impact of concrete, about 1000 concrete mix designs were surveyed. As shown in Figure 2, the range of compressive strength is 18 MPa~80 MPa. Among them, 800 are normal strength (below 18~40 MPa) concrete mix, and 200 are high-strength (40~80 MPa) concrete mix.

Furthermore, there are 76 concrete mixes that use only OPC (Ordinary Portland Cement) as binder, 546 concrete mixes in which GGBS (Ground Granulated Blast furnace Slag) is mixed, 253 concrete mixes where fly ash is mixed, and 125 concrete mixes where both GGBS and fly ash are mixed. The range of unit binder amount was distributed the most in 300~400 (kg/m^3), and the least in 500~600 (kg/m^3). The effect of the admixture mixing ratio on the unit binder amount was not very significant. Also, based on the Korean statute regarding the mandatory amount of recycled aggregate to be used, the recycled aggregate mixing ratio was divided into 0%, 10%, 20%, and 30%.

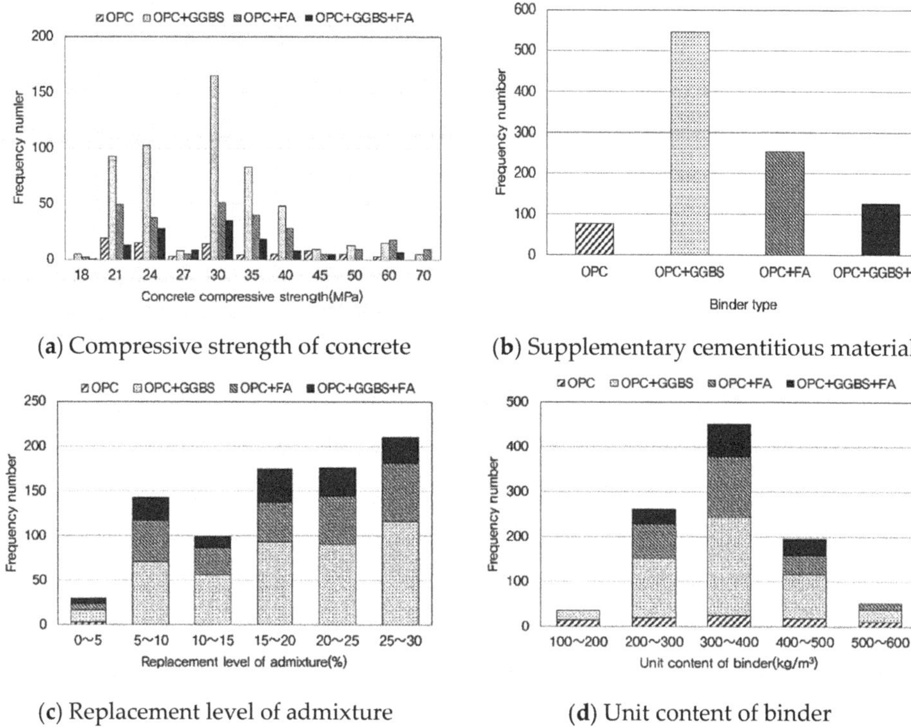

(**a**) Compressive strength of concrete (**b**) Supplementary cementitious materials

(**c**) Replacement level of admixture (**d**) Unit content of binder

Figure 2. Distribution of main parameters in the database.

3.2. Environmental Impact According to Concrete Strength

The stronger the concrete was, the greater tendency it had to increase the global warming potential (GWP), photochemical ozone creation potential (POCP), and abiotic depletion potential (ADP). According to existing research results, it was found that the mixing amount of ordinary cement had the greatest impact on the increase of GWP, POCP, and ADP [24,25]. As shown in Figure 3a, when extracting limestone and iron ore, which are the main raw materials for ordinary cement, sulfur dioxide and sulfuric acid are emitted due to the use of dynamite, which is composed of sulfuric acid, nitric acid, and sulfur substances. Also, due to the energy used in extracted ore and clinker crushing plants, NO_x and PO_4^{3-} are emitted. During the production of ordinary cement, the pyro process is the stage where maximum energy is inputted and the most materials with an environmental impact are emitted [26]. This is because, in the process of increasing the rotary kiln temperature up to 1000~1450 °C in order to produce clinkers, the input of fuels such as Bunker-C Oil, bituminous coal, waste tires, and waste plastics emits environmental impact substances such as carbon dioxide (CO_2) and ammonia (NH_3). This is because most fuels used are mainly composed of carbon and hydrogen, and are made from crude oil that contains oxygen and sulfur [27]. Also, the amount of cement put in high strength concrete (40~80 MPa) was about 20% more than in normal strength concrete (below 18~40 MPa). Therefore, as natural resources such as limestone, iron ore, and gypsum, which are the major elements of cement, were used, abiotic depletion potential (ADP) increased. However, the higher the strength was, the acidification potential (AP) and eutrophication potential (EP) indices showed a tendency to become slightly decreased, as shown in Figure 4d,f. It was found that the coarse aggregate mixing amount of concrete had the greatest impact on AP and EP. Generally, as coarse aggregate mixing amount of concrete decreased to 850~880 kg/m^3 when its strength was increased to 890~910 kg/m^3 on an average, it was found that environmental impact substances produced in coarse aggregate production process also decreased. As shown in Figure 3b, lubricating oil and dynamite, which were used in logging and blasting processes to produce aggregate, were mainly made up of coal-type mineral and sulfuric acid. Hence, sulfur dioxide (SO_2), sulfuric acid (H_2SO_4), and nitrate (NO_3^-), which have an impact on acidification and eutrophication, are emitted. Also, when extracting and crushing the blasted rocks, the use of light oil emits ammonia (NH_3), ammonium (NH_4), phosphate (PO_4^{3-}), and nitrogen oxides (NO_x).

3.3. Environmental Impact According to Admixture

The higher the admixture mixing ratio of concrete was, the lower the global warming potential (GWP), acidification potential (AP), ozone depletion potential (ODP), abiotic depletion potential (ADP), and photochemical ozone creation potential (POCP) became. But the eutrophication potential (EP) index showed a tendency to increase slightly [28,29].

As shown in Figure 5a,b,e, as the admixture mixing ratio increased to 10%, 30%, and 50%, compared to OPC (mixing ratio 0%), GWP, POCP and ADP were found to be decreased to as much as about 10% to 28%. This was because, when producing GGBS and FA (Fly ash), the effect of by-products of other industrial products, such as CO_2, CH_4, N_2O, CO, S, soft coal, hard coal, and crude oil on global warming potential (GWP), the photochemical ozone creation potential (POCP), and abiotic depletion potential (ADP) were very small compared to the process of ordinary cement production. As shown in Figure 3c, GGBS is produced by crushing, mixing, and cooling blast-furnace slag, the by-product of iron ore, and natural gypsum together [30]. As shown in Figure 3d, FA is produced by saving, selecting, and scaling fly ash, the by-product. Manufacturing plants of GGBS and FA used electricity and light oil as their energy sources, and it was found that substances emitted due to the use of light oil and electricity were of 45 types including carbon dioxide (CO_2), methane (CH_4), sulfuric acid (S), ammonia (NH_3), and nitrogen oxide (NO_x).

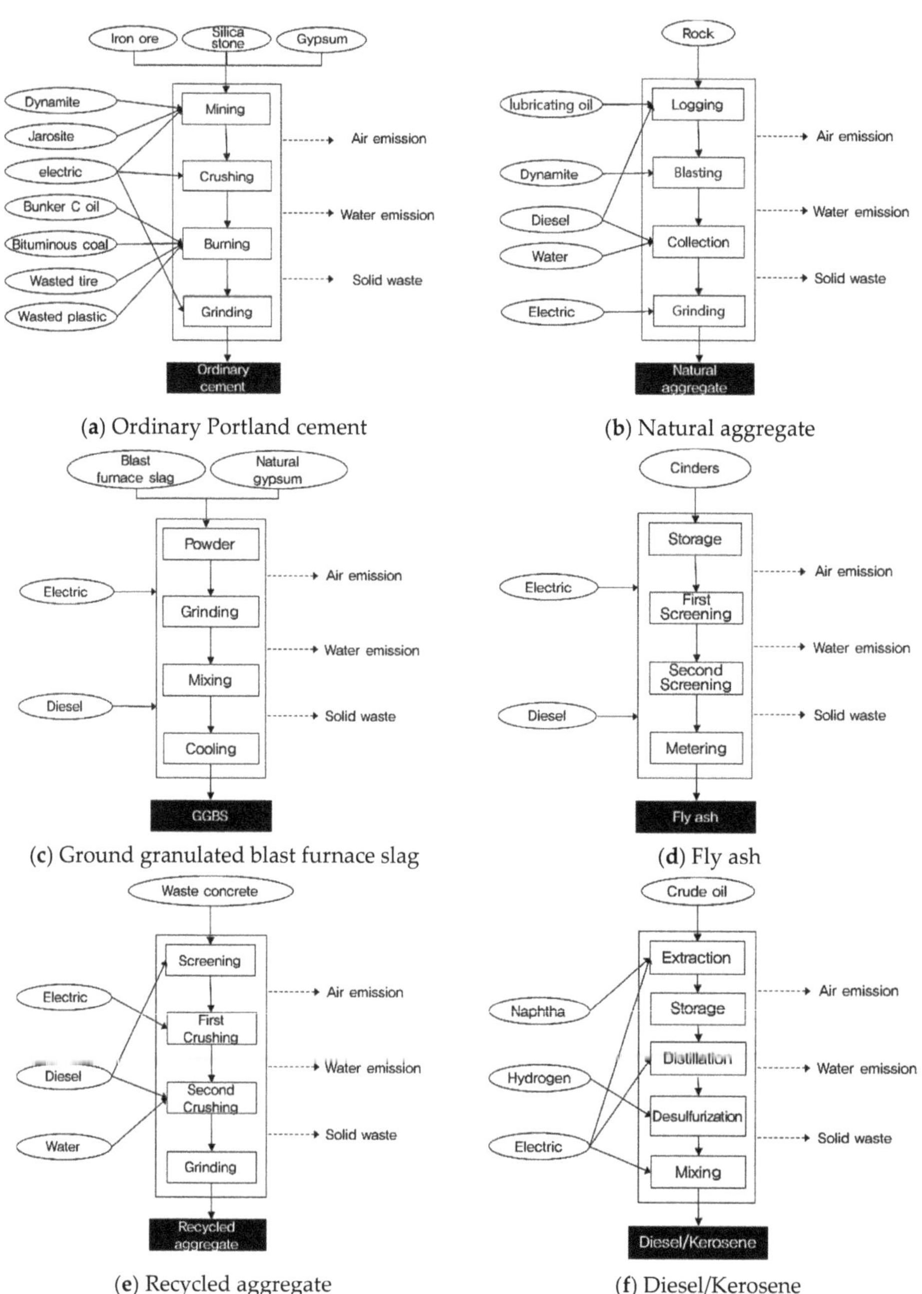

Figure 3. Production process of raw material and oil.

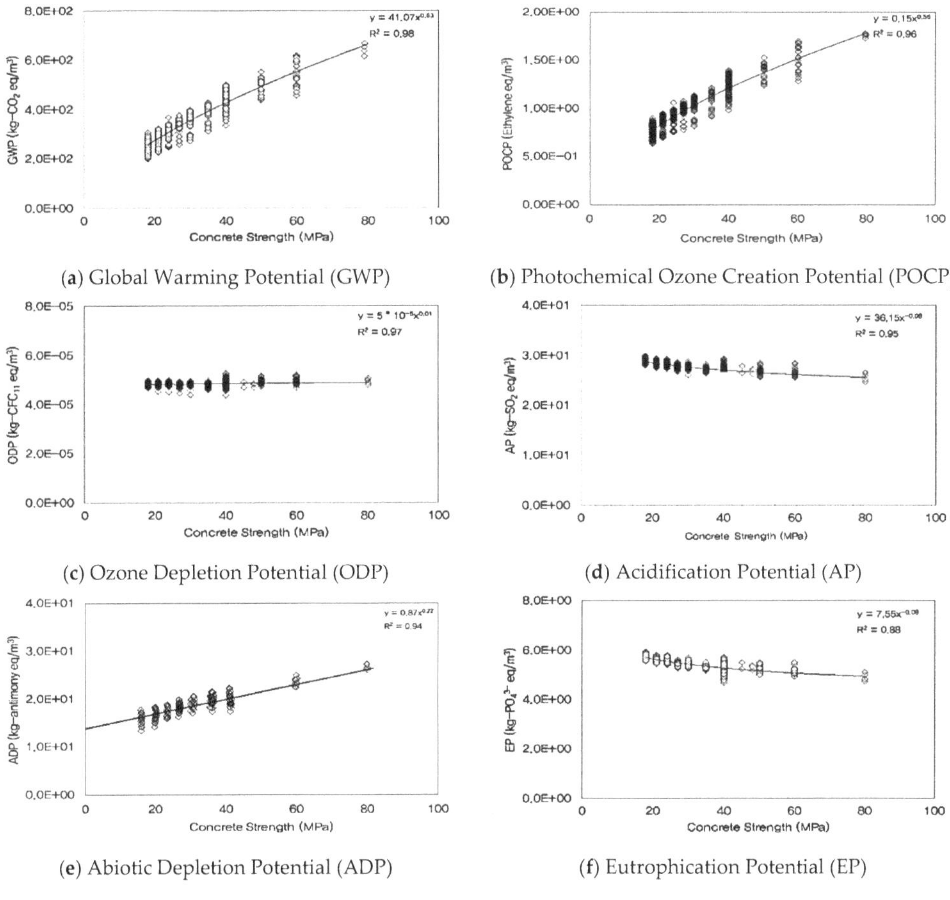

(a) Global Warming Potential (GWP) (b) Photochemical Ozone Creation Potential (POCP) (c) Ozone Depletion Potential (ODP) (d) Acidification Potential (AP) (e) Abiotic Depletion Potential (ADP) (f) Eutrophication Potential (EP)

Figure 4. Environmental impact analysis by concrete strength.

As shown in Figure 5c,d, it was found that, compared to OPC, AP and ODP were reduced to about 1%~3% as the admixture mixing ratio increased. In the process of GGBS production, nitrogen oxide (NO_x), sulfur dioxide (SO_2), halon, and CFC, the major impact materials of AP and ODP, were emitted in lesser amounts compared to the production process of ordinary cement, but there was no considerable difference. As shown in Figure 5f, it was found that compared to OPC, EP increased up to about 3%~9% as the admixture mixing ratio increased. This was due to the fact that emissions of NH_4, NH_3, NO_3, N_2, and PO_4^{3-}, the main materials that have an impact on EP in the production process of GGBS and FA, were greater than when ordinary cement was produced [31].

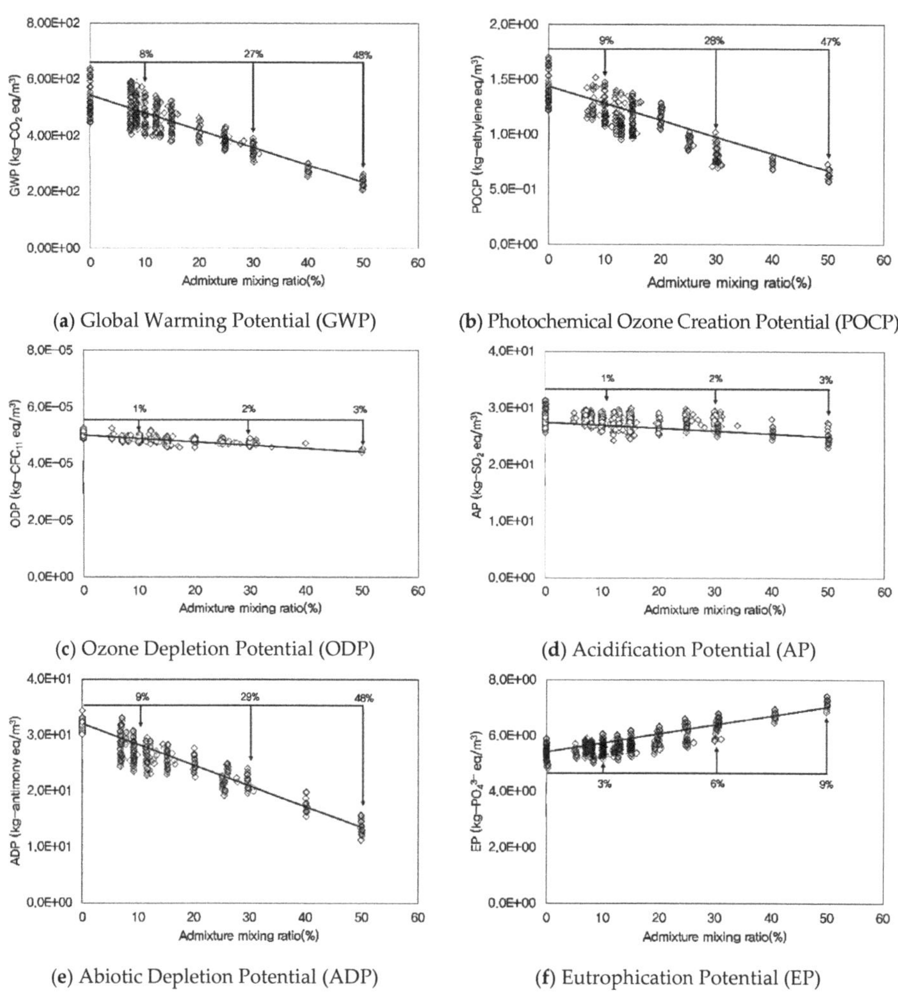

(**a**) Global Warming Potential (GWP)

(**b**) Photochemical Ozone Creation Potential (POCP)

(**c**) Ozone Depletion Potential (ODP)

(**d**) Acidification Potential (AP)

(**e**) Abiotic Depletion Potential (ADP)

(**f**) Eutrophication Potential (EP)

Figure 5. Environmental impact analysis by admixture (GGBS + (fly ash) FA) mixing ratio.

3.4. Environmental Impact According to Recycled Aggregate

As the recycled aggregate mixing ratio of concrete increased, acidification potential (AP), eutrophication potential (EP), ozone depletion potential (ODP), and abiotic depletion potential (ADP) decreased, but the global warming potential (GWP) and photochemical ozone creation potential (POCP) increased [32].

As shown in Figure 6a, when the recycled aggregate mixing ratio was increased, GWP was found to have increased to up to about 14%~29% compared to concrete in which only natural aggregate was mixed. This was because, in the production process of recycled aggregate, major impact materials in terms of global warming potential (GWP) such as CO_2, CH_4, and N_2O, were emitted more than in the case of natural aggregate production process. As shown in Figure 3e, recycled aggregate is produced by crushing, separating, and selecting waste concrete from the demolition of building structures. The amount of light oil and electricity used in the complicated process of making 1 ton of recycled aggregate thus becomes greater than the amount of energy used to produce 1 ton of natural aggregate. As shown in Figure 6c–f, when the mixing ratio of recycled aggregate was increased to 10%, 20%, and 30%, compared to the concrete in which only natural aggregate was mixed, AP, EP, ODP, and ADP were reduced to as much as about 9%~29%. Analysis revealed that it was because, in the process of producing recycled aggregate, NO_x, NH_3, SO_2, NH_4, halon, and CFC, the major impact materials of AP, EP, ODP, and ADP, were emitted less compared to the production process of natural aggregate. In particular, as natural resources were not used in recycling waste concrete which was construction waste, it was found that soft coal, hard coal, and crude oil, the major impact materials of abiotic depletion potential (ADP), were significantly reduced. As shown in Figure 6b, as the recycled aggregate mixing ratio was increased, compared to OPC, POCP was found to be reduced to about 2%~9%. CH_4, CO, S, and $C_4H_{10,}$ the major impact materials of photochemical ozone creation potential (POCP) in recycled aggregate production process, were emitted less than in the case of natural aggregate production process, but there was not much difference.

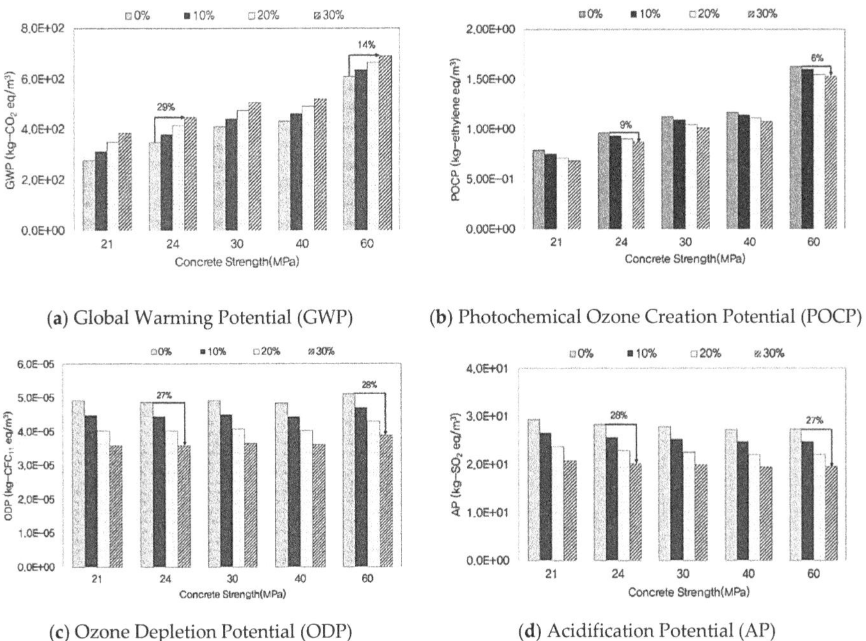

(**a**) Global Warming Potential (GWP)

(**b**) Photochemical Ozone Creation Potential (POCP)

(**c**) Ozone Depletion Potential (ODP)

(**d**) Acidification Potential (AP)

Figure 6. *Cont.*

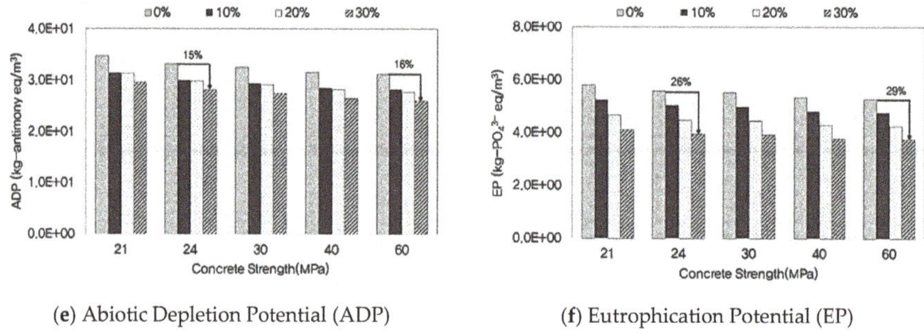

(**e**) Abiotic Depletion Potential (ADP) (**f**) Eutrophication Potential (EP)

Figure 6. Environmental impact analysis by recycled aggregate mixing ratio.

4. Conclusions

(1) By using the concrete life cycle assessment (LCA) technique suitable for the Korean situation, the effects of increase in strength, admixture material mixing, recycled aggregate mixing, and the amount of binder on environmental impact (global warming, acidification, eutrophication, ozone depletion, photochemical ozone creation, and abiotic depletion) were evaluated in a quantitative manner.

(2) It was found that the higher the strength of the concrete, the higher were the indices of global warming potential (GWP), photochemical ozone creation potential (POCP), and abiotic depletion potential (ADP). However, the acidification potential (AP), and eutrophication potential (EP) indices showed a tendency to decrease slightly.

(3) As admixture mixing ratio of concrete increased, the global warming potential (GWP), acidification potential (AP), ozone depletion potential (ODP), abiotic depletion potential (ADP), and photochemical ozone creation potential (POCP) decreased, but the eutrophication potential (EP) index showed a tendency to increase slightly.

(4) As the recycled aggregate mixing ratio of concrete increased, the acidification potential (AP), eutrophication potential (EP), ozone depletion potential (ODP), and abiotic depletion potential (ADP) decreased, but the global warming potential (GWP), and photochemical ozone creation potential (POCP) increased.

(5) GWP and POCP per unit compressive strength (1 MPa) of high-strength concrete (60 MPa) were found to be about 13% lower than that of normal strength (24 MPa). Also, in the case of AP, EP, ODP, and ADP per unit compressive strength (1 MPa), high-strength concrete (60 MPa) was found to be about 10%~25% lower than that of normal strength (24 MPa).

(6) Among the six environmental impact categories, it was found that ordinary cement had the greatest impact on global warming potential (GWP), photochemical ozone creation potential (POCP), and abiotic depletion potential (ADP), and

aggregate had the most effect on acidification potential (AP), eutrophication potential (EP), and ozone depletion potential (ODP).

Acknowledgments: This research was supported by Basic Science Research Program through the National Research Foundation of Korea (NRF) funded by the Ministry of Science, ICT & Future Planning (No.2015R1A5A1037548) and the National Research Foundation of Korea (NRF) grant funded by the Korea government (MSIP) (No.20110028794).

Author Contributions: All authors contributed substantially to all aspects of this article.

Conflicts of Interest: The authors declare no conflict of interest.

References

1. Dayoung, O.; Takafumi, N.; Ryoma, K.; Wonjun, P. CO_2 emission reduction by reuse of building material waste in the Japanese cement industry. *Renew. Sustain. Energy Rev.* **2014**, *38*, 796–810.
2. Federica, C.; Idiano, D.; Massimo, G. Sustainable management of waste-to-energy facilities. *Renew. Sustain. Energy Rev.* **2014**, *33*, 719–728.
3. Global BRE. *Methodology for Environmental Profiles of Construction Products-Product Category Rules for Type 3 Environmental Product Declaration of Construction Product*; BRE Press: Watford, UK, 2011.
4. International EPD system. Product category rules-Concrete. 2013. Available online: http://www.environ dec.com/PCR/ (accessed on 13 April 2016).
5. ISO/DIS 13315-2. In *Environmental Management for Concrete and Concrete Structures)-Part 2: System Boundary and Inventory Data*; ISO: Geneva, Switzerland, 2013.
6. Ministry of government legislation. Law of Low-Carbon on Green growth. 2013. Available online: http://www.moleg.go.kr/english/korLawEng?pstSeq=54792 (accessed on 13 April 2016).
7. Sakai, K.; Kawai, K. Recommendation of Environmental Performance Verification for Concrete Structures, JSCE Guidelines for Concrete No. 7. Available online: http://www.jsce.or.jp/committee/concrete/e/newsletter/newsletter06/newsletter06.asp (accessed on 13 April 2016).
8. Van den Heede, P.; De Belie, N. Environmental impact and life cycle assessment (LCA) of traditional and "green" concretes: Literature review and theoretical calculations. *Cem. Concr. Compos.* **2012**, *34*, 431–442.
9. Organización Internacional de Normalización. *ISO 14025: Environmental Labels and Declarations-Principles and Procedures*; ISO: Geneva, Switzerland, 2006.
10. Organización Internacional de Normalización. *ISO 14044: Life Cycle Assessment-Requirements and Guidelines*; ISO: Geneva, Switzerland, 2006.
11. Youngche, H.; Sangwon, S.; Sangwon, H.; Keonmo, L. Determination of normalization values for korean eco-indicator. *Korean Soc. Life Cycle Assess.* **2000**, *2*, 69–78.
12. Korea Concrete Institute. *Concrete and Environment*; Kimoondang Publishing Company: Seoul, Korea, 2011; pp. 16–30.

13. MOLIT. *National Database for Environmental Information of Building Products*; Ministry of Land, Transport and Maritime Affairs of the Korean Government: Sejong-si, Korea, 2008.
14. National Life Cycle Index Database Information Network. Available online: http://www.edp.or.kr (accessed on 11 August 2015).
15. The Ecoinvent Database. Available online: http://www.ecoinvent.org/database (accessed on 11 August 2015).
16. Eco-Labeling. Available online: http://www.edp.or.kr/edp/edp_intro.asp (accessed on 11 August 2015).
17. IPCC guidelines for national greenhouse gas inventories. 2006. Available online: http://www.ipcc- nggip.iges.or.jp/public/2006gl/ (accessed on 13 April 2016).
18. Guinee, J.B. Development of a Methodology for the Environmental Life Cycle Assessment of Products: With a Case Study on Margarines. Ph.D. Thesis, Leiden University, Leiden, The Netherlands, 1995.
19. Heijungs, R.; Guinée, J.B.; Huppes, G.; Lamkreijer, R.M.; Udo de Haes, H.A.; Wegener Sleeswijk, A.; Ansems, A.M.M.; Eggels, P.G.; van Duin, R.; de Goede, H.P. *Environmental Life Cycle Assessment of Products. Guide (Part1) and Background (Part 2)*; CML Leiden University: Leiden, The Netherlands, 1992.
20. World Metrological Organization (WMO). *Scientific Assessment of Ozone Depletion: Global Ozone Research and Monitoring Project*; WHO: Geneva, Switzerland, 1991; p. 25.
21. Derwent, R.G.; Jenkin, M.E.; Saunders, S.M.; Piling, M.J. Photochemical ozone creation potentials for organic compounds in Northwest Europe calculated with a master chemical mechanism. *Atmos. Environ.* **1998**, *32*, 2429–2441.
22. Jenkin, M.; Hayman, G. Photochemical Ozone Creation Potentials for oxygenated volatile organic compounds: Sensitivity to variation is in kinetic and mechanistic parameters. *Atmos. Environ.* **1999**, *33*, 1275–1293.
23. Ministry of Government Legislation. Details of Laws and enforcement ordinance and regulations on recycling promotion of construction wastes. 2010. Available online: https://www.google.com/url?sa =t&rct=j&q=&esrc=s&source=web&cd=1&ved=0ahUKEwiB1u7mwJDMAhULjiwKHRZTDXgQFggdMAA&url=http%3a%2f%2fwww.meti.go.jp%2fenglish%2finformation%2fdownloadfiles%2fcRecycle3R20403e.pdf&usg=AFQjCNE3tzbP17a8V2s7OVD1biXT4gN6KQ&sig2=8AccAo4Gl6buk49UI46BKw (accessed on 11 August 2015).
24. Taehyoung, K.; Sungho, T.; Seongjun, R. Assessment of the CO_2 emission and cost reduction performance of a low-carbon-emission concrete mix design using an optimal mix design system. *Renew. Sustain. Energy Rev.* **2013**, *25*, 729–741.
25. Sungho, T.; Sungwoo, S.; Ganghee, L. The Environmental Load Reduction Effect of the Reinforced Concrete Structure Using High-Strength Concrete. *Korea Inst. Ecol. Archit. Environ.* **2008**, *8*, 61–66.
26. Junghoon, P.; Sungho, T.; Taehyoung, K. Life Cycle CO_2 Assessment of Concrete by compressive strength on Construction site in Korea. *Renew. Sustain. Energy Rev.* **2012**, *16*, 2940–2946.

27. Gazquez, M.J.; Bolivar, J.P.; Vaca, F.; Garcia-Tenorio, R.; Caparros, A. Evaluation of the use of TiO_2 industry red gypsum waste in cement production. *Cem. Concr. Compos.* **2013**, *37*, 76–81.
28. Owaid, H.M.; Hamid, R.B.; Taha, M.R. A review of sustainable supplementary cementitious materials as an alternative to all-Portland cement mortar and concrete. *Aust. J. Basic Appl.* **2012**, *6*, 2887–3303.
29. Taehyoung, K.; Sungho, T.; Seongjun, R. Life Cycle Assessment for Carbon Emission Impact Analysis of Concrete Mixing Ground Granulated Blast-furnace Slag (GGBS). *Archit. Inst. Korea* **2013**, *29*, 75–82.
30. Yang, K.H.; Seo, E.A.; Jung, Y.B.; Tae, S.H. Effect of Ground Granulated Blast-Furnace Slag on Life-Cycle Environmental Impact of Concrete. *Korea Concr. Inst.* **2014**, *26*, 13–21.
31. Korea Ground Granulated Furnace Slag Association. Technical Report. 2015. Available online: http://www. asa-inc.org.au/news-and-events/connections (accessed on 11 August 2015).
32. Jongsuk, J.; Jaesung, L.; Yangjin, A.; Kyunghee, L.; Kisun, B.; Myunghun, J. An Analysis of Emission of Carbon Dioxide from Recycling of Waste Concrete. *Archit. Inst. Korea* **2008**, *24*, 109–116.

The Use of MIVES as a Sustainability Assessment MCDM Method for Architecture and Civil Engineering Applications

Oriol Pons, Albert de la Fuente and Antonio Aguado

Abstract: Environmental and sustainability assessment tools have an important role in moving towards a better world, bringing knowledge and raising awareness. In the architecture and civil engineering sector, these assessment tools help in moving forward to constructions that have less economic, environmental and social impacts. At present, there are numerous assessment tools and methods with different approaches and scopes that have been analyzed in numerous technical reviews. However, there is no agreement about which method should be used for each evaluation case. This research paper synthetically analyzes the main sustainability assessment methods for the construction sector, comparing their strengths and weaknesses in order to present the challenges of the Spanish Integrated Value Model for Sustainability Assessment (MIVES). MIVES is a Multi-Criteria Decision Making method based on the value function concept and the Seminars of experts. Then, this article analyzes MIVES advantages and weak points by going through its methodology and two representative applications. At the end, the area of application of MIVES is described in detail along with the general application cases of the main types of assessment tools and methods.

Reprinted from *Sustainability*. Cite as: Pons, O.; de la Fuente, A.; Aguado, A. The Use of MIVES as a Sustainability Assessment MCDM Method for Architecture and Civil Engineering Applications. *Sustainability* **2016**, *8*, 460.

1. Introduction

Environmental impacts caused by humans have been increasing for decades with serious consequences such as climate change [1]. During the last decade, some countries have seriously tackled these ecological problems [2] but their measures have been insufficient partly due to the high rate of development, resources consumption and environmental impact in developing countries [3].

Nowadays, more than 50 years after Life Cycle Assessment originated [4], more than 40 years after the 1973 oil crisis and almost 30 years after the 1987 Brundtland Commission report [5], the importance of sustainability and its assessment is accepted worldwide. Brundtland and other reports that followed [6] promoted sustainable development in a holistic point of view taking into account economic, environmental and social impacts.

The building sector causes an important part of these impacts during its materials production, the buildings usage phase and the buildings demolition or end of life among other phases [7]. For this sector, there are several tools, database and methods available to assess sustainability and environmental aspects within the architectural and civil engineering area [8]. However, there is no unanimity yet as to which criteria and indicators or which method is better to use in each case [9].

These tools differ both in their scope and approach [10]. Numerous methods focus on measuring disaggregated indicators in detail [11]. An important group of well-known tools have been broadly used during decades to assess the environmental impacts within the construction sector. This is the case of the Life Cycle Assessment (LCA) [12–16] in which environmental inputs, outputs and impacts are evaluated through the life cycle of a building or part of it. Some important LCA partial steps are: the Life Cycle Inventory (LCI) [17], which catalogues and quantifies a specific product's inputs and outputs during its life cycle, and the Life Cycle Impact Assessment (LCIA), which evaluates impacts for a building system throughout the life cycle of a construction product [18].

On the other hand, Life Cycle Cost (LCC) [19] is a tool that permits analysts to carry out an economic analysis focusing on the purchasing and operating phases of a building over a period of time; while Life Cycle Energy Analysis allows assessing energy inputs during the life cycle of a building [8]. Material Flow Analysis (MFA) is a related method that can measure and analytically quantify flows and stocks of construction materials [20]. Material and Energy Flow Analysis (MEFA) also incorporates the flow of energy [21]. These tools are exemplary in using scientific methods to measure and apply criteria to assess environmental impacts [9]. However, numerous researchers agree that is not feasible to apply these LCA related methods for covering the analysis of complete buildings [22–24].

Other tools and methods include more than one sustainability requirement. In this regard, methods that add economic and/or social indicators to environmental requirements have increased in the last decade [25,26], as well as standards for social indicators [27]. More recently, some of these tools and assessments also include technical, functional and governance requirements as well [28]. In this category, there are the certification tools for the building sector [29]. Table 1 presents eight of these certification tools, which are a representative sample of more than 30 methodologies studied in review papers [9,30,31].

Most of these exemplify this recent tendency since social and economic aspects are incorporated. The acceptance of these methods has shown to be variable, as only two have been internationally applied for decades, while the others are mainly used in the country of origin. The methods gathered within Table 1 also differ in being credits or percentage based rating tools, the application's complexity and the outcomes resulting from each method, which is in most of the cases, either a certification with a qualification of satisfaction or a graphic sustainability index.

Table 1. Different sustainability assessment tools for buildings.

Name	Institution	Origin	Use	C	E	S	CR	PR	C	R	
BEAM	BEAM	Hong Kong 1996	N	L	X	L	X	-	L	CQ	[32]
BREEAM	BRE	UK 1990	I	-	X	L	X	-	L	CQ	[33]
DGNB	DGNB	Germany 2008	N	X	X	X	-	X	H	CQ	[34]
EcoEffect	KTH	Sweden 2000	N	X	X	-	X	-	M	GI	[35]
Green Star	GBCA	Australia 2003	N	-	X	X	X	-	M	CQ	[36]
HQE	AssoHQE	France 1996	N	X	X	X	-	X	M	CQ	[37,38]
LEED	USGBC	USA 2000	I	L	X	L	X	-	L	CQ	[39,40]
VERDE	GBCE	Spain 2010	N	X	X	X	-	X	M	CQ	[41,42]

Legend: Use: I = internationally consolidated, N = nationally consolidated; C. Economic requirements (cost, time, *etc.*). X means included, L means low consideration; E. Environmental requirements (energy consumption, CO_2 emissions, *etc.*). X means included; S. Social requirements (health, safety, quality, *etc.*). X means included, L means low consideration; CR. X means credits based rating tool that gives credits to carry out the assessment; PR. X means percentage based rating tool that assesses the percentage of satisfaction of each indicator; C. Tool complexity of application. H means high, M means medium, L means low; R. Result. CQ means certification with a qualification of satisfaction; I means graphic index.

Most of the mentioned tools and methods can be applied during the design, construction and use phases of a building and mainly rely on experimental and quantitative data. On the other hand, some sustainability assessment studies and research focus on the post occupancy period like Post Occupancy Evaluations [43], others focus on data from participatory processes [44,45] and other studies are advancing in order to incorporate both [46]. All the aforementioned tools have contributed to advance towards a more sustainable construction sector and to raise awareness of this issue within the sector [9]. However, most of them are specialized in quantifying specific branches of sustainability, such as the environmental or the economic branch. Only a few methods are capable of quantifying all the different social, economic and environmental requirements that permit researchers to derive a global sustainability index. Finally, when a sustainability assessment requires a specialized tool for a particular study case, the aforementioned tools are scarcely representative.

In this regard, this research paper presents a detailed analysis of the MIVES method (from the Spanish Integrated Value Model for the Sustainability Assessment). MIVES is a Multi-Criteria Decision Making (MCDM) method capable of defining specialized and holistic sustainability assessment models to obtain global sustainability indexes. There have already been numerous applications of MCDM in engineering [47], most focusing on economic aspects [48–50] and fewer about environmental issues [51–53] or social aspects [54,55]. Some MCDM tools incorporate different sustainability branches [56,57] and some are specialized in the construction sector [58]. The MIVES method is a unique MCDM based on the use of value functions [59] to assess the satisfaction of the different stakeholders involved in the decision-making process. The use of these functions allows minimizing the subjectivity in the assessment. So far, MIVES has already been used for industrial buildings [60–63], underground infrastructures [64], hydraulic structures [65,66], wind towers [67], sewage systems [68], post-disaster sites and housing selection [69,70] and construction projects [71,72]. It should be highlighted that in the current Spanish Structural Concrete Code [73], MIVES method is proposed for assessing the sustainability of concrete structures [74]. Finally, it must be added that the MIVES method has even been expanded to include the uncertainties involved in the process of analysis [75].

This present research paper describes the main advantages and weak points of this method. It also shows these features by presenting two representative examples of MIVES sustainability assessment tools and their potential applications. Finally, this research paper concludes providing the area of application for this MCDM and the main types of sustainability assessment tools.

2. Methodology

MIVES is a methodology that was developed at the start of the new Millennium [53]. As previously said, it is a unique MCDM because MIVES combines: (a) a specific holistic discriminatory tree of requirements; (b) the assignation of weights for each requirement, criteria and indicator; (c) the value function concept [59] to obtain particular and global indexes; and (d) seminars with experts using Analytic Hierarchy Process (AHP) [76] to define the aforementioned parts.

Experts from the field of research and from the different institutions and companies involved in each research project participate in these seminars. These experts use AHP to define: the requirements tree, their criteria and indicators, their weights and their value functions. They bring knowledge and expertise and also take into account related previous research projects and technical bibliography. These seminars by experts bring objectivity, reality and complexity to the resulting assessment tool. For example, when defining the requirements tree this objectivity is crucial in order to obtain the correct assessment. Each case of study needs its own tree that incorporates exclusively its most significant and discriminatory indicators. It is also of great importance that the amount of indicators is not excessive.

2.1. MIVES Process

The assessment of the sustainability index by using the MIVES method should be carried out following these steps: (S1) define the problem to be solved and the decisions to be made; (S2) produce a basic diagram of the decision model, establishing all those aspects that will be part of a requirements tree that may include qualitative and quantitative variables; (S3) establish the value functions to convert the qualitative and quantitative variables into a set of variables with the same units and scales; (S4) define the importance or relative weight of each of the aspects to be taken into account in the assessment; (S5) define the various design alternatives that could be considered to solve the previously identified problem; (S6) evaluate and assess those alternatives by using the previously created model; and (S7) make the right decisions and choose the most appropriate alternative.

Examples of step S1 are explained in Section 3. The requirements tree in S2 (Figure 1) is a hierarchical diagram in which the various characteristics of the product or processes to be evaluated are organized, normally at three levels: indicators, criteria, and requirements. At the final level, the specific requirements are defined and the previous levels (criteria and indicators) are included in order to desegregate the requirements; thus permitting: (1) having a global view of the problem; (2) organizing the ideas; and (3) facilitating the comprehension of the model to any stakeholder involved in the decision process. There are also examples of requirements trees in Section 3.

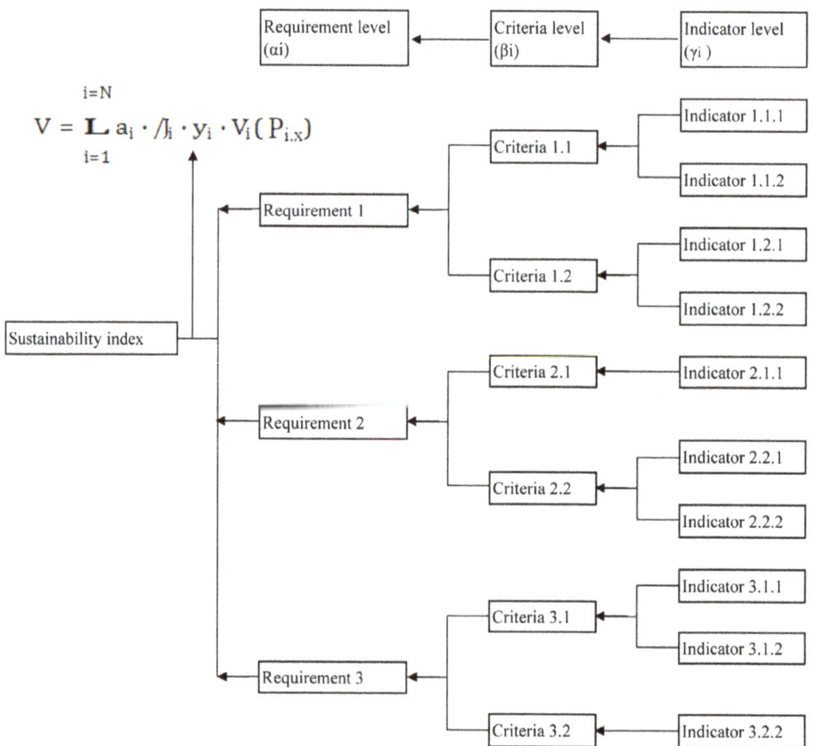

Figure 1. General requirements tree.

Afterwards, in S3 the value functions are used to formalize a method to convert the different criteria magnitudes and units into a common, non-dimensional unit that will be called value. In this sense, it should be noticed that this method accounts for both qualitative and quantitative variables related with the indicators.

In any multi-criteria decision problem, the decision maker has to choose between a group of alternatives [77], these being discrete or continuous. Thus, when the preferences (x) are known with respect to a set of design alternatives (X), a value function $V{:}P \rightarrow R$ can be fixed such that $P_x > P'_x$ so that $V(P_x) > V(P'_x)$, P being equal to a set of criteria to be evaluated for alternative x. The problem consists of generating a non-dimensional value function $V(P_x)$ that reflects the preferences of the decision maker for each alternative while integrating all the criteria $P_x = (P_{1,x}; P_{2,x}; \ldots ; P_{N,x})$. The solution is a function V consisting of the sum of N value functions Vi corresponding to the N criteria which comply with $V_i{:}P \rightarrow R$ so $P_{i,x} > P'_{i,x} \leftrightarrow V_i(P_{i,x}) > V_i(P'_{i,x})$. For the case of problems structured in the form of a requirements tree, the resulting Sustainability Index (SI) can be assessed using Equation (1).

$$SI = V(P_x) = \sum_{i=1}^{i=N} \alpha_i \cdot \beta_i \cdot \gamma_i \cdot V_i(P_{i,x}) \tag{1}$$

In Equation (1), $V(P_x)$ measures the degree of sustainability (value) of the alternative x evaluated with respect to various criteria $P_x = (P1,x;P_{2,x}; \ldots ;P_{N,x})$ considered. α_i are the weights of each requirement i, β_i are the weights of each criteria i and γ_i are the weights of the different indicators i. These weights are the preference, respectively, of these requirements, criteria and indicators. $V_i(P_{i,x})$ are the value functions used to measure the degree of sustainability of the alternative x with respect to a given criterion i. Finally, N is the total number of criteria considered in the assessment.

A main objective of the V_i functions is to homogenize the different indicators values, which have different measurement units, so to obtain a global sustainability index. In this regard, it is also highly recommended to delimit the values that these functions can generate. In this way, all the criteria have one single scale of assessment, normally between 0 and 1. These values represent the minimum and maximum degree of sustainability, respectively. A second main objective is to make it possible to weigh the V_i functions by weights α_i, β_i, and γ_i. It also makes it easier to obtain these weights (α_i, β_i, and γ_i) since it will only be necessary to establish the relative priority of certain requirements, criteria, or indicator with respect to other ones, regardless of whether some may present different scales of quantification.

Once the value functions have been defined, it is necessary to calculate weights α_i, β_i, and γ_i for each branch of the requirements tree (step S4 of MIVES). To this end, numerical values established by experts in the field are used. These weights are obtained using AHP in seminars as explained in the previous section. First, the weights of each requirement (α_i) are calculated. Then, within each requirement, weights β_i for the several criteria are calculated, and finally, the same process is done for each criteria to obtain the indicator weights γ_i. AHP is useful when the resulting initial trees are excessively complex, or discrepancies occur among the experts, or, simply, it is desirable to carry out an organized process to avoid difficulties in establishing the weights. Afterward, to compensate for possible subjective bias because of the use of semantic labels in AHP, a subsequent process of analyzing, comparing and, in case of being necessary, modifying the resulting weights is recommended.

The various alternatives x are defined in the following stage S5. After that, these alternatives are evaluated (step S6), and the sustainability index associated with each of them is calculated using Equation (1).

2.2. Definition of the Value Functions

Defining value functions is also crucial to add homogeneity to different indicators, which have different measurement units in order to obtain sustainability indexes. There is a value function (V_{Ii}) for each indicator. Defining each value function requires measuring preference or the degree of satisfaction produced by a certain alternative. Each measurement variable may be given in different units; therefore, it is necessary to standardize these into units of value or satisfaction, which is basically what the value function does. The method proposed a scale for which 0.0 reflects minimum satisfaction (P_{min}) and 1.0 reflects maximum satisfaction (P_{max}).

To determine the satisfaction value for an indicator [59], the MIVES model outlines a procedure consisting in the definition of: (1) the tendency (increase or decrease) of the value function; (2) the points corresponding to P_{min} and P_{max}; (3) the shape of the value functions (linear, concave, convex, S-shaped); and (4) the mathematical expression of the value function.

The general expression of the value function V_i used in MIVES to assess the satisfaction of the stakeholders for each indicator corresponds to Equation (2).

$$V_i = K_i \cdot \left[1 - e^{-m_i \cdot \left(|P_{i,x} - P_{i,min}|/n_i\right)^{A_i}}\right] \tag{2}$$

In Equation (3), variable *Ki* is a factor that ensures that the value function will remain within the range of 0.0–1.0 and that the best response is associated with a value equal to 1.0:

$$K_i = \frac{1}{1 - e^{-m_i \cdot \left(|P_{i,max} - P_{i,min}|/n_i\right)^{A_i}}} \tag{3}$$

In both Equations (2) and (3): (a) $P_{i,max}$ and $P_{i,min}$ are the maximum and minimum values of the indicator assessed. (b) *Pi,x* is the score of alternative x

that is under assessment, with respect to indicator i under consideration, which is between $P_{i,min}$ and $P_{i,max}$. This score generates a value that is equal to $V_i(P_{i,x})$, which has to be calculated. (c) A_i is the shape factor that defines whether the curve is concave ($A_i < 1$), a straight line ($A_i \approx 1$) or whether it is convex or S-shaped ($A_i > 1$). (d) n_i is the value used, if $A_i > 1$, to build convex or S-shaped curves. (e) m_i defines the value of the ordinate for point n_i, in the former case where $A_i > 1$.

The geometry of the functions V_i allows establishing greater or lesser exigency when complying with the requisites needed to satisfy a given criterion. For example, the convex functions experience a great increase in value for scores that are close to the minimum value, and the increase in value diminishes as the score approaches the maximum. This type of function is used when one wishes to encourage compliance with minimum requirements. That may be the case, for instance, with sufficiently exacting standards in which mere compliance is highly satisfactory. Another instance may be when the aim is to reward the use of new technologies, and their implementation is seen as very positive (even when it is a partial or a minor one), with a view to encouraging better practices. The maximum and minimum value criteria can be 0, the greatest or worst value of the studied alternatives, *etc.* depending on each study case.

It can be seen that the shape of the function depends on the values that the parameters A_i, n_i and m_i. The interpretation of these parameters facilitates the understanding and the use of Equation (2). Tables 2 and 3 give characteristic values of these parameters for the definition of increasing and decreasing value functions, respectively. These parameters may vary according to the preferences of the decision maker.

Table 2. Typical values of n_i, m_i and A_i for increasing value functions.

Function	n_i	m_i	A_i
Linear	$n_i \approx P_{i,min}$	≈0.0	≈1.0
Convex	$P_{i,min} + \frac{P_{i,max}-P_{i,min}}{2} < n_i < P_{i,min}$	<0.5	>1.0
Concave	$P_{i,min} < n_i < P_{i,min} + \frac{P_{i,max}-P_{i,min}}{2}$	>0.5	<1.0
S-shaped	$P_{i,min} + \frac{P_{i,max}-P_{i,min}}{5} < n_i < P_{i,min} + \left(\frac{P_{i,max}-P_{i,min}}{2}\right)\frac{4}{5}$	0.2–0.8	>1.0

Table 3. Typical values of n_i, m_i and A_i for increasing and decreasing functions.

Function	n_i	m_i	A_i
Linear	$n_i \approx P_{i,min}$	$\approx A_0$	≈1.0
Convex	$P_{i,max} + \frac{P_{i,min}-P_{i,max}}{2} < n_i < P_{i,max}$	<0.5	>1.0
Concave	$P_{i,max} < n_i < P_{i,max} + \frac{P_{i,min}-P_{i,max}}{2}$	>0.5	<1.0
S-shaped	$P_{i,max} + \frac{P_{i,min}-P_{i,max}}{5} < n_i < P_{i,max} + \left(\frac{P_{i,min}-P_{i,max}}{2}\right)\frac{4}{5}$	0.2–0.8	>1.0

When the shape of the value function for an indicator is unclear, this may be defined by a working group. In these cases, several value functions (discrete or continuous) may be defined according to the members of the group. Therefore, a family of functions is obtained as can be seen in Figure 2.

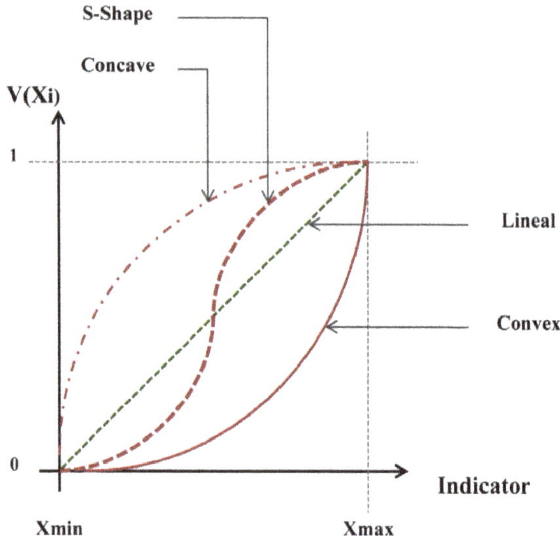

Figure 2. Value function generated by a working group composed of different decision makers.

The simplest way to solve these differences consists in taking the mean of the different values (after excluding extreme cases). The parameters A_i, A_i, n_i and m_i can then be estimated through a minimum squares approach. It is also possible to work with a range of values in such a way that two values correspond to each y-value (the mean and the standard deviation). This would require a statistical approach in the subsequent decision process.

This is the explanation that can be applied to study cases that have limited complexity and, therefore, a deterministic approach or for studies with homogeneous alternatives like a same building typology such as school buildings [71]. Nevertheless, for large and complex cases and assessments with uncertainty there are already developed probabilistic approaches [75] using MIVES. There are also MIVES methods for heterogeneous alternatives like public infrastructures, which can vary from metro lines to a health center building and require a previous homogenization phase [65,66].

3. Applications to the Building Sector

As explained in Section 2, MIVES has already been successfully applied to study the sustainability in numerous cases within the building sector [55–69]. Table 4 presents a representative sample of cases classified in four levels: energy, urban, edifices and building systems and elements. The diversity of these cases, some at an energy or urban level and some about building elements, some assessing broad samples in general and some carrying out analysis in detail, *etc.*, shows MIVES versatility. This table incorporates the reference where each sustainability assessment is explained thoroughly.

These sustainability assessments have their own particularities. In the following sections, the specific study Cases 6 and 9 are explained in detail. These two cases are representative of the particularities of the different cases shown in Table 4 and they show the main differentiate characteristics of MIVES. Case 6 assesses a large sample of more than 400 educational edifices while Case 9 analyzes a specific construction element but considering all its constructive and structural characteristics. They both follow all the methodology steps described in Section 2 but their requirements tree, weights and value functions are particular for each case, as shown in the following sections.

Table 4. Different sustainability assessments within the construction sector that have used MIVES.

Level	Sustainability Assessment	Ref.
Energy	1. Sustainability index of wind-turbine support systems	[67]
	2. Electricity generation systems	[78]
Urban	3. Sustainable site location of post-disaster temporary housing in urban areas	[69]
	4. Sustainability assessment of sewerage pipe systems	[68]
Edifices	5. Sustainability of post-disaster temporary housing units technologies	[70]
	6. Sustainable assessment applied to technologies used to build schools	[71]
	7. Environmental analysis of industrial buildings	[61]
Building systems and elements	8. Sustainability assessment of concrete structures	[74]
	9. Sustainability assessment method applied to structural concrete columns	[72]
	10. Sustainability assessment of concrete flooring systems	[79]

3.1. *Sustainable Assessment Applied to Technologies Used to Build Schools*

This assessment tool and its application are explained in detail in technical bibliography [71]. It was a tool designed and applied to assess the sustainability of more than 400 schools built in Spain in the early 2000s [80]. This tool focused on the analysis of these educational edifices construction processes and the technologies used to build them. These schools were public kindergarten and primary centers for 200 to 400 students, constructed from 2000 to 2014 in short time frames and tight budgets due to an extreme need for new educational centers in short time. They each had a surface from 1500 to 3000 m^2 in which there were: (a) classes and auxiliary spaces for the kindergarten and the primary area; and (b) common areas like a lunchroom, gym, administration and teachers' area, *etc.*

These educational edifices were constructed using different technologies. The most representative of which were: on site concrete structure system (NC) [81], off-site concrete framed structure technology (FC) [82], off-site steel modules structure system (FS) [83] and off-site timber structure system (FT) [84]. Table 5 presents some important characteristics of them.

Table 5. Important characteristics of the main building technologies.

	NC	FC	FS	FT
Structural material	Concrete	Concrete	Steel	Timber
Structural technology	Frames	Frames	Modules of frames	Load-bearing walls
Average building speed (m^2/month)	150	250	250	250
Average distance factory-site (km)	85	150	900	1600
Average weight (kg/m^2)	1464	946	408	507
Scaffolding required	High	Low	None	Low
Disassembly possibilities	None	Low	High	High

The definition and application of this MIVES tool solved the endemic necessity of evaluating the economic, social and environmental impact of these school centers and their construction as well as new schools in the future. Therefore, this tool incorporates a simplified Life Cycle Assessment (LCA) [7]. Table 6 shows the requirements tree of this tool, with its weights and value functions shapes, which were defined during seminars attended by experts from all the involved parties [71].

The main results of this assessment were the global and the partial sustainability indexes for the construction alternatives. With the global indexes, it was possible to qualify the assessed technologies from more to less sustainable and partial indexes were useful to give advice to their industries in order to improve. These global indexes were: 0.35 for NC; 0.72 for FC; 0.71 for FS and 0.59 for FT. For example, to the studied timber technology, this had an unexpectedly low sustainable index although being a high performance environmental building system.

The reason was its unsuitability for this particular study case due to the unsustainable distance between the production center and the building site. This conclusion demonstrated that technologies are not excellent by themselves but depend on their application. Finally, the authors recommended building future educational edifices following this study sustainability requirement.

Table 6. Requirements tree, weights and value functions shapes for the schools technologies assessment.

Requirements	Criteria	Indicators
R1. Economic (50%)	C1. Cost (52%)	I1. Production and assembly cost (30%, DCx) I2. Cost deviation probability (25%, DS)
		I3. Maintenance cost (45%, DS)
	C2. Time (48%)	I4. Production and assembly timeframe (38%, DCx) I5. Timeframe deviation probability (62%, DCx)
R2. Environmental (30%)	C3. Phase 1: extraction and fabrication of materials (30%)	I6. Water consumption (22%, DCx) I7. CO_2 emissions (40%, DCx) I8. Energy consumption (38%, DCx)
	C4. Phase 2: transport (10%)	I9. CO_2 emissions (100%, DCx)
	C5. Phase 3: building and assembly (15%)	I10. CO_2 emissions (58%, DCv) I11. Solid waste (42%, DCv)
	C6. Phase 4: use and maintenance (30%)	I12. CO_2 emissions (Weight 100%, DCx)
	C7. Phase 5: demolition (15%)	I13. Solid waste (Weight 100%, DS)
R3. Social (20%)	C8. Adaptability to changes (35%)	I14. Neither adaptable nor disassemble building percentage (theoretical) (50%, DS) I15. Deviation of neither adaptable nor disassemble building percentage (50%, DS)
	C9. Users' safety (65%)	I16. Labor risk of accidents during building and assembly (40%, DCv) I17. Users risk of accidents during buildings enlargements (60%, DCv)

Legend: weights are in percentage between brackets; value functions shapes: DCx stands for decrease convexly, DCv decrease concavely, DS decrease like an S.

3.2. Sustainability Assessment Method Applied to Structural Concrete Columns

This columns sustainability assessment tool was designed to evaluate the sustainability of this single structural component and it is described in detail in technical literature [72]. Columns are crucial for the mechanical functionality and safety of most buildings, in which they are the structural elements that transmit loads from each floor to the floor below and down to the foundation components. Columns can total up to 25% of the concrete and steel consumption of a building. Therefore, they can significantly reduce a building's environmental impact by being designed and constructed with the optimum geometry, materials and construction process.

In this sense, this sustainability tool has been applied to analyze alternatives that use fast hardening, self-compacting and high strength concretes. With these advanced concretes it is possible to build columns with smaller cross-sections and higher load capacity. Thus, they permit the building to consume less material, to reduce the columns section and achieve a more optimum profit in available edifice space. These concretes also increase the work performance with shorter construction timeframes. In consequence, several social impact factors such as construction noises and special transportation methods are reduced.

This tool has already been applied to evaluate the sustainability of several alternatives of *in situ* reinforced concrete columns for medium size buildings with a maximum of six levels and 500 m^2 per floor. The analysis focuses on the third and fourth floor columns inside the building. These columns are 3 m high and distributed in a 6 m by 6 m structural net. These have mainly moderated compression stresses and not have an excessive reinforcement ratio due to bending loads. Nevertheless, uncertainties such as initial imperfections and building faults have been considered. To do so, as suggested in Model Code 2010 [85], a minimum eccentricity (e_{min}) of a value of h/30 has been taken into account, h being the maximum cross-section dimension. These alternatives differ in their cross-section shapes and dimensions, concrete compressive strengths and construction processes. This MIVES tool has also been able to analyze the influence of these variables in the sustainability index of the assessed alternatives, which are presented in Table 7. The requirements tree, weights and value functions shapes of this tool are shown in Table 8.

Table 7. Alternatives of columns assessed in this study case.

Alternative	C.Ch. Strength (N/mm^2)	Cross-Section (cm)	Construction Process
Circular 1 Circular 2	25	Ø 30	S-C V
Circular 3 Circular 4	50	Ø 35	S-C V
Circular 5 Circular 6	75	Ø 50	S-C V
Square 1 Square 2	25	25 × 25	S-C V
Square 3 Square 4	50	30 × 30	S-C V
Square 5 Square 6	75	40 × 40	S-C V

Legend: C.Ch. strength: Concrete characteristic strength; S-C: Using self-compacting concrete; V: vibrating the concrete.

As a result of this assessment we concluded that the most sustainable columns are those with smaller cross-sections and are built using high characteristic compressive strength concretes. This is the alternative Circular 1 with a global sustainable index of 0.85. Analyzing the aforementioned variables we also concluded that: columns executed using self-compacting concretes have a higher sustainability index than those which require being vibrated; circular columns are more sustainable than those square or rectangular shaped due to aesthetic and functional reasons;

circular columns have a higher index when using high performance concrete and having small cross-section areas; and square and rectangular alternatives are more sustainable when using conventional concretes and having bigger cross-sections.

Table 8. Requirements tree, weights and value functions shapes for structural concrete columns.

Requirements	Criteria	Indicators
R1. Economic (50%)	C1. Construction costs (67%)	I1. Building costs (85%, DS) I2. Non acceptance costs (15%, IL)
	C2. Efficiency (33%)	I3. Maintenance (60%, DS) I4. Habitability (40%, DCv)
R2. Environmental (33%)	C3. Emissions (67%)	I5. CO_2 emissions (100%, DS)
	C4. Resources consumption (33%)	I6. Concrete consumption (90%, DCv) I7. Steel consumption (10%, DCx)
R3. Social (17%)	C5. Negative effects on the producer industry (80%)	I8. Workers' inconveniences (20%, DS) I9. Workers' safety (80%, IL)
	C6. Effects to third party (20%)	I10. Environment nuisances (100%, IL)

Legend: weights are in percentage between brackets; value functions shapes: DCx stands for decrease convexly, DCv decrease concavely, DS decrease like an S, IL increase lineally.

4. Discussion

MIVES can generate useful sustainability assessment tools in a broad range of cases within the building sector [55–69]. As seen in previous sections, it can define holistic tools for specific samples and study cases. In some of these cases, the suitability of MIVES application has been assessed and this MCDM has proved to be the best sustainable tool to use in those studies [71]. To use this tool, a rigorous and complete process must be carried out following the seven steps described in Section 2. This process is the origin of the main weak points of MIVES. These drawbacks are the experts' time and dedication needed to define each specialized assessment tool following the aforementioned steps. These weaknesses can be overcome with outstanding results, as proven in tight framed cases in which MIVES has already been applied [69–71]. Their results prove that these weaknesses exclusively difficult the methodology process but do not have any effect on the resulting tool or its sustainability assessment. Therefore, solutions to these weaknesses are based on advancing the seven steps of this methodology process before the application of the assessment tool. However, in some assessment cases, these drawbacks will result in it not being feasible to apply this MCDM.

This is the case of samples that can be assessed successfully using existing environmental and sustainability assessment tools. The main sample of these tools has been presented in the introduction. As has been explained, there are two main groups with their own field of application:

(a) Environmental evaluating tools for detailed specific and particular studies. This is the case of LCA, LCC and similar tools, which are the best options to assess a specific issue during part or the whole life cycle of a construction defined entity. This issue can be the environmental impact, economic impact, energetic consumption, materials flow, *etc.* The defined entity can be a construction element like a brick, a construction material like concrete, *etc.*
(b) Certification tools that give a qualification of satisfaction of different sustainability aspects that are evaluated of a building. A representative sample of these tools and their main features are presented in the introduction. It can be seen that they differ in their area of application, rating system, complexity, result, *etc.* Choosing between one tool and another is a complex task but internationally recognized tools have broader experience and reputation while local tools are more sensible to local features.

There is a third more heterogeneous group comprised of specific research projects and studies. Some cases could be solved by following the steps of a previous similar study using MCDM, POE or participatory processes, *etc.* However, any new study case should be similar to the previous one in which this new study relies on. Moreover, any required changes and adaptations would be important time consuming drawbacks.

Finally, in other cases not described in the previous paragraphs, no other tools have been found that could be used to carry out the assessment.

5. Conclusions

A thorough review of sustainability assessment tools confirms that there is a lack of a general method capable of covering all assessment cases. Contrarily, it has been highlighted that each tool is rather oriented to specific purposes: (a) environmental evaluating tools for detailed specific and particular studies; (b) certification tools for sustainability certifications of buildings; (c) MIVES for holistic sustainability assessments of specific cases; and (d) a more heterogeneous group comprised of particular sustainability research projects and study cases. It has also pointed out that the sustainability assessment of some cases cannot be dealt representatively with the existing tools. This means that from now on there is still work to do to cover all sustainable assessments, either increasing the application capacity of the existing tools or defining new ones.

In this research paper, a complete analysis of MIVES has been carried out. This methodology has been studied in detail by showing and discussing its phases, its area of application and two study cases. This analysis concludes that MIVES can define complete, objective and easy to apply sustainability assessment methods for most samples within the construction sector. These methods are specific for each case, can be deterministic or probabilistic, assess homogenous or heterogeneous alternatives and give integrated sustainability indexes. These MCDM advantages rely on a rigorous defining process that requires time and dedication from multidisciplinary experts. In consequence, MIVES is not applicable to urgent assessments that need a new specific tool. This limitation is common to first time applied tools that need a definition process prior to the assessment, like previously mentioned MCDM, POE, *etc.* Nevertheless, as shown in previous sections, several MIVES study cases have already solved this limitation anticipating the defining process of each new tool.

Acknowledgments: The authors acknowledge all their colleagues and Professors who have collaborated in developing MIVES method, as well as the 16 PhDs who have carried out their theses using this methodology. Acknowledgements are also given to the economic support provided by the Spanish Ministry of Economy and Competitiveness to several MIVES research projects such as the recent project BIA2013-49106-C2-1-R.

Author Contributions: All authors designed the research, applications, discussion, conclusions and read and approved the final manuscript. Oriol Pons wrote the research paper; Antonio Aguado and Oriol Pons did the state of the Art; and Antonio Aguado and Albert de la Fuente did the methodology.

Conflicts of Interest: The authors declare no conflict of interest.

References

1. UNEP. Annual Report 2014. Available online: http://www.unep.org/annualreport/2014/en/index.html (accessed on 6 May 2016).
2. Zhang, Z.; Xue, B.; Chen, X.; Pang, J. The decoupling of resource consumption and environmental impact from economic growth in china: Spatial pattern and temporal trend. *Sustainability* **2016**.
3. Environmental Performance Index, Global Metrics for Environment. Report 2016. Available online: http://epi.yale.edu/reports/2016-report (accessed on 6 May 2016).
4. Hunt, R.G.; Franklin, W.E. LCA—How it came about? *Int. J. Life Cycle Assess* **1996**, *1*, 147–150.
5. Brundtland, G.H. Our Common Future: Report of the World Commission on Environment and Develepoment. 1987. Available online: http://www.un-documents.net/ (accessed on 6 May 2016).
6. ICLEI. *Towards Sustainable Cities & Towns: Report of the First European Conference on Sustainable Cities & Towns*; International Council for Local Environmental Initiatives: Freiburg, Germany, 1994.
7. Pons, O.; Wadel, G. Environmental impacts of prefabricated school buildings in Catalonia. *Habitat Int.* **2011**.

8. Cabeza, F.; Rincón, L.; Vilariño, V.; Castell, G.P.A. Life Cycle Assessment (LCA) and life Cycle Energy Analysis (LCEA) of buildings and the building sector: A review. *Renew. Sustain. Energy Rev.* **2014**.
9. Marjaba, G.E.; Chidiac, S.E. Sustainability and resiliency metrics for buildings—Critical review. *Build. Environ.* **2016**.
10. Todd, J.A.; Crawley, D.; Lindsey, S.G.G. Comparative assessment of environmental performance tools and the role of the Green Building Challenge. *Build. Res. Inf.* **2001**.
11. Gundes, S. The Use of Life Cycle Techniques in the Assessment of Sustainability. *Procedia Soc. Behav. Sci.* **2016**.
12. Buyle, M.; Audenaert, J.B.A. Life cycle assessment in the construction sector: A review. *Renew. Sustain. Energy Rev.* **2013**.
13. Ortiz, O.; Castells, F.; Sonnemann, G. Sustainability in the construction industry: A review of recent developments based on LCA. *Construct. Build. Mater.* **2009**.
14. ISO. *14040: 2006. Environmental Management. Life Cycle Assessment. Principles and Framework*; International Organization for Standardization: Geneva, Switzerland, 2006.
15. Joint Research Centre, European Commission, Institute for Environment and Sustainability. International reference life cycle data system (ILCD) handbook: General guide for life cycle assessment—Detailed guidance. 2010. Available online: http://eplca.jrc.ec.europa.eu/ (accessed on 6 May 2016).
16. Glass, J.; Dyer, T.; Georgopoulos, C.; Goodier, C.; Paine, K.; Parry, T.; Baumann, H.; Gluch, P. Future use of life-cycle assessment in civil engineering. *ICE Construct. Mater.* **2013**, *166*, 204–212.
17. Gursel, A.P.; Masanet, E.; Horvath, A.; Stadel, A. Life-cycle inventory analysis of concrete production: A critical review. *Cement. Concr. Comp.* **2014**, *51*, 38–48.
18. Grosso, M.; Thiebat, F. Life cycle environmental assessment of temporary building constructions. In Proceedings of the 6th International Building Physics Conference, Torino, Italy, 14–17 June 2015.
19. Asiedu, Y.; Gu, P. Product life cycle cost analysis: State of the art review. *Ind. J. Prod. Res.* **1998**, *36*, 883–908.
20. Allesch, A.; Brunner, P. Material flow analysis as a decision support tool for waste management: A literature review. *J. Ind. Ecol.* **2015**.
21. Liu, L.; Issam, S.; Hermreck, C.; Chong, W. Integrating G2G, C2C and resource flow analysis into life cycle assessment framework: A case of construction steel's resource loop. *Resour. Conserv. Recycl.* **2015**, *102*, 143–152.
22. Quale, J.; Eckelman, M.; Williams, K.; Slodits, G. Construction matters: Comparing environmental impacts of building modular and conventional homes in the United States. *J. Ind. Ecol.* **2012**, *16*, 243–253.
23. Malin, N. Life cycle assessment for whole buildings: Seeking the holy grail. *Build. Des. Constr.* **2005**, *46*, 6–11.
24. Kohler, N.; Moffatt, S. Life-cycle analysis of the built environment. *Sustain. Build. Construct.* **2003**, *26*, 17–21.

25. Ali, H.; al Nsairat, S. Developing a green building assessment tool for developing countries—Case of Jordan. *Build. Environ.* **2009**, *44*, 1053–1064.
26. Cole, J. Building environmental assessment methods: Redefining intentions and roles. *Build. Res. Inf.* **2005**, *35*, 455–467.
27. ISO. *ISO 26000:2010 Guidance on Social Responsibility*; International Organization for Standardization: Geneva, Switzerland, 2010.
28. Salzer, C.; Wallbaum, H.; Lopez, L.; Kouyoumji, J. Sustainability of Social Housing in Asia: A Holistic Multi-Perspective Development Process for Bamboo-Based Construction in the Philippines. *Sustainability* **2016**.
29. Sustainable Building Alliance. 2016. Available online: http://www.sballiance.org/our-work/libraries/sbtool/ (accessed on 6 May 2016).
30. Haapio, A.; Viitaniemi, P. A critical review of building environmental assessment tools. *Environ. Impact Assess.* **2008**, *28*, 469–482.
31. Reed, R.; Bilos, A.; Wilkinson, S.; Schulte, K. International Comparison of Sustainable Rating Tools. *J. Sustain. Real State* **2009**, *1*, 1–22.
32. BEAM. BEAM Assessment Tool—BEAM Plus New Buildings, Version 1.2 (2012.07), Building Environmetn Assessment. 2012. Available online: http://www.beamsociety.org.hk/en_beam_assessment_project_1.php (accessed on 6 May 2012).
33. BREAM. BREAM International New Construction 2016. Technical Manual. SD233-1.0:2016. Available online: http://www.breeam.com/BREEAMInt2016SchemeDocument/ (accessed on 6 May 2016).
34. DGNB. The DGNB Certification System. Available online: http://www.dgnb-system.de/en/system/certification_system/ (accessed on 6 May 2016).
35. Assefa, G.; Glaumann, M.; Malmqvist, T.; Eriksson, O. Quality *versus* impact: Comparing the environmental efficiency of building properties using the EcoEffect tool. *Build. Environ.* **2010**, *45*, 1095–1103.
36. Green Building Council of Australia. Green Star Design & As Built. Available online: https://www.gbca.org.au/green-star/green-star-design-as-built/ (accessed on 6 May 2016).
37. HQE. Haute Qualité Environnementale (HQE). Available online: http://www.sballiance.org/our-work/libraries/haute-qualite-environnementale/ (accessed on 6 May 2016).
38. King Sturge. European Property Sustainability Matters (EPSM)—Delivering a Sustainable Future. 2011. Available online: http://www.propertyweek.com/Journals/44/Files/2011/2/24/EPSM%202011.pdf (accessed on 6 May April 2016).
39. USGBC. LEED. Available online: http://www.usgbc.org/leed (accessed on 6 May 2016).
40. USGBC. LEED V4 for Building Design and Construction: New Construction. 2015. Available online: http://greenguard.org/uploads/images/LEEDv4forBuildingDesignandConstructionBallotVersion.pdf (accessed on 6 May April 2016).
41. Green Building Council España (GBCe). VERDE. Available online: http://www.gbce.es/en/node/2075 (accessed on 6 May 2016).
42. Macías, M.; García, J. Metodología y herramienta VERDE para la evaluación de la sostenibilidad en edificios (VERDE, a methodology and tool for a sustainable building assessment). *Inf. Constr.* **2010**, *517*, 87–100.

43. Bonde, M.; Ramirez, J. A post-occupancy evaluation of a green rated and conventional on-campus residence hall. *Int. J. Sustain. Built Environ.* **2015**, *4*, 400–408.
44. Purtik, H.; Zimmerling, E.; Welpe, I. Cooperatives as catalysts for sustainable neighborhoods—A qualitative analysis of the participatory development process toward a 2000-Watt Society. *J. Clean. Prod.* **2016**.
45. Van den Burg, S.; Marian, S.; Norrman, J.; Garção, R.; Söderqvist, T.; Röckmann, C. Participatory Design of Multi-Use Platforms at Sea. *Sustainability* **2016**, *8*.
46. Brkovic, M.; Pons, O.; Parnell, R. Where sustainable school meets the 'third teacher': Primary school case study from Barcelona, Spain. *Archnet IJAR* **2015**, *9*, 77–97.
47. Kazimieras, E.; Antucheviciene, J.; Turskis, Z.; Adeli, H. Hybrid multiple-criteria decision-making methods: A review of applications in engineering. *Sci. Iran.* **2016**, *23*, 1–20.
48. Brauers, W.K. *Optimization Methods for a Stakeholder Society—A Revolution in Economic Thinking of Multi-Objective Optimization*; Springer US: New York, NY, USA, 2004.
49. Kazimieras, E.; Turskis, Z. Multiple criteria decision making (MCDM) methods in economics: An overview. *Technol. Econ. Dev. Ecol.* **2011**, *17*, 397–427.
50. Liou, J.; Tzeng, G. Multiple criteria decision making (MCDM) methods in economics: An overview. *Technol. Econ. Dev. Ecol.* **2012**, *18*, 672–695.
51. Chen, Y.; Lien, H.; Tzeng, G. Measures and evaluation for environment watershed plans using a novel hybrid MCDM model. *Expert. Syst. Appl.* **2010**, *37*, 926–938.
52. Chithambaranathan, P.; Subramanian, N.; Gunasekaran, A.; Palaniappan, P. Service supply chain environmental performance evaluation using grey based hybrid MCDM approach. *Int. J. Prod. Econ.* **2015**, *166*, 163–176.
53. Ilangkumaran, M.; Karthikeyan, M.; Ramachandran, T.; Boopathiraja, M.; Kirubakaran, B. Risk analysis and warning rate of hot environment for foundry industry using hybrid MCDM technique. *Saf. Sci.* **2015**, *72*, 133–143.
54. Tavana, M.; Momeni, E.; Rezaeiniya, N.; Mirhedayatian, S.; Rezaeiniya, H. A novel hybrid social media platform selection model using fuzzy ANP and COPRAS-G. *Expert. Syst. Appl.* **2013**, *40*, 5694–5702.
55. Chang, K. A hybrid program projects selection model for nonprofit TV stations. *Math. Probl. Eng.* **2015**.
56. Azadnia, A.; Saman, M.M.; Wong, K. Sustainable supplier selection and order lot-sizing: An integrated multi-objective decision-making process. *Int. J. Prod. Res.* **2015**, *53*.
57. Govindan, K.; Diabat, A.; Shankar, K. Analyzing the drivers of green manufacturing with fuzzy approach. *J. Clean. Prod.* **2014**, *96*, 182–193.
58. Smith, R.; Ferrebee, E.; Ouyang, Y.; Roesler, J. Optimal staging area locations and material recycling strategies. Comput-Aided Civ Inf. Special Issue: Sustainability and Resilience of Spatially Distributed Civil Infrastructure Systems. *Comput. Aided Civil Inf.* **2014**, *29*, 559–571.
59. Alarcón, B.; Aguado, A.; Manga, R.; Josa, A. A Value Function for Assessing Sustainability: Application to Industrial Buildings. *Sustainability* **2011**, *3*, 35–50.

60. San-Jose, J.; Garrucho, I. A system approach to the environmental analysis industry of buildings. *Build. Environ.* **2010**, *45*, 673–683.
61. San-Jose, J.; Cuadrado, J. Industrial building design stage based on a system approach to their environmental sustainability. *Constr. Build. Mater.* **2010**, *24*, 438–447.
62. Reyes, J.; San-Jose, J.; Cuadrado, J.; Sancibrian, R. Health & Safety criteria for determining the value of sustainable construction projects. *Saf. Sci.* **2014**.
63. San-José, J.; Losada, R.; Cuadrado, J.; Garrucho, I. Approach to the quantification of the sustainable value in industrial buildings. *Build. Environ.* **2007**, *42*, 3916–3923.
64. Ormazabal, G.; Viñolas, B.; Aguado, A. Enhancing value in crucial decisions: Line 9 of the Barcelona Subway. *Manag. Eng. J.* **2008**, *24*, 265–272.
65. Pardo, F.; Aguado, A. Investment priorities for the management of hydraulic structures. *Struct. Infrastruct. E Maint. Manag. Life Cycl. Design Perform.* **2014**, *11*, 1338–1351.
66. Pardo, F.; Aguado, A. Sustainability as the key to prioritize investments in public infrastructures. *Environ. Impact. Assess.* **2016**.
67. de la Fuente, A.; Armengou, J.; Pons, O.; Aguado, A. Multi-criteria decision-making model for assessing the sustainability index of wind-turbine support systems. Application to a new precast concrete alternative. *J. Civil Eng. Manag.* **2015**.
68. De la Fuente, A.; Pons, O.; Josa, A.; Aguado, A. Multi-Criteria Decision Making in the sustainability assessment of sewerage pipe systems. *J. Clean. Prod.* **2016**, *112*, 4762–4770.
69. Hosseini, S.; de la Fuente, A.; Pons, O. Multi-criteria decision-making method for sustainable site location of post-disaster temporary housing in urban areas. *J. Constr. Eng. M ASCE* **2016**.
70. Hosseini, S.; de la Fuente, A. Multi-criteria decision-making method for assessing the sustainability of post-disaster temporary housing units' technologies: A case study in Bam. *Sustain. Cities Soc.* **2016**, *20*, 38–51.
71. Pons, O.; Aguado, A. Integrated value model for sustainable assessment applied to technologies used to build schools in Catalonia. Spain. *Build. Environ.* **2012**, *53*, 49–58.
72. Pons, O.; de la Fuente, A. Integrated sustainability assessment method applied to structural concrete columns. *Constr. Build. Mater.* **2013**, *49*, 882–893.
73. CPH. Spanish Structural Concrete Standard. Annex 14: Recommendations for the use of fiber reinforced concrete. Madrid. 2008. Available online: http://www.fomento.gob.es/NR/rdonlyres/AF90A2BE-9F34- 4ACB-82DA-C1ACD58F3879/37440/Anejo14 borde.pdf (accessed on 6 May 2016).
74. Aguado, A.; del Caño, A.; de la Cruz, P.; de la Cruz, P.; Gómez, P.; Josa, A. Sustainability assessment of concrete structures. *J. Constr. Eng. M ASCE* **2012**, *138*, 268–276.
75. Caño, A.; Gómez, D.; de la Cruz, M. Uncertainty analysis in the sustainable design of concrete structures: A probabilistic method. *Constr. Build. Mater.* **2012**, *37*, 865–873.
76. Saaty, T. How to make a decision: The analytic hierarchy process. *Eur. J. Oper. Res.* **1990**, *48*, 9–26.
77. Saaty, T. *Fundamentals of Decision Making and Priority Theory with the Analytic Hierarchy Process*; RWS Publications: Pittsburgh, PA, USA, 2006.

78. Cartelle, J.; Lara, M.; de la Cruz, M.; del Caño, A. Assessing the global sustainability of different electricity generation systems. *Energy* **2015**, *89*, 473–489.
79. Ballester, M.; Vea, F.; Yepes, V. *Análisis multivariante para la estimación de la contribución a la sostenibilidad de los forjados reticulares*; de V ACHE Congress: Barcelona, Spain, 2011. (In Spanish)
80. Pons, O. Arquitectura Escolar Prefabricada a Catalunya. Ph.D. Thesis, UPC—Barcelona Tech, Spain, March 2009.
81. Ministerio de la Vivienda. Código Técnico de la Edificación (CTE). Madrid. 2006. Available online: http://www.codigotecnico.org/index.php (accessed on 6 May 2016).
82. CEB-FIB. *Structural Connections for Precast Concrete Buildings—Guide to Good Practice Prepared by Task Group 6.2*; CEB-FIB: Lausanne, Switzerland, 2008.
83. Staib, G.; Dörrhöfer, A.; Rosenthal, M. *Components and Systems*; Birkhäuser Architecture: Munich, Germany, 2008.
84. EOTA (European Organization for Technical Approvals). European Technical Approval ETA-06/0138. 2006. Available online: http://www.klhuk.com/media/9379/en%20eta-06%200138%20klh%20electronic%20copy.pdf (accessed on 6 May 2016).
85. CEB. *Model Code 2010*; Comité Euro-International du Beton-Federation International de la Precontrainte: Paris, France, 2010.

Building Simplified Life Cycle CO_2 Emissions Assessment Tool (B-SCAT) to Support Low-Carbon Building Design in South Korea

Seungjun Roh and Sungho Tae

Abstract: Various tools that assess life cycle CO_2 ($LCCO_2$) emissions are currently being developed throughout the international community. However, most building $LCCO_2$ emissions assessment tools use a bill of quantities (BOQ), which is calculated after starting a building's construction. Thus, it is difficult to assess building $LCCO_2$ emissions during the early design phase, even though this capability would be highly effective in reducing $LCCO_2$ emissions. Therefore, the purpose of this study is to develop a Building Simplified $LCCO_2$ emissions Assessment Tool (B-SCAT) for application in the early design phase of low-carbon buildings in South Korea, in order to facilitate efficient decision-making. To that end, in the construction stage, the BOQ and building drawings were analyzed, and a database of quantities and equations describing the finished area were conducted for each building element. In the operation stage, the "Korea Energy Census Report" and the "Korea Building Energy Efficiency Rating Certification System" were analyzed, and three kinds of models to evaluate CO_2 emissions were proposed. These analyses enabled the development of the B-SCAT. A case study compared the assessment results performed using the B-SCAT against a conventional assessment model based on the actual BOQ of the evaluated building. These values closely approximated the conventional assessment results with error rates of less than 3%.

Reprinted from *Sustainability*. Cite as: Roh, S.; Tae, S. Building Simplified Life Cycle CO_2 Emissions Assessment Tool (B-SCAT) to Support Low-Carbon Building Design in South Korea. *Sustainability* **2016**, *8*, 567.

1. Introduction

Since CO_2 reduction has been globally established as a paradigm of sustainable development, governments all over the world are competitively announcing mid- to long-term goals for the reduction of CO_2 emissions [1,2]. The USA has set its INDC (Intended Nationally Determined Contributions) to reduce CO_2 emissions by 26%–28% (compared with the baseline year 2005) by the year 2025. The EU has set its INDC to reduce CO_2 emissions by 40% (compared with the year 1990) by the year 2030. South Korea has set its INDC to reduce CO_2 emissions by 37% (compared with Business as Usual) by the year 2030.

The building industry, which is a large-scale energy consumer accounting for more than 30% of all CO_2 emissions, poses a major obstacle in CO_2 reductions for all countries [3–7]. Accordingly, a realistic policy to reduce CO_2 emissions in this industry is required [8–10]. Techniques for assessing life cycle CO_2 ($LCCO_2$) emissions of buildings are gaining attention [11–14], and many countries are performing diverse studies to assess and reduce building $LCCO_2$ emissions befitting their respective national circumstances [15–19]. Moreover, tools for evaluating $LCCO_2$ emissions of buildings starting in the early design phase are being developed to reduce these emissions [20–22], given that a building's CO_2 emissions determined during the early design phase continue to affect the building for the entirety of its life cycle [23,24]. A number of programs to address this have already been implemented throughout the world, e.g., an impact estimator for buildings developed by the ASBI in Canada, Envest2 developed by BRE in the UK, and LISA (LCA in Sustainable Architecture) developed in Australia [17,25].

South Korea has also developed diverse building CO_2 emissions assessment tools such as SUSB-LCA [26], K-LCA [27], BEGAS [28], and BEGAS 2.0 [29], in order to meet global requirements. However, research reveals that previous tools have two limitations. First, most current CO_2 emissions assessment tools focus on assessing operational CO_2 emissions based on energy consumption during the operation stage [30–34]. Second, most of the $LCCO_2$ emissions assessment tools directly use the bill of quantities (BOQ) calculated after the construction of a building begins [35,36]. These constraints complicate assessments made during the early design phase, when $LCCO_2$ emissions can be efficiently reduced [37,38].

The purpose of this study is to develop a Building Simplified $LCCO_2$ emissions Assessment Tool (B-SCAT) that is applicable in the early design phase for the facilitation of efficient decision-making of low-carbon buildings in South Korea. To that end, this study consists of the following steps: (1) proposal of a simplified $LCCO_2$ emissions assessment model for buildings; (2) development of a B-SCAT; and (3) a case study comparing the assessment results of an evaluated building using a B-SCAT and a conventional assessment model based on the building's actual BOQ.

2. Proposal for Simplified $LCCO_2$ Assessment Model for Buildings

The building $LCCO_2$ emissions represent the total CO_2 emissions in all stages from construction, operation, to end-of-life [39,40], as described in Equation (1):

$$LCCO_2 = CO_2{}^{CS} + CO_2{}^{OS} + CO_2{}^{ES}, \quad (1)$$

where $LCCO_2$ represents the life cycle CO_2 emissions (kg-CO_2) of the evaluated building; $CO_2{}^{CS}$ represents the CO_2 emissions (kg-CO_2) in the construction stage;

CO_2^{OS} represents the CO_2 emissions (kg-CO_2) in the operation stage; and CO_2^{ES} represents the CO_2 emissions (kg-CO_2) in the end-of-life stage.

This section proposes a simplified CO_2 emissions assessment model for each stage (*i.e.*, construction, operation, and end-of-life) that can evaluate the CO_2 emissions of an apartment complex, office building, and mixed-use building during the early design phase. Figure 1 shows the framework for simplifying building $LCCO_2$ emissions assessment in this study.

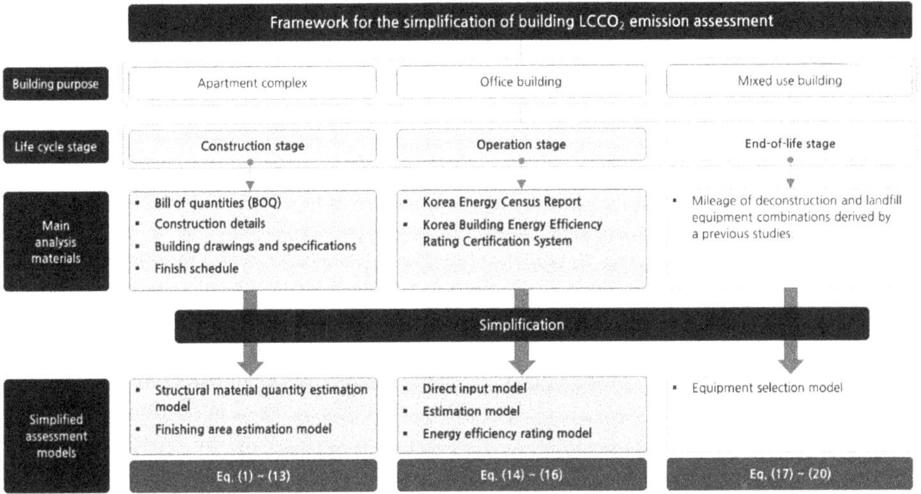

Figure 1. Framework of the simplification of building $LCCO_2$ emissions assessment.

2.1. Construction Stage

Construction stage can be subdivided into the material production process and construction process, as represented in Equation (2):

$$CO_2^{CS} = CO_2^{PP} + CO_2^{CP}, \quad (2)$$

where CO_2^{CS} is the CO_2 emissions (kg-CO_2) in the construction stage; CO_2^{PP} is the CO_2 emissions (kg-CO_2) of the manufacturing of building materials; and CO_2^{CP} is the CO_2 emissions (kg-CO_2) of construction process.

2.1.1. Material Production Process

In the material production process, CO_2 emitted during the manufacturing of building materials generally producing 30% of building $LCCO_2$ emissions [29] are evaluated. The CO_2 emissions of this process include those released during the production of structural materials and finishing materials, as represented in Equation (3):

$$CO_2{}^{PP} = CO_2{}^{SM} + CO_2{}^{FM}, \tag{3}$$

where $CO_2{}^{PP}$ is the CO_2 emissions (kg-CO_2) in the material production process, mostly produced by building materials; $CO_2{}^{SM}$ is the CO_2 emissions (kg-CO_2) of structural materials; and $CO_2{}^{FM}$ is the CO_2 emissions (kg-CO_2) of finishing materials.

This study categorized the assessment criteria for building elements, which are included in the structural materials and finishing materials, as shown in Figure 2, to assess the CO_2 emissions of the material production process while considering the function of the building. In other words, the apartment complex was subdivided into a residential building, annexed building, and underground parking lot; while the office building was subdivided into an office building, annexed building, and underground parking lot. Finally, the mixed-use building was divided into a residential building, office building, annexed building, and underground parking lot. In addition, the interior and exterior finishing materials were analyzed according to the finish schedule, and building elements were divided into the following categories: wall, wall opening, roof, exclusive space, elevator hall, and staircase.

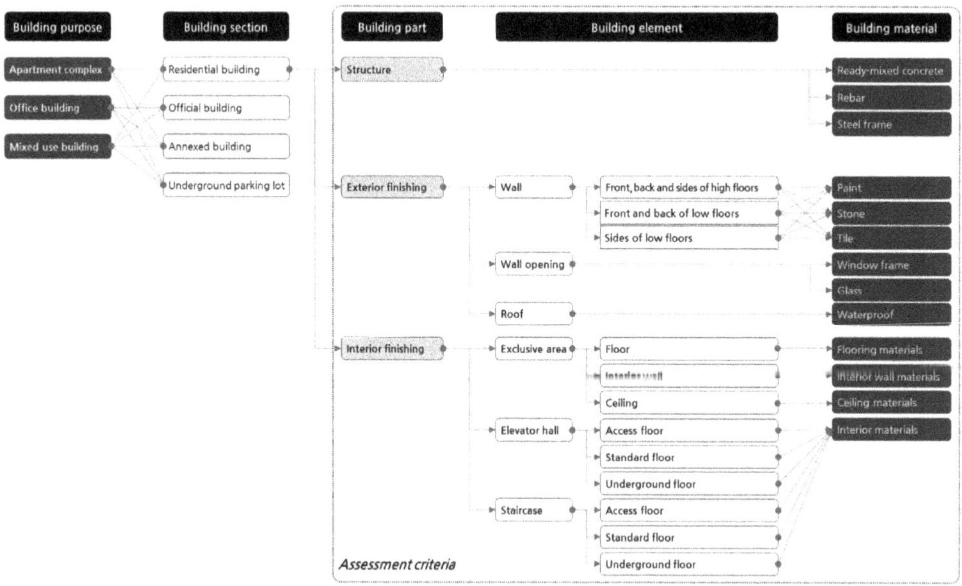

Figure 2. Assessment criteria of building elements.

(1) Structural Materials

To calculate the CO_2 emissions of structural materials, such as ready-mixed concrete, rebar, and steel frames, the supply quantities of these materials were determined after analyzing 60 types of BOQ and construction details of recently constructed buildings. Table 1 lists the average supply quantities of structural materials per unit area by building section.

Table 1. Average supply quantities of structural materials per unit area.

Building Section	Structure Type	Structure Form	Plane Type	Structural Material		
				Ready-Mixed Concrete (m^3/m^2)	Rebar (kg/m^2)	Steel Frame (kg/m^2)
Residential building	RC [1]	Wall	Flat-type	0.66	60.00	-
			Tower-type	0.59	62.20	-
			Mixed-type	0.63	61.10	-
		Column	Flat-type	0.65	63.52	-
			Tower-type	0.57	75.56	-
			Mixed-type	0.61	69.54	-
		Flat slab	Flat-type	0.62	82.34	-
			Tower-type	0.56	77.50	-
			Mixed-type	0.58	79.92	-
	SRC [2]	Column	Flat-type	0.35	37.67	74.98
			Tower-type	0.32	29.01	74.98
			Mixed-type	0.33	33.34	74.98
Office building	SRC	Wall	-	0.46	63.00	59.07
		Curtain wall	-	0.30	41.58	59.07
Annexed building	RC	Wall	-	0.74	87.00	-
Underground parking lot	RC	Column	-	1.46	157.00	-

[1] RC: Reinforced concrete; [2] SRC: Steel framed reinforced concrete.

For each assessment item, the supply quantities of structural materials can be determined from the floor area, number of stories, and supply quantities coefficient, as described in Equations (4)–(6). In the ready-mixed concrete (refer to Equation (4)), the modification factor was applied in order to consider the decrease in supply quantity of the vertical members according to use of high-strength concrete [41]. Table 2 lists the modification factor of the supply quantity for high-strength concrete.

The CO_2 emissions of the structure materials were then assessed using Equation (7) as follows:

$$SQ_i^{RMC} = FA_i^{STD} \times NS_i \times QC_i^{RMC} \times \alpha, \quad (4)$$

$$SQ_i^{RB} = FA_i^{STD} \times NS_i \times QC_i^{RB}, \quad (5)$$

$$SQ_i^{SF} = FA_i^{STD} \times NS_i \times QC_i^{SF}, \quad (6)$$

and

$$CO_2{}^{SM} = \sum_i (SQ_i^{RMC} \times CF_j^{RMC}) + \sum_i (SQ_i^{RB} \times CF_j^{RB}) + \sum_i (SQ_i^{SF} \times CF_j^{SF}), \quad (7)$$

where $SQ_i{}^{RMC}$ is the supply quantity (m^3) of ready-mixed concrete in vertical zone i; $FA_i{}^{STD}$ is the floor area (m^2) of a standard floor in vertical zone *i*; and NS_i is the number of stories in vertical zone i. Furthermore, $QC_i{}^{RMC}$ is the supply quantity coefficient (m^3/m^2) of ready-mixed concrete in vertical zone i (refer to Table 1); α is the modification factor of the ready-mixed concrete (refer to Table 2); $SQ_i{}^{RB}$ is the supply quantity (kg) of rebar in vertical zone i; $QC_i{}^{RB}$ is the supply quantity coefficient (kg/m^2) of rebar in vertical zone *i* (refer to Table 1); $SQ_i{}^{SF}$ is the supply quantity (kg) of steel frame in vertical zone i; $QC_i{}^{SF}$ is the supply quantity coefficient (kg/m^2) of steel frame in vertical zone *i* (refer to Table 1); $CO_2{}^{SM}$ is the CO_2 emissions (kg-CO_2) of structure materials; $CF_i{}^{RMC}$ is the CO_2 emissions factor (kg-CO_2/m^3) of ready-mixed concrete j (refer to Table 3); $CF_j{}^{RB}$ is the CO_2 emissions factor (kg-CO_2/kg) of rebar j; and $CF_j{}^{SF}$ is the CO_2 emissions factor (kg-CO_2/kg) of steel frame j.

Table 2. Modification factors of the ready-mixed concrete.

Strength (MPa)	Reduction Ratio (%)	Modification Factor
21	-	1.000
24	-	1.000
27	4.77	0.952
30	9.70	0.903
35	16.84	0.852
40	22.61	0.774
50	30.08	0.699
60	32.11	0.679

(2) Finishing Materials

The CO_2 emissions of the interior and exterior finishing materials for each building function and section were calculated using only the limited information available during the early design phase [42–44]. The assessment items were categorized according to building element, as shown in Figure 2. The models to determine the area of the finishing materials for each building element were developed after analyzing the 60 types of drawings and finish schedules. These models use the provisional perimeter formula developed in this study to calculate the element in which a particular finishing material was used for each building element, encompassing the interior and exterior perimeters of the standard floor for each major plane type and using the variables of numbers of units and cores, unit area, and exclusive use area, as well as the basic information entered during the first process of the assessment. Table 4 presents provisional perimeter formulas of a standard floor.

Table 3. CO_2 emissions factors of concrete.

Strength (MPa)	Admixture Material	Mixture Composition (%)		CO_2 Emissions Factor (kg-CO_2/m^3)
		Blast Furnace Slag	Fly-Ash	
21	-	-	-	346.0
	Blast furnace slag	10	0	328.5
		20	0	297.2
		30	0	266.0
		40	0	230.7
	Fly-ash	0	10	328.3
		0	20	296.8
		0	30	265.3
		0	40	229.8
	Blast furnace slag + Fly-ash	10	10	297.0
		10	20	265.5
		10	30	234.0
		20	10	265.7
		20	20	234.2
		30	10	234.5
27	-	-	-	364.0
	Blast furnace slag	10	0	329.7
		20	0	294.1
		30	0	258.5
		40	0	226.7
	Fly-ash	0	10	329.4
		0	20	293.6
		0	30	257.8
		0	40	225.6
	Blast furnace slag + Fly-ash	10	10	293.9
		10	20	258.0
		10	30	222.2
		20	10	258.3
		20	20	222.5
		30	10	222.7

Table 4. Provisional perimeter formulas of a standard floor.

Classification			Flat-Type	Tower-Type
			Types 2 and 4	Types 3 and 4
Exterior material	Exterior wall	Front, back, and side walls on high floors	$(2J + K + 2)\sqrt{A}$	$(3J + 1)\sqrt{A}$
		Front and back on low floors	$(2J + K)\sqrt{A}$	$(2J + 1)\sqrt{A}$
		Side wall on low floors	$2\sqrt{A}$	$J\sqrt{A}$
Interior material	Interior wall	Residential exclusive area	$(4J + K)\sqrt{a}$	$(4J + 1)\sqrt{a}$
		Elevator hall/Staircase	$4K\sqrt{a}$	$4\sqrt{a}$

J: Number of units; K: Number of cores; A: Floor area; a: Exclusive area.

The walls, which are considered exterior finishing, were divided into the following categories according to the typical finishing execution: front, back, and sides of high floors; front and back of low floors; and sides of low floors. The area of finishing materials can be calculated as the product of exterior perimeter of the standard floor of the building calculated in Table 4, number of stories, story height, and wall surface rate as described in Equation (8). For wall openings, such as window frames and glass, as well as for the exterior walls, the area can be calculated as the product of exterior perimeter of the building standard floor, number of stories, story height, and window surface rate (1-the wall surface rate) as described in Equation (9). In addition, for the interior finishing, such as interior walls of the residential building, elevator hall, and staircases, the area can be calculated as the product of interior wall perimeter, which is calculated using the formula presented in Table 4, number of stories, story height, and number of units as described in Equation (10). The areas of floor and ceiling of the residential unit (exclusive area), access floor, and staircases in the building were determined as the area of the locations where the materials were applied, calculated from the unit area and building area determined in the first step of the assessment.

The CO_2 emissions of the finishing materials can be assessed using the product of the area of the interior and exterior materials for each building element and the CO_2 emissions factor for each material type, as described in Equation (11):

$$FA_i^{EW} = EP_i^{STD} \times NS_i \times SH_i \times \beta_i, \quad (8)$$

$$FA_i^{EO} = EP_i^{STD} \times NS_i \times SH_i \times \gamma_i, \quad (9)$$

$$FA_i^{IW} = IP_i^{STD} \times NS_i \times SH_i, \quad (10)$$

and

$$CO_2{}^{FM} = \sum_i \left(FA_i^{EW} \times CF_j^{FM}\right) + \sum_i \left(FA_i^{EO} \times CF_j^{FM}\right) + \left(FA^{ER} \times CF_j^{FM}\right) + \sum_i \left(FA_i^{IW} \times CF_j^{FM}\right) + \sum_i \left(FA_i^{IF} \times CF_j^{FM}\right) + \sum_i \left(FA_i^{IC} \times CF_j^{FM}\right), \quad (11)$$

where $FA_i{}^{EW}$ is the area (m^2) of the finishing material for the exterior wall in vertical zone i; $EP_i{}^{STD}$ is the exterior perimeter (m) of a standard floor in vertical zone i (refer to Table 4); NS_i is the number of stories in vertical zone i; and SH_i is story height (m) in vertical zone i. Furthermore, β_i is the wall surface rate of the exterior wall in vertical zone i; $FA_i{}^{EO}$ is the area (m^2) of finishing material for the exterior wall opening in vertical zone *i*; γ_i is the window surface rate (1-the wall surface rate) of the exterior wall in vertical zone i; $FA_i{}^{IW}$ is the area (m^2) of finishing material for the interior wall in vertical zone i; $IP_i{}^{STD}$ is the interior perimeter (m) of a standard floor in vertical zone *i* (refer to Table 4); $CO_2{}^{FM}$ is the CO_2 emissions (kg-CO_2) of finishing materials; FA^{ER} is the area (m^2) of finishing material for the roof; $FA_i{}^{IF}$ is the area (m^2) of finishing material for the floor in vertical zone *i*; $FA_i{}^{IC}$ is the area (m^2) of finishing material for the ceiling in vertical zone *i*; and $CF_j{}^{FM}$ is the CO_2 emissions factor (kg-CO_2/m^2) of finishing material j (refer to Table 5).

Table 5. CO_2 emissions factors of finishing materials.

Classification	Element	Finishing Material	Units	CO_2 Emissions Factor (kg-CO_2/Unit)
Exterior material	Exterior wall	Water-based paint	m^2	0.36
		Silicone-based paint	m^2	0.32
		Stone coat	m^2	11.22
		Granite with stone molding	m^2	13.43
		Tile	m^2	7.06
	Window frame	PVC window frame	m^2	5.91
		Aluminum window frame	m^2	7.57
		Curtain wall window frame	m^2	4.65
	Glass	Plate glass	m^2	9.86
		Insulating glass	m^2	22.43
		Tempered glass	m^2	13.35

(3) CO_2 Emissions Factors of Building Materials

This study determined the CO_2 emissions factors for each type of building material using an individual integration method and the South Korean carbon emissions factor [45] established by the South Korean Ministry of the Environment. In particular, even though the CO_2 emissions factor depends on concrete strength, the current South Korean carbon emissions factor and South Korean LCI DB [46] include only some of the types of concrete and their strengths. This study used the CO_2 emissions factor determined with the individual integration method for each type of concrete strength and admixture material obtained from a previous study [47,48]. Furthermore, for consistency in the assessment of the CO_2 emissions factor and assessment results, this study used the South Korean carbon emissions factor as the CO_2 emissions factors of all building materials, excluding ready-mixed concrete. Tables 3 and 5 present the CO_2 emissions factors of concrete and finishing materials.

2.1.2. Construction Process

In the construction process, the CO_2 emissions can be evaluated in terms of energy consumption by freight vehicles transporting building materials to the building site, in addition to emissions produced by construction machinery, field offices, and other facilities involved in the construction of the building. However, it is difficult to produce a detailed construction schedule in the early design phase. Moreover, this stage makes up less than 3% of the building $LCCO_2$ emissions. Hence, this study used the average energy consumption by unit area (*i.e.*, diesel consumption: 5.24 ℓ/m^2, gasoline consumption: 0.05 ℓ/m^2, electricity consumption: 10.47 kWh/m^2) derived by a previous study [42]. Equations (12) and (13) represent the CO_2 emissions in the construction stage:

$$CO_2^{CP} = \left(5.24 \times CF_d^{EN} + 0.05 \times CF_g^{EN} + 10.47 \times CF_e^{EN}\right) \times GA, \tag{12}$$

and

$$CO_2^{CS} = 18.44 \times GA, \tag{13}$$

where CO_2^{CP} is the CO_2 emissions (kg-CO_2) in the construction stage; CF_d^{EN} is the CO_2 emissions factor of diesel (2.58 kg-CO_2/ℓ); CF_g^{EN} is the CO_2 emissions factor of gasoline (2.08 kg-CO_2/ℓ); CF_e^{EN} is the CO_2 emissions factor of electricity (0.46 kg-CO_2/kWh); and GA is the gross area (m^2) of a building.

2.2. Operation Stage

The operation stage considers the CO_2 emissions due to energy consumed during the service life of the building. This is a major stage responsible for about 70% of the building's $LCCO_2$ emissions [29]. The emissions from this stage can be assessed using the service life of the building, amount of energy consumed, and the CO_2 emissions factor as described in Equation (14).

$$CO_2^{OS} = \sum_{n=1}^{SL} (1+RR)^{n-1} \times \sum_k (EC_k \times CF_k^{EN}), \tag{14}$$

where CO_2^{OS} is the CO_2 emissions (kg-CO_2) in the operation stage; SL is the service life of the building (years); RR is the annual reduction rate of operational energy effectiveness; EC_k is the annual energy consumption of the energy source k; and CF_k^{EN} is the CO_2 emissions factor of energy source k (refer to Table 6).

This study proposed three kinds of assessment models (*i.e.*, direct input model, estimation model, and energy efficiency rating model) based on analysis of the "South Korea Energy Census Report" [49] and the "South Korea Building Energy Efficiency Rating System" [50] in order to efficiently assess energy consumption depending on the timing of the assessment and available data. Moreover, the "2006 IPCC Guidelines for National Greenhouse Gas Inventories" [51] has been analyzed to evaluate CO_2 emissions during the operation stage, and the corresponding database of CO_2 emissions factors has been created, as shown in Table 6. The measured CO_2 emissions factors for electricity and district heating as determined by the Korea Power

Exchange and Korea District Heating Corporation should be applied [52,53]. Gas and kerosene utilize the basic CO_2 emissions factor of the 2006 IPCC Guidelines [51].

Table 6. CO_2 emissions factors of energy sources.

Classification	CO_2 Emissions Factor	Unit	Source
Kerosene	2.441	kg-CO_2/ℓ	2006 IPCC Guidelines for National Greenhouse Gas Inventory [51]
Medium quality heavy oil	3.003	kg-CO_2/ℓ	
Diesel	2.580	kg-CO_2/ℓ	
Gasoline	2.080	kg-CO_2/ℓ	
Propane	2.889	kg-CO_2/kg	
Gas	2.200	kg-CO_2/Nm^3	
Electricity	0.495	kg-CO_2/kWh	Korea Power Exchange
District heating	0.051	kg-CO_2/MJ	Korea District Heating Corporation

2.2.1. Direct Input Model

The direct input model uses the annual amount of energy from various sources consumed by a building (refer to Equation (14)). This method is used when annual energy consumption data are available, e.g., if the energy consumption can be predicted based on computer simulations during the early design phase.

2.2.2. Estimation Model

The estimation model predicts the energy consumption pattern of a building using an analysis of previously accumulated survey data. The calculated result is typically in the form of annual energy consumption and depends on the utility and gross area of the building. To ensure the reliability of the estimation model, this study investigated and analyzed the average energy consumption based on the heating system used by the apartment building and the average energy consumption of the office building determined from the Energy Census Report (2014) [49], which is published every three years by the Korea Ministry of Trade, Industry, and Energy. The mixed-use building, which was not specified in the Energy Census Report, was categorized as part apartment and part office building and, therefore, utilized the average energy consumption values of both an apartment and office building. Table 7 lists the average energy consumption for the apartment building analyzed in this study. Equation (15) represents the estimation model for evaluating the CO_2 emissions during the operation stage.

$$CO_2{}^{OS} = \sum_{n=1}^{SL} (1 + RR)^{n-1} \times GA \times \sum_{k} (EC_k^{EM} \times CF_k^{EN}), \tag{15}$$

where $CO_2{}^{OS}$ is the CO_2 emissions (kg-CO_2) in the operation stage; SL is the service life of the building (years); RR is the annual reduction rate of operational energy effectiveness; GA is the gross area (m^2) of the building; $EC_k{}^{EM}$ is the annual energy consumption per unit area based on the estimation model (refer to Table 7); and $CF_k{}^{EN}$ is the CO_2 emissions factor of energy source k (refer to Table 6).

2.2.3. Energy Efficiency Rating Model

The energy efficiency rating model is the one used by the South Korea Building Energy Efficiency Rating Certification System for the construction of an apartment building or commercial building. The annual CO_2 emissions per exclusive area due to air-conditioning, heating, hot water, lighting, and ventilation were inputted into the model based upon the Building Energy Efficiency Rating Certification System [50]. Equation (16) represents the energy efficiency rating model for evaluating the CO_2 emissions during the operation stage:

$$CO_2{}^{OS} = \sum_{n=1}^{SL} (1 + RR)^{n-1} \times EA \times \sum_{l} CE_l^{EERM}, \quad (16)$$

where $CO_2{}^{OS}$ represents the CO_2 emissions (kg-CO_2) in the operation stage; SL is the service life of the building (years); RR is the annual reduction rate of operational energy effectiveness; EA is the exclusive area (m^2) of the building; and $CE_l{}^{EERM}$ is the annual CO_2 emissions of energy consumption part l, according to the energy efficiency rating model.

Table 7. Average energy consumption values of the apartment building components.

Classification		Kerosene (ℓ/year/m^2)	Medium Quality Heavy Oil (ℓ/year/m^2)	Propane (kg/year/m^2)	City Gas-Cooking (Nm^3/year/m^2)	City Gas-Heating (Nm^3/year/m^2)	Electricity (kWh/year/m^2)	Heat Energy (Mcal/year/m^2)	Hot Water (Mcal/year/m^2)
Heating System	Heat Source								
Individual heating	Petroleum	6.801	-	1.189	0.008	-	30.785	-	-
	LPG	-	-	5.529	-	-	31.355	-	-
	Electricity	0.045	-	1.346	0.021	-	37.099	-	-
	City Gas	-	-	0.013	1.141	7.934	35.287	-	-
Central heating	Ordinary	-	2.567	0.181	1.039	5.793	33.458	-	0.587
	Petroleum	-	10.492	0.649	0.567	-	29.277	-	0.484
	City Gas	-	-	0.030	1.191	7.670	34.813	-	0.621
District heating	Ordinary	-	-	0.054	1.376	-	37.990	94.360	0.750

2.3. End-of-Life Stage

The CO_2 emissions of the end-of-life stage include those released during the building's demolition process, transportation of the waste building materials, and the landfill gas produced by the waste building materials, as described in Equation (17). The demolition process includes an evaluation of the CO_2 emissions from the equipment used to demolish the building. Waste transport emissions include CO_2 emitted during the transport of the generated waste to the landfill. Once in landfill, an evaluation is performed on the CO_2 emissions generated by the waste building materials as landfill gas. However, it is difficult to obtain detailed disposal information in the early design phase. Hence, in this study, the oil consumption for each combination of demolition equipment and landfill equipment was organized into a database and adapted using CO_2 emissions assessment methods based on an analysis of the results of previous studies [20,54,55]. Table 8 lists the equipment mileage used during the demolition and landfill processes, and Equations (18)–(20) represent CO_2 emissions in each process of the end-of-life stage:

$$CO_2{}^{ES} = CO_2{}^{DP} + CO_2{}^{TP} + CO_2{}^{LP}, \tag{17}$$

$$CO_2{}^{DP} = QW \times EM_m^{DP} \times CF_d^{EN}, \tag{18}$$

$$CO_2{}^{TP} = QW \times DT \times CF^{TR}, \tag{19}$$

and

$$CO_2{}^{LP} = QW \times EM_m^{LP} \times CF_d^{EN}, \tag{20}$$

where $CO_2{}^{ES}$ represents the CO_2 emissions (kg-CO_2) in the end-of-life stage; $CO_2{}^{DP}$ is the CO_2 emissions (kg-CO_2) in the demolition process based on demolition equipment; $CO_2{}^{TP}$ is the CO_2 emissions (kg-CO_2) in the transportation process based on transportation vehicles; $CO_2{}^{LP}$ is the CO_2 emissions (kg-CO_2) in the disposal process based on disposal equipment; QW is the quantities of wasted building materials (ton); $EM_m{}^{DP}$ is the mileage (ℓ/ton) of demolition equipment m (refer to Table 8); $CF_d{}^{EN}$ is the CO_2 emissions factor of diesel (2.58 kg-CO_2/ℓ); DT is the distance (km) that waste building materials are transported to the landfill site; CF^{TR} is the CO_2 emissions factor of a truck (0.249 kg-CO_2/ton$\cdot$ km); and $EM_m{}^{LP}$ is the mileage (ℓ/ton) of landfill equipment m (refer to Table 8).

Table 8. Mileage of demolition and landfill equipment.

Usage	Equipment Combination and Dimensions	Mileage (ℓ/ton)
Demolition	Backhoe (1.0 m^3) + Giant Breaker (0.7 m^3)	3.642
	Pavement Breakers (25-kg grade) 2 units + Air Compressor (3.5 m^3/min)	2.385
	Backhoe (1.0 m^3) + Hydraulic Breaker (1.0 m^3) + Giant Breaker (0.7 m^3)	4.286
	Backhoe (0.4 m^3) + Breaker (0.4 m^3)	4.760
Landfill	Dozer (D8N, 15 PL, 6 PL) + Compactor (32 tons)	0.150

3. Development of a B-SCAT

This section describes the development of a B-SCAT for supporting low-carbon building design and efficient decision-making processes in the early design phase of a building. This tool divides the assessment procedure into basic information, construction, operation, and end-of-life steps. In particular, it facilitates assessment by making simple selections of supply materials for each building area in the construction stage. This process enables diverse alternative assessments to be made within a limited timeframe. Default values calculated from the database were provided for the construction process, operation stage, and end-of-life stage in order to reduce the time and labor required for the assessment.

3.1. Step 1: Basic Information

The basic information includes the architectural scheme data of the evaluated building. Items, such as site location and zone, are entered; the function and structural form of the evaluated building are selected; and the gross area, building-to-land ratio, and floor area ratio within the complex profile are calculated. In addition, the details of the evaluated building are set, establishing details, such as standard floor area, exclusive area, number of units, number of stories, structural type, plane type, and wall surface rate. Figure 3 illustrates the interface of the basic information in the B-SCAT.

3.2. Step 2: Construction Stage

During the construction stage, the CO_2 emissions resulting from the production of building materials are assessed, and the input interface is established depending on the function of the building. To assess the CO_2 emissions for an apartment complex, data on the residential building, annexed building, underground parking lot, and landscaping were entered. To assess the emissions for an office building, data on the office building, annexed building, underground parking lot, and landscaping were entered. To assess the emissions for a mixed-use building, data on the residential building, office building, annexed building, underground parking lot, and landscaping were entered. In addition, the CO_2 emissions were assessed by selecting the type of materials supplied as structural and finishing materials for each assessment item. Figure 4 illustrates the interface of the construction stage.

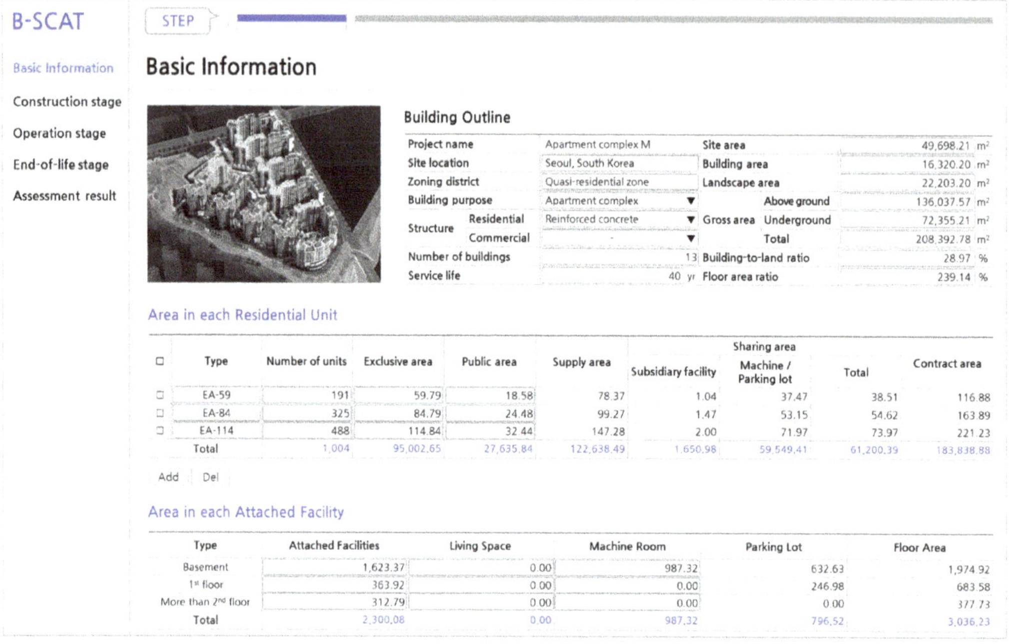

Figure 3. Interface of the basic information.

3.3. Step 3: Operation Stage

The assessment method of the operation stage is divided into three types. In the direct input model, the annual energy consumption of the evaluated building is entered and assessed directly. The estimation model assesses the CO_2 emissions based on annual energy consumption per unit area, which depends on the building function and heating system. This model utilizes the database included in the tool and can be useful when energy consumption data is unavailable for the building of interest. The energy efficiency rating model assesses the CO_2 emissions by directly inputting the assessment results of the CO_2 emissions of a building, utilizing the Energy Efficiency Rating Certification System of the evaluated building or the energy simulation program provided by the Korea Energy Management Corporation. Figure 5 illustrates the interface of the operation stage.

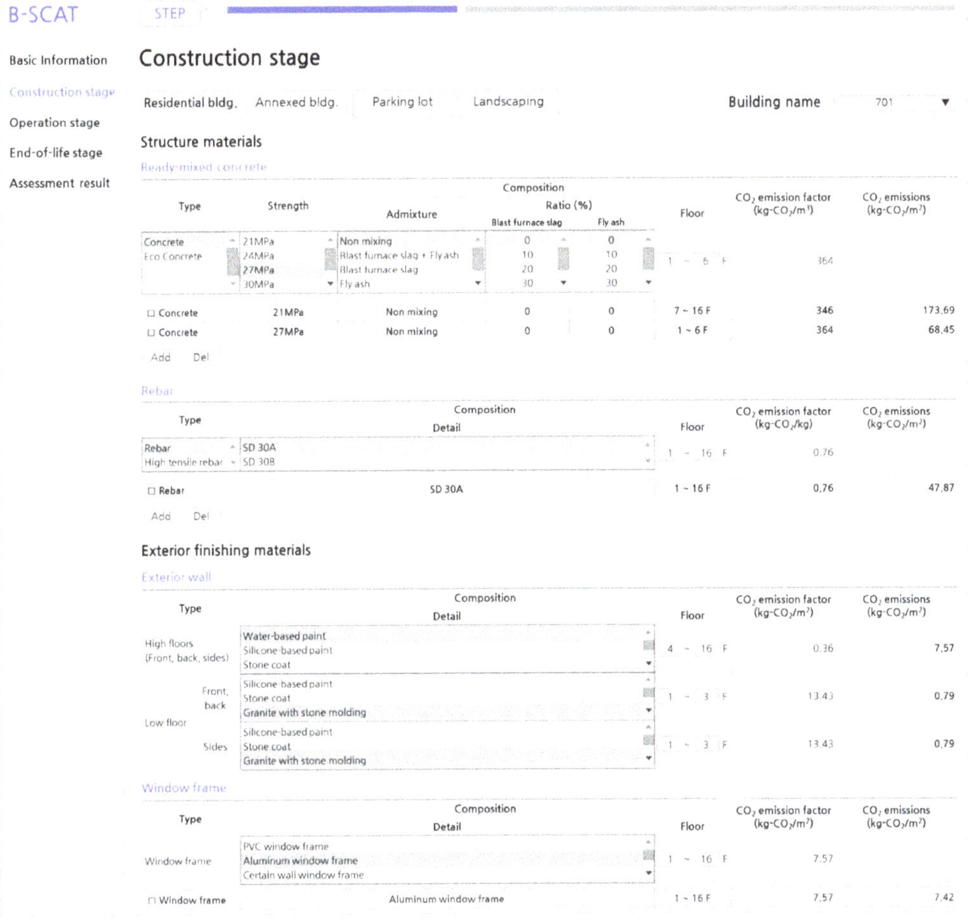

Figure 4. Interface of the construction stage.

3.4. Step 4: End-of-Life Stage

The end-of-life stage involves an assessment of the CO_2 emissions produced at the end of a building's life cycle, when structures are demolished and waste building material is generated and processed. The assessment includes analysis of the equipment used in the building demolition and waste landfill process. Figure 6 illustrates the interface of the end-of-life stage.

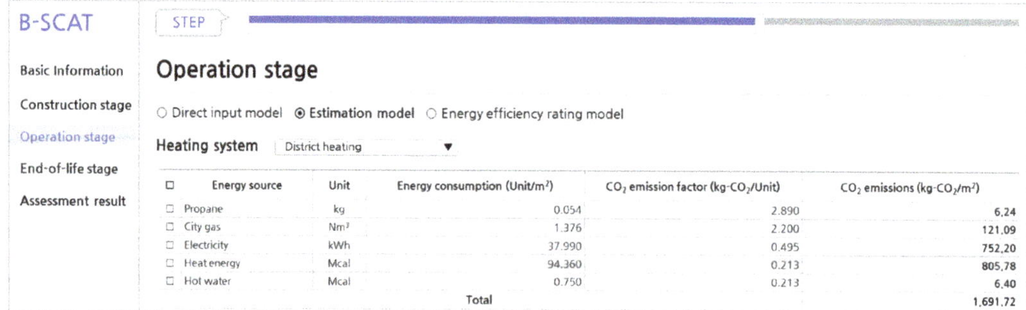

Figure 5. Interface of the operation stage.

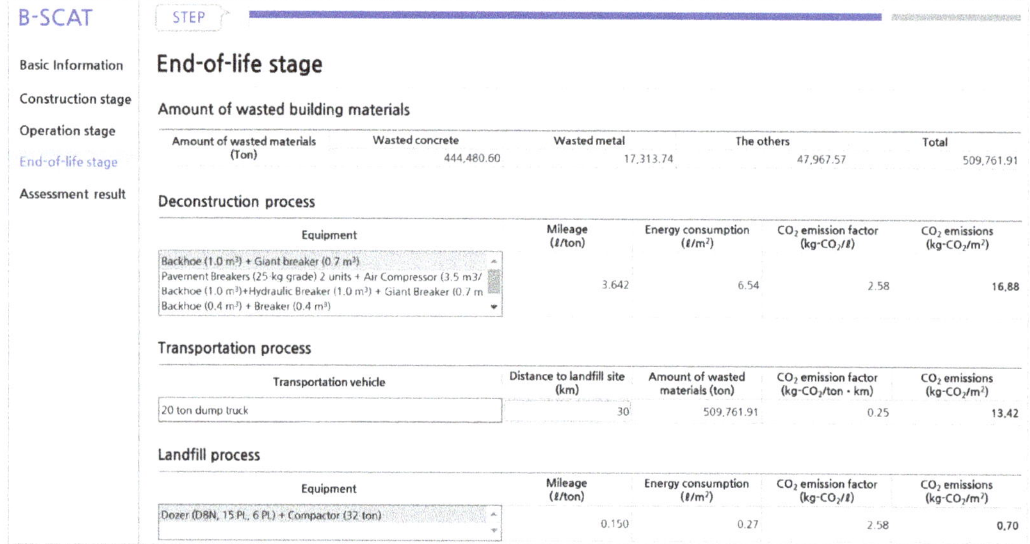

Figure 6. Interface of the end-of-life stage.

3.5. Step 5: Assessment Results

The assessment results, as shown in Figure 7, are displayed on one screen that includes all of the details of the assessment of the $LCCO_2$ emissions. The upper region of the comprehensive assessment view displays the profile of the building of interest, the assessment method used for each stage, the details of the database used, and the basis for the calculations. The lower region presents a comparative analysis of the CO_2 emissions assessment results in each stage according to the standard building type selected during the assessment.

Figure 7. Interface of the assessment result.

4. Case Study

To review the applicability of the B-SCAT, an assessment was conducted using the basic data for a building that was recently completed. For comparison with the assessment results, the finishing materials used during the production process of construction stage were selected based on the same basic drawings and specifications drafted during the early design phase used for those results.

4.1. Evaluated Building

The project's evaluated building comprised Apartment Complex M, which contains 13 residential buildings. Table 9 presents the architectural scheme of the analyzed building.

4.2. Assessment Conditions

As shown in Table 10, the assessment conditions were selected according to the input items for each assessment stage, which were based on the plan, drawings, and specifications of the apartment complex.

Table 9. Architectural scheme of the analyzed building.

Project Name	Apartment Complex M				
Zoning district	Quasi-residential area	Site area		49,698.21	m^2
Structure	Reinforced concrete structure	Building area		16,320.20	m^2
Number of buildings	13	Landscape area		22,203.20	m^2
Unit type	Types 2, 4, and 6	Gross area	Above ground	136,037.57	m^2
Plane type	Flat type, Tower type		Underground	72,355.21	m^2
Service life	40 years		Total	208,392.78	m^2
Heating system	Local heating	Building-to-land ratio		28.97	%
Construction period	25 months	Floor area ratio		239.14	%

Table 10. Assessment conditions.

Classification		B-SCAT	Conventional Assessment Model
Construction stage		Basic drawing and specification	BOQ
		Default value (=18.44 kg-CO_2/m^2)	
Operation stage		Estimation model (local heating) (Reduction rate of operational energy effectiveness: 0%, 1%, 1.5%)	
End-of-life stage	Demolition process	Backhoe (1.0 m^3) + giant breaker (0.7 m^3)	
	Transportation process	20-ton dump truck (distance: 30 km)	
	Landfill process	Dozer (D8N, 15 PL, 6 PL) + compactor (32 tons)	

B-SCAT, and the construction and design provisions of the evaluated building, were analyzed according to the input items of the residential and annexed buildings. The plane type and structural form of the residential building were determined to be the flat-type and tower-type, reinforced concrete structure, and wall type, respectively, and the wall surface ratio was set at 55%. In addition, the superintendent office, holding facilities, and sports center were identified as annexes in the analysis, and their wall surface ratio was also set to 60%. In the construction stage, the materials used for each assessment item in each building element were analyzed based on an analysis of the plan of the apartment complex and the table of interior and exterior finishing materials. In particular, the use of 27 MPa ordinary concrete was assumed for the first to the sixth floors of the residential buildings, in the interest of structural stability, while the use of 21 MPa concrete was assumed for the seventh floors and higher, to achieve economic efficiency. In addition, the exterior walls were assumed to use granite and stone moldings for the first three floors and water-based paint for the fourth floors and higher. Aluminum window frames and insulating glass were assumed for all 13 buildings of the apartment complex. The annexed buildings, low-rise buildings with 1 to 3 stories, which comprised the superintendent office, holding facilities, and sports center, were assumed to use 21 MPa concrete. Given the function of those buildings, it was assumed the exterior walls were marble and granite, and the interior walls had terrazzo and water-based paint. In the operation stage, given the absence of results from a simulation of the energy consumption of the apartment complex or from the preliminary Energy Efficiency Rating Certification System, the estimation model was used for analysis. The local heating system, which is the actual heating system of the evaluated building, was selected to calculate CO_2 emissions. The service life of the evaluated building was set to 40 years, according to the building durability period of the South Korean Corporate Tax Act [56]. The reduction rate of operational energy effectiveness was assumed as 0%, 1%, and 1.5% in the end-of-life stage, the equipment selected for demolition included a backhoe (1.0 m^3) and a giant breaker (0.7 m^3). Also included was the 30 km distance between the building site and the landfill processing site. A bulldozer (D8N, 15 PL, 6 PL) and compactor (32 tons) were selected as the equipment used in the landfill process.

4.3. Assessment Results

Figure 8 presents the results of the $LCCO_2$ emissions assessment of the apartment complex. The CO_2 emissions produced during the construction stage were assessed as 502.76 kg-CO_2/m^2 using the tool developed in this study and 515.71 kg-CO_2/m^2 based on the actual BOQ, yielding an error rate of 2.51%. The CO_2 emissions of the operation stage, which applied 0% of the reduction rate of operational energy effectiveness, were assessed as 1691.72 kg-CO_2/m^2. In addition, the $LCCO_2$ emissions were assessed as 2225.48 kg-CO_2/m^2 and 2238.43 kg-CO_2/m^2, respectively, yielding an error rate of approximately 0.58%.

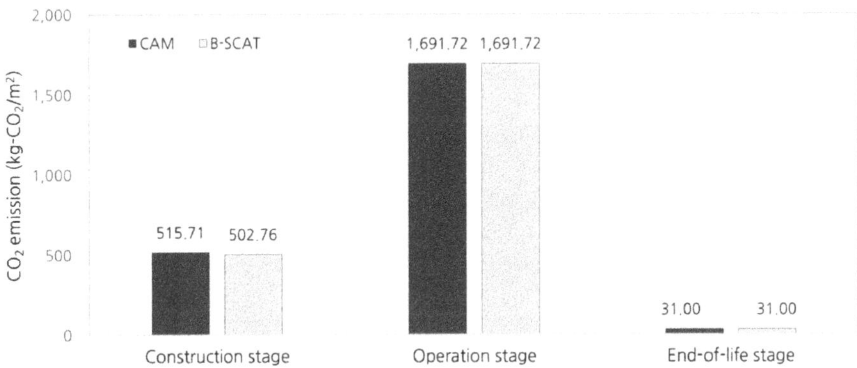

Figure 8. Assessment results.

4.4. *Comparative Analysis of Assessment Results of Construction Stage*

From the assessment results from the previously conducted building $LCCO_2$ emissions assessment tool and from the drawings and specifications, this study conducted a comparative analysis of the assessment results of the production stage after subdividing the results into residential buildings, annexed buildings, and underground parking lots.

4.4.1. Residential Buildings

As shown in Figure 9, this study conducted a comparative analysis of the CO_2 emissions per unit area of the supply materials for each residential building region calculated using this tool. The assessment items (Buildings 701, 702, 703, and 704) and the average CO_2 emissions per unit area of the residential buildings were calculated using the BOQ. Consequently, the results calculated with the tool for Buildings 701, 702, 703, and 704 were 443.74 kg-CO_2/m^2, 437.13 kg-CO_2/m^2, 438.42 kg-CO_2/m^2, and 445.16 kg-CO_2/m^2, respectively. Compared with the value of 449.23 kg-CO_2/m^2 assessed from the BOQ, these values yielded error rates of 1.22%, 2.69%, 2.41%, and 0.91%, respectively. In addition, the average assessment result of the tool was 441.59 kg-CO_2/m^2, which closely approximated the BOQ assessment results with an error rate of 1.70%.

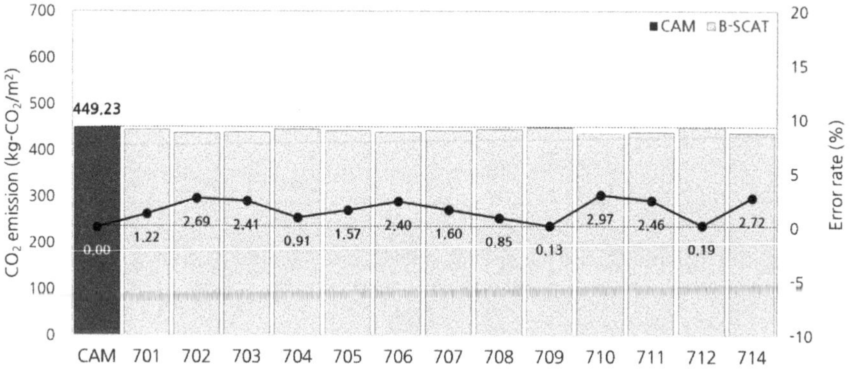

Figure 9. Assessment results for each residential building.

4.4.2. Annexed Building

For the annexed buildings, as shown in Figure 10, a comparative analysis was conducted on the CO_2 emissions per unit area of supply materials for each building part in the superintendent office (SO), holding facilities (HF), and sports center (SC). The annexed buildings' average CO_2 emissions per unit area were calculated from the BOQ. Consequently, the results assessed using this tool for the SO, the HF, and the SC were 427.46 kg-CO_2/m^2, 445.65 kg-CO_2/m^2, and 432.54 kg-CO_2/m^2, respectively;

these are valid results compared with the value of 442.52 kg-CO_2/m^2 obtained from the BOQ. In addition, the error rates were 3.40%, 0.71%, and 2.26%, respectively, and the average error rate was 1.65%.

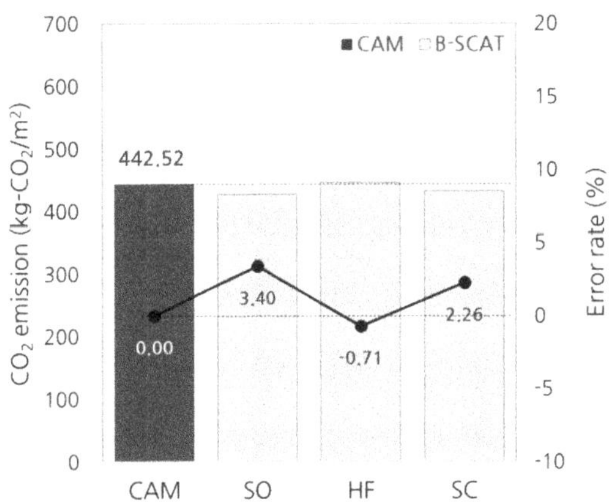

Figure 10. Assessment results for each annexed building.

4.4.3. Underground Parking Lot

As shown in Figure 11, a comparative analysis was conducted on the CO_2 emissions per unit area of supply materials for each building part of the underground parking lot (PL). The average CO_2 emissions per unit area of the underground parking lot was calculated from the BOQ. Consequently, the results assessed using this tool for the PL was 676.52 kg-CO_2/m^2, respectively; this is a valid result compared with the value of 654.27 kg-CO_2/m^2 obtained from the BOQ. In addition, the error rate was 3.40%, respectively.

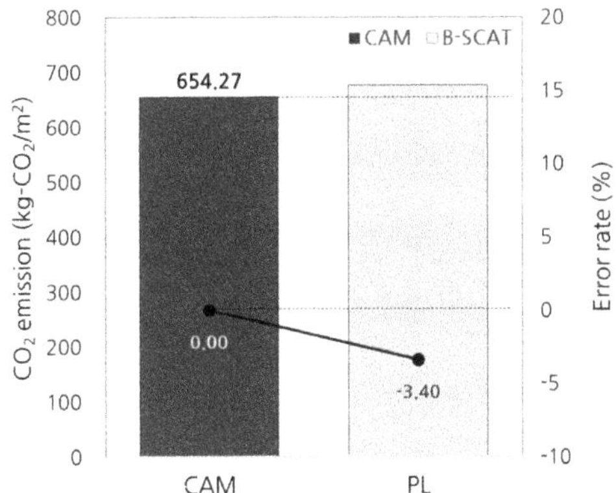

Figure 11. Assessment results for each underground parking lot.

4.5. Comparative Analysis of Assessment Results of Operation Stage

As shown in Figure 12, this study conducted a comparative analysis of the CO_2 emissions per unit area of operation stage by the reduction rate of operational energy effectiveness. The assessment results applied 0%, 1%, and 1.5% of the reduction rate of operational energy effectiveness were 1691.72 kg-CO_2/m^2, 2493.80 kg-CO_2/m^2, and 3023.46 kg-CO_2/m^2, respectively. Through this evaluation result, it confirmed that the evaluation result of the operational stage changed according to whether or not the annual reduction rate of operational energy effectiveness and size of this value was applied. That is, even if 1% of the annual reduction rate of operational energy effectiveness was applied, 47% of energy consumption increased, and 79% of energy consumption increased in 1.5% application during the service life of the building (40 years). Therefore, in order to achieve the low-carbon building, the selection of energy equipment, which have low reduction rates of operational energy effectiveness, is very important.

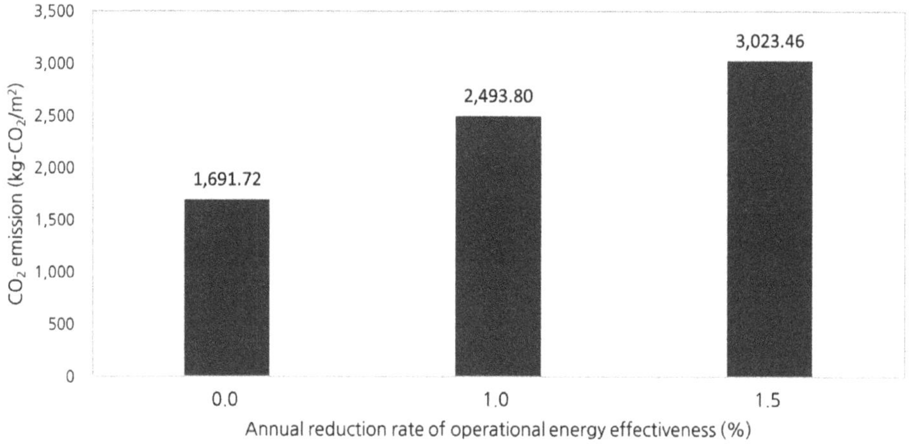

Figure 12. Assessment results by the annual reduction rate of operational energy effectiveness.

5. Conclusions

The purpose of this study was to develop a B-SCAT that is applicable in the early design phase for low-carbon building design. The conclusions of this study are as follows:

(1) After separating the life cycle of a building into various stages, including construction, operation, and end-of-life, a simplified $LCCO_2$ emissions assessment model and B-SCAT were developed for application to the early design phase of buildings.

(2) In the construction stage, the supply quantities coefficient of structural materials for each building function and section were analyzed, and the equations were constructed based on an analysis of the types and areas of the finishing materials used for each building element.
(3) In the operation stage, the model of assessment was identified using models for direct input, estimation, and energy efficiency rating in order to provide a proactive assessment according to the time of the assessment and the available data. An assessment method was subsequently proposed.
(4) The average of the CO_2 emissions assessment results for residential buildings tested during the case study of the B-SCAT was 441.59 kg-CO_2/m^2 per unit area; this is close to the assessment result of 449.23 kg-CO_2/m^2 based on the BOQ, yielding an error rate of 1.70%.
(5) According to the analysis of the annexed buildings and underground parking lots using the B-SCAT, the average CO_2 emissions were determined to be 435.22 kg-CO_2/m^2 and 676.52 kg-CO_2/m^2 per unit area, respectively, which closely approximates the results of 442.52 kg-CO_2/m^2 and 654.27 kg-CO_2/m^2, respectively, based on the BOQ, with error rates of 1.65% and 3.40% respectively.

The B-SCAT developed by this study for use in the early design phase is expected to predict the environmental performance of future construction projects and alternative assessments, leading to low-carbon building designs.

Currently, according to application of the mainly-constructed database in Korea, it is considered to broaden the range of the B-SCAT database in order that other countries utilize B-SCAT. Especially, it is considered to be possible to apply identical building life cycle CO_2 emission assessment methods in the early stage of a project, which is suggested in this paper, to other countries.

Acknowledgments: This research was supported by Basic Science Research Program through the National Research Foundation of Korea (NRF) funded by the Ministry of Science, ICT & Future Planning (No. 2015R1A5A1037548) and the National Research Foundation of Korea (NRF) grant funded by the Korea government (MSIP) (No. 20110028794).

Author Contributions: All authors contributed substantially to all aspects of this article.

Conflicts of Interest: The authors declare no conflict of interest.

Abbreviations

The following abbreviations are used in this manuscript:

$LCCO_2$	Life Cycle CO_2
BOQ	Bill of Quantities
B-SCAT	Building Simplified $LCCO_2$ emissions Assessment Tool
INDC	Intended Nationally Determined Contributions

References

1. Wang, Q.; Chen, X. Energy policies for managing China's carbon emission. *Renew. Sustain. Energy Rev.* **2015**, *50*, 470–479.
2. Gorobets, A. Eco-centric policy for sustainable development. *J. Clean. Prod.* **2014**, *64*, 654–655.
3. Mahapatra, K. Energy use and CO_2 emission of new residential buildings built under specific requirements—The case of Växjö municipality, Sweden. *Appl. Energy* **2015**, *152*, 31–38.
4. Shi, Q.; Yu, T.; Zuo, J. What leads to low-carbon buildings? A China study. *Renew. Sustain. Energy Rev.* **2015**, *50*, 726–734.
5. Li, J.; Colombier, M. Managing carbon emissions in China through building energy efficiency. *J. Environ. Manag.* **2009**, *90*, 2436–2447.
6. Stojiljkovic, M.M.; Ignjatovic, M.G.; Vuckovic, G.D. Greenhouse gases emission assessment in residential sector through buildings simulations and operation optimization. *Energy* **2015**, *92*, 420–434.
7. International Energy Agency (IEA). *Energy Technology Perspectives OECD/IEA*; IEA: Paris, France, 2010.
8. Zuo, J.; Zhao, Z.Y. Green building research-current status and future agenda: A review. *Renew. Sustain. Energy Rev.* **2014**, *30*, 271–281.
9. Urge-Vorsatz, D.; Koeppel, S.; Mirasgedis, S. Appraisal of policy instruments for reducing buildings' CO_2 emissions. *Build. Res. Inf.* **2007**, *35*, 458–477.
10. Annunziata, E.; Frey, M.; Rizzi, F. Towards nearly zero-energy buildings: The state-of-art of national regulations in Europe. *Energy* **2013**, *57*, 125–133.
11. Rashid, A.F.A.; Yusoff, S. A review of life cycle assessment method for building industry. *Renew. Sustain. Energy Rev.* **2015**, *45*, 244–248.
12. Bribián, I.Z.; Usón, A.A.; Scarpellini, S. Life cycle assessment in buildings: State-of-the-art and simplified LCA methodology as a complement for building certification. *Build. Environ.* **2009**, *44*, 2510–2520.
13. Wen, T.J.; Siong, H.C.; Noor, Z.Z. Assessment of embodied energy and global warming potential of building construction using life cycle analysis approach: Case studies of residential buildings in Iskandar Malaysia. *Energy Build.* **2015**, *93*, 295–302.
14. Basbagill, J.; Flager, F.; Lepech, M.; Fischer, M. Application of life-cycle assessment to early stage building design for reduced embodied environmental impacts. *Build. Environ.* **2013**, *60*, 81–92.
15. Kovacic, I.; Zoller, V. Building life cycle optimization tools for early design phases. *Energy* **2015**, *92*, 409–419.
16. Asdrubail, F.; Baldassarri, C.; Fthenakis, V. Life cycle analysis in the construction sector: Guiding the optimization of conventional Italian buildings. *Energy Build.* **2013**, *64*, 73–89.
17. The American Institute of Architects (AIA). *AIA Guide to Building Life Cycle Assessment in Practice*; AIA: Washington, DC, USA, 2010.
18. Haapio, A.; Viitaniemi, P. A critical review of building environmental assessment tools. *Environ. Impact Assess. Rev.* **2008**, *28*, 469–482.

19. Srinivasan, R.S.; Ingwersen, W.; Trucco, C.; Ries, R.; Campbell, D. Comparison of energy-based indicators used in life cycle assessment tools for buildings. *Build. Environ.* **2014**, *79*, 138–151.
20. Roh, S.; Tae, S.; Shin, S.; Woo, J. Development of an optimum design program (SUSB-OPTIMUM) for the life cycle CO_2 assessment of an apartment house in Korea. *Build. Environ.* **2014**, *73*, 40–54.
21. Tsai, W.; Lin, S.; Liu, J.; Lin, W.; Lee, K. Incorporating life cycle assessments into building project decision-making: An energy consumption and CO_2 emission perspective. *Energy* **2011**, *36*, 3022–3029.
22. Castellano, J.; Castellano, D.; Ribera, A.; Ciurana, J. Developing a simplified methodology to calculate CO_2/m^2 emissions per year in the use phase of newly-built, single-family houses. *Energy Build.* **2015**, *109*, 90–107.
23. Islam, H.; Jollands, M.; Setunge, S. Life cycle assessment and life cycle cost implication of residential buildings—A review. *Renew. Sustain. Energy Rev.* **2015**, *42*, 129–140.
24. Alshamrani, O.S.; Galal, K.; Alkass, S. Integrated LCA-LEED sustainability assessment model for structure and envelope systems of school buildings. *Energy Build.* **2014**, *80*, 61–70.
25. BRE. Envest2 and IMPACT. Available online: http://www.bre.co.uk/page.jsp?id=2181 (accessed on 8 May 2016).
26. Lee, K.; Tae, S.; Shin, S. Development of a life cycle assessment program for building (SUSB-LCA) in South Korea. *Renew. Sustain. Energy Rev.* **2009**, *13*, 1994–2002.
27. Jeong, Y.S.; Choi, G.S.; Kang, J.S.; Lee, S.E. Development of Life Cycle Assessment Program (K-LCA) for Estimating Environmental Load of Buildings. *Archit. Inst. Korea* **2008**, *24*, 259–266.
28. Roh, S.; Tae, S.; Shin, S. Development of building materials embodied greenhouse gases assessment criteria and system (BEGAS) in the newly revised Korea Green Building Certification System (G-SEED). *Renew. Sustain. Energy Rev.* **2014**, *35*, 410–421.
29. Roh, S.; Tae, S.; Suk, S.J.; Ford, G.; Shin, S. Development of a building life cycle carbon emissions assessment program (BEGAS 2.0) for Korea's green building index certification system. *Renew. Sustain. Energy Rev.* **2016**, *53*, 954–965.
30. Luo, Z.; Yang, L.; Liu, J. Embodied carbon emissions of office building: A case study of China's 78 office buildings. *Build. Environ.* **2016**, *95*, 365–371.
31. Dong, Y.H.; Ng, S.T. A life cycle assessment model for evaluating the environmental impacts of building construction in Hong Kong. *Build. Environ.* **2015**, *89*, 183–191.
32. Radhi, H.; Sharples, S. Global warming implications of facade parameters: A life cycle assessment of residential buildings in Bahrain. Environ. *Impact Assess. Rev.* **2013**, *38*, 99–108.
33. Taborianski, V.M.; Prado, R.T.A. Methodology of CO_2 emission evaluation in the life cycle of office building façades. *Environ. Impact Assess. Rev.* **2012**, *33*, 41–47.
34. Jang, M.; Hong, T.; Ji, C. Hybrid LCA model for assessing the embodied environmental impacts of buildings in South Korea. *Environ. Impact Assess. Rev.* **2015**, *50*, 143–155.

35. Kim, C.J.; Kim, J.; Hong, T.; Koo, C.; Jeong, K.; Park, H.S. A program-level management system for the life cycle environmental and economic assessment of complex building projects. *Environ. Impact Assess. Rev.* **2015**, *54*, 9–21.
36. Roh, S.; Tae, S.; Baek, C.; Shin, S.; Lee, J.; Lee, J.; An, J. The development of object-oriented building life cycle CO_2 assessment system (LOCAS). *Archit. Inst. Korea* **2012**, *28*, 101–108.
37. Baek, C.; Park, S.H.; Suzuki, M.; Lee, S.H. Life cycle carbon dioxide assessment tool for buildings in the schematic design phase. *Energy Build.* **2013**, *61*, 175–187.
38. Tae, S.; Shin, S. Current work and future trends for sustainable buildings in South Korea. *Renew. Sustain. Energy Rev.* **2009**, *13*, 1910–1921.
39. ISO. *ISO 21931-1: Sustainability in Building Construction—Framework for Methods of Assessment of the Environmental Performance of Construction Works—Part 1: Buildings*; ISO: Geneva, Switzerland, 2010.
40. Sharma, A.; Saxena, A.; Sethi, M.; Shree, V. Life cycle assessment of buildings: A review. *Renew. Sustain. Energy Rev.* **2011**, *15*, 871–875.
41. Tae, S.; Baek, C.; Shin, S. Life cycle CO_2 evaluation on reinforced concrete structures with high-strength concrete. *Environ. Impact Assess. Rev.* **2011**, *31*, 253–260.
42. Tae, S.; Shin, S.; Woo, J.; Roh, S. The development of apartment house life cycle CO_2 simple assessment system using standard apartment houses of South Korea. *Renew. Sustain. Energy Rev.* **2011**, *15*, 1454–1467.
43. Kim, J.; Koo, C.; Kim, C.; Hong, T.; Park, H. Integrated CO_2, cost, and schedule management system for building construction projects using the earned value management theory. *J. Clean. Prod.* **2015**, *103*, 275–285.
44. Huedo, P.; Mulet, E.; Lopez-Mesa, B. A model for the sustainable selection of building envelope assemblies. *Environ. Impact Assess. Rev.* **2016**, *57*, 63–77.
45. Korea Environmental Industry & Technology Institute (KEITI). Korea Carbon Emission Factor. 2010. Available online: http://www.edp.or.kr/lci/co2.asp (accessed on 8 May 2016). (In Korean)
46. Korea Environmental Industry & Technology Institute (KEITI). Korea Life Cycle Inventory Database. 2004. Available online: http://www.edp.or.kr/lci/lci_db.asp (accessed on 8 May 2016). (In Korean)
47. Park, J.; Tae, S.; Kim, T. Life cycle CO_2 assessment of concrete by compressive strength on construction site in Korea. *Renew. Sustain. Energy Rev.* **2012**, *16*, 2490–2496.
48. Kim, T.; Tae, S.; Roh, S. Assessment of the CO_2 emission and cost reduction performance of a low-carbon-emission concrete mix design using an optimal mix design system. *Renew. Sustain. Energy Rev.* **2013**, *25*, 729–741.
49. Ministry of Trade, Industry and Energy (MOTIE): South Korea. *2014 Energy Consumption Survey*; Ministry of Trade, Industry and Energy: Sejong, Korea, 2015. (In Korean)
50. Korea Energy Management Corporation (KEMCO). South Korea Building Energy Efficiency Rating System. 2010. Available online: http://www.kemco.or.kr/building/v2/buil_cert/buil_cert_4_1.asp (accessed on 8 May 2016). (In Korean)
51. IPCC. *IPCC Guidelines for National Greenhouse Gas Inventories*; IPCC: Geneva, Switzerland, 2006.

52. Korea Power Exchange (KPX). CO_2 Emissions Factors for Electricity. 2013. Available online: http://www.kpx.or.kr/KOREAN/htdocs/popup/pop_1224.html (accessed on 8 May 2016). (In Korean)
53. Korea District Heating Corporation (KDHC). CO_2 Emissions Factors for District Heating. 2013. Available online: http://www.kdhc.co.kr/content.do?sgrp=S10&siteCmsCd=CM3650&topCmsCd=CM3655&cmsCd=CM4018&pnum=1&cnum=9 (accessed on 8 May 2016). (In Korean)
54. Korea Institute of Civil Engineering and Building Technology (KICT). *Standard Estimating System of the Construction Work*; Korea Institute of Civil Engineering and Building Technology: Goyang, Korea, 2014. (In Korean)
55. Korea Institute of Civil Engineering and Building Technology (KICT). *Standard of Estimate for Construction*; Korea Institute of Civil Engineering and Building Technology: Goyang, Korea, 2014. (In Korean)
56. Korea Legislation Research Institute (KLRI). Korea Corporate Tax Act: Korea Ministry of Strategy and Finance. Available online: http://elaw.klri.re.kr/kor_service/lawView.do?hseq=28577&lang=ENG (accessed on 8 May 2016).

Environmental Impact Analysis of Acidification and Eutrophication Due to Emissions from the Production of Concrete

Tae Hyoung Kim and Chang U. Chae

Abstract: Concrete is a major material used in the construction industry that emits a large amount of substances with environmental impacts during its life cycle. Accordingly, technologies for the reduction in and assessment of the environmental impact of concrete from the perspective of a life cycle assessment (LCA) must be developed. At present, the studies on LCA in relation to greenhouse gas emission from concrete are being carried out globally as a countermeasure against climate change. However, the studies on the impact of the substances emitted in the concrete production process on acidification and eutrophication are insufficient. As such, assessing only a single category of environmental impact may cause a misunderstanding about the environmental friendliness of concrete. The substances emitted in the concrete production process have an impact not only on global warming but also on acidification and eutrophication. Acidification and eutrophication are the main causes of air pollution, forest destruction, red tide phenomena, and deterioration of reinforced concrete structures. For this reason, the main substances among those emitted in the concrete production process that have an impact on acidification and eutrophication were deduced. In addition, an LCA technique through which to determine the major emissions from concrete was proposed and a case analysis was carried out. The substances among those emitted in the concrete production process that are related to eutrophication were deduced to be NO_x, NH_3, NH_4^+, COD, NO_3^-, and PO_4^{3-}. The substances among those emitted in the concrete production process that are related to acidification, were found to be NO_x, SO_2, H_2S, and H_2SO_4. The materials and energy sources among those input into the concrete production process, which have the biggest impact on acidification and eutrophication, were found to be coarse aggregate and fine aggregate.

Reprinted from *Sustainability*. Cite as: Kim, T.H.; Chae, C.U. Environmental Impact Analysis of Acidification and Eutrophication Due to Emissions from the Production of Concrete. *Sustainability* **2016**, *8*, 578.

1. Introduction

Concrete is a major construction material that emits a large amount of substances with environmental impacts during its entire life cycle (production process, construction, maintenance, dismantlement, and scrapping).

Accordingly, technologies for the assessment of and reduction in environmental impacts of concrete from the perspective of life cycle assessment (LCA) must be developed. At present, the studies on LCA of greenhouse gas emission from concrete are being carried out globally as a countermeasure against climate change. However, the studies on the impact of the substances emitted in the concrete production process on acidification and eutrophication are insufficient. The substances emitted into the air and water during concrete production have impacts not only on global warming, but also on acidification and eutrophication [1]. Acidification is an environmental problem caused by acidified rivers/streams and soil due to anthropogenic air pollutants such as SO_2, NH_3, and NO_x. Acidification increases mobilization and leaching behavior of heavy metals in soil and exerts adverse impacts on aquatic and terrestrial animals and plants by disturbing the food web. Eutrophication is a phenomenon in which inland waters are heavily loaded with excess nutrients due to chemical fertilizers or discharged wastewater, triggering rapid algal growth and red tides.

Such acidification and eutrophication are the main causes of air pollution, red tide phenomena, and deterioration of reinforced concrete structures. Emissions such as NO_x and SO_2 come down to the ground as acid rain, mist, or snow to be absorbed into lakes, rivers, and soil. As a result, surface water, ground water, and soil are acidified in ways that cause devastation of forests and many shelled animals. The increasing damage to reinforced concrete structures, which are highly resistant to alkali, is the result of chemical attack by nitrogen oxide (NO_x) and sulfite gas (SO_2) contained in acid rain and snow. More specifically, the water pollution caused by the very large amount of concrete used for the four large river refurbishment projects in Korea, started in 2008 under the justification of river ecosystem restoration [2], has become a serious social problem.

Accordingly, in this study, an LCA technique was proposed to assess the impact of the substances emitted during the concrete production process on acidification and eutrophication, in accordance with the standards of ISO 14044 [3], 21930 [4], and 13315 [5].

The substances emitted during the concrete production process were analyzed. Derived through the analysis, the main substances that had an impact on acidification and eutrophication were determined. Acidification and eutrophication were analyzed for 24 MPa concrete using the main substances deduced to have such impacts. In addition, the analysis was carried out by evaluating acidification and eutrophication according to increase in the mixing ratios of ground-granulated blast-furnace slag (GGBS) and recycled aggregate (RA) during concrete mixing. Based on these findings, a way to reduce acidification and eutrophication from the concrete production process was developed and presented here.

2. Review of Environmental Impact According to Concrete Production

Concrete is a construction material comprised of normal cement, mixing water, and admixture. For normal cement (the main raw material) in particular, a large amount of energy is used during collection of limestone and clay and production of clinker, the processes during which environmental impact substances are emitted. Also, soil erosion and destruction of ecosystems is caused by collection (surface mining) of natural aggregate.

Substances with environmental impacts are also emitted due to the energy consumed by the equipment used in the process of transporting the raw materials (ordinary cement and aggregate) to a concrete factory. In particular, air pollution and water pollution may be caused by combustion of the energy sources used in batch plants and concrete manufacturing facilities, as well as by the discharged sludge waste and wastewater. Figure 1 shows the mechanism for the impact of the substances emitted in the concrete production process on acidification and eutrophication. NO_x and SO_2 emitted into the atmosphere returns to the ground in precipitation (rain, snow, mist, *etc.*). NO_x and SO_2 increase the hydrogen ion concentration in soil, streams, and oceans, and reduce their pH (they become more acidic). This increases leaching of heavy metals and adverse impacts on ecosystems, such as the food supply and nutrition of algae, plants, and fishes, and the body coverings (shells and exoskeletons) of many other animals. Moreover, NH_3 and PO_4^{3-} flow into ground water through the sewage systems of concrete factories. As NH_3 and PO_4^{3-} increase in an ecosystem, the activities of microorganisms also rise, causing increased consumption of oxygen. As a result, rapid spikes in nutritive substances in underwater ecosystems cause red tides due to rapid reproduction of algae.

Also, the nitric acid, sulfuric acid, ammonia, and phosphorus discharged into the air and water systems gradually generate calcium carbonate ($CaCO_3$) by chemically reacting with the alkaline calcium hydroxide in concrete. As the pH falls due to neutralization, the normally alkaline concrete becomes neutral. This makes the alkaline passive state coating unstable and allows corrosion of the rebar. Corrosion of the rebar is accelerated by water and air, which flow into the concrete as cracks form. The weakening of the tensile strength of the rebar results in deterioration of the durability of the reinforced concrete structure. For concrete, a construction material that is sensitive to the substances emitted that have such environmental impacts, categories of environmental impact assessment need to be selected from diverse perspectives. Assessing a single environmental impact may cause a misunderstanding about the environmental friendliness of concrete. The substances emitted during the concrete production process have impacts on the environment that include air pollution, water pollution, and generation of waste.

From these arise environmental problems that include not only global warming, but also acidification and eutrophication.

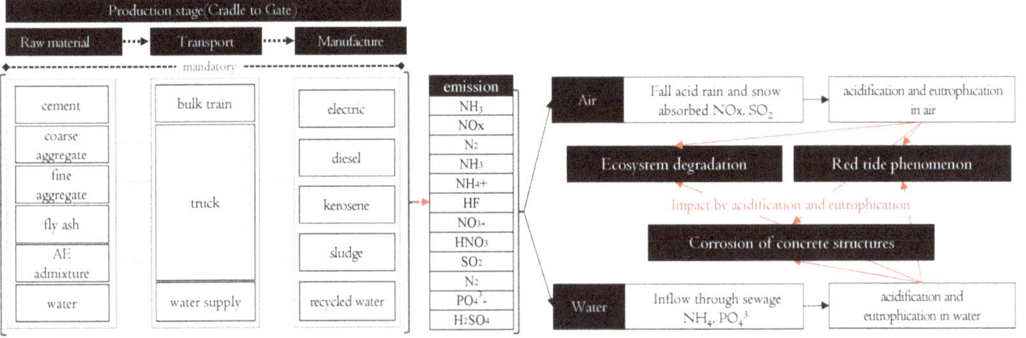

Figure 1. Mechanism of acidification and eutrophication potential in the production of concrete.

The designers are forced to make estimates of CO_2 emissions for concrete in ESD (environmentally sustainable design) based on conjecture rather than data.

Flower *et al.* presented hard data collected from a number of quarries and concrete manufacturing plants so that accurate estimates can be made for concretes in ESD [6].

Tait *et al.* analyzed the overall environmental impact, with a particular focus on carbon dioxide (CO_2) emissions of three concrete mix designs: CEM I (100% PC content), CEM II/B-V (65% PC content, 35% Fly Ash (FA) content) and CEM III/B (30% PC content, 70 % ground-granulated blast-furnace slag (GGBS) content) [7].

Huntzinger *et al.* evaluated the environmental impact of four cement manufacturing processes: (1) the production of traditional Portland cement, (2) blended cement (natural pozzolans), (3) cement where 100% of waste cement kiln dust is recycled into the kiln process, and (4) Portland cement produced when cement kiln dust (CKD) is used to sequester a portion of the process-related CO_2 emissions. Also, Huntzinger *et al.* presented a cradle-to-gate life cycle assessment of several cement products [8].

To make clear the environmental damages and potential improvements of the Chinese cement industry, Chen *et al.* conducted the detailed life cycle inventory (LCI) of cement manufacture with direct input and output in the boundary of the cement plant as well as corresponding transport [9].

Chen *et al.* proposed a hybrid life cycle assessment method based on national and provincial statistics to study pollutants generated by the cement industry in China, the impacts of these pollutants, and the potential for environmental improvement. Results showed that the key factors that contribute to overall environmental burden are the direct emissions of nitrogen oxides (NO_x), particulates, and carbon dioxide (CO_2) into the atmosphere, as well as the use of coal during cement production [10].

Scott *et al.* presented a cradle-to-gate framework for design engineers and concrete ready-mix producers to implement in an effort to optimize mixture designs across economic, environmental, and mechanical performance criteria. The framework was assessed through the examination of a newly constructed highway in South Georgia [11].

Serres *et al.* analyzed environmental impacts associated with mixing compositions of concrete made of waste materials by using LCA. Environmental performances of natural, recycled and mixed 20 mm concrete samples, formulated with the same mechanical strength regarding the functional unit, were evaluated. The LCA results are presented using various impact assessment methods, according to both EN 15804 and NF P 01-010 standards. Recycled samples present good environmental behavior, even if recycled materials (sand and aggregates) involve different operations (crushing against extraction, *etc.*) [12].

3. Methodology of Life Cycle Acidification and Eutrophication for Concrete

3.1. Goal and Scope Definition

This study proposed a concrete LCA (life cycle assessment) method including the definition of the goal and scope of concrete, inventory analysis, and impact analysis in compliance with the LCA method meeting the ISO standards. For the life cycle impact assessment (LCIA), the environmental impact categories of acidification potential (AP) and eutrophication potential (EP) were selected.

The 1 m^3 concrete was set as the functional unit on the basis of the main function to facilitate data management and application. As the system boundary for the concrete LCA, the product stage of concrete was selected, as shown in Figure 2. In addition, concrete production steps were divided into raw material extraction, transportation, and manufacturing steps, and environmental impacts of the elements involved in each step on air and water systems were assessed [13].

3.2. Inventory Analysis

Based on the life cycle assessment ranges (system boundary) of concrete, input factors and output factors such as energy, raw material, product, and waste were analyzed. To this end, as can be seen from Table 1, LCI DB (life cycle index database) on each of the input materials and energy sources in concrete production was investigated.

LCI DB on the input materials and energy sources used in this life cycle assessment utilized the existing data of Korea's Ministry of Land, Infrastructure, and Transport [14] and Ministry of Environment [15]. As LCI DB is different for each country, DB offered in one's own country should be used. However, LCI DB on ground-granulated blast-furnace slag, fly ash, and admixture in Korea's LCI DB has not been established yet. Therefore, the DB of ecoinvent [16], an overseas LCI DB, was used.

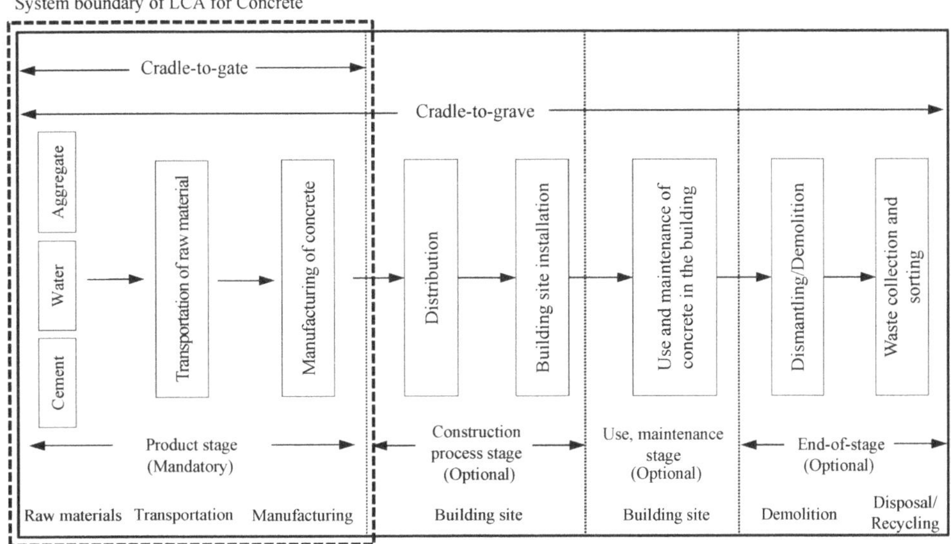

Figure 2. System boundary of the life cycle assessment (LCA) for concrete.

Table 1. Classification value of composition material for concrete. The six environmental impact categories are as follow: global warming potential (GWP) kg-CO_2 eq/kg; abiotic depletion potential (ADP) kg-Antimony eq/kg; acidification potential (AP) kg-SO_2 eq/kg; eutrophication potential (EP) kg-PO_4^{3-} eq/kg; ozone depletion potential (ODP) kg-CFC_{11} eq/kg; and photochemical ozone creation potential (POCP) kg-Ethylene eq/kg.

Inventory List	Environmental Impact Categories						Composition Material	
	GWP	ADP	AP	EP	ODP	POCP	Cement	Aggregate
Ammonia (NH_3)	-	-	■	■	-	-	-	6.95×10^{-7}
Carbon Dioxide (CO_2)	■	-	-	-	-	-	9.31×10^{-1}	3.40×10^{-1}
CFC-11	-	-	-	-	■	-	2.05×10^{-9}	4.02×10^{-13}
Methane (CH_4)	■	-	-	-	-	■	1.71×10^{-2}	5.57×10^{-4}
Sulfur Dioxide (SO_2)	-	-	■	-	-	■	1.27×10^{-2}	4.42×10^{-4}
Phosphate (PO_4^{3-})	-	-	-	■	-	-	-	4.22×10^{-8}

■: included, -: not included

3.3. Life Cycle Impact Assessment

This paper presented the input and output elements of the energy, raw materials, products, and waste to the scope of concrete LCA as shown in Figure 3.

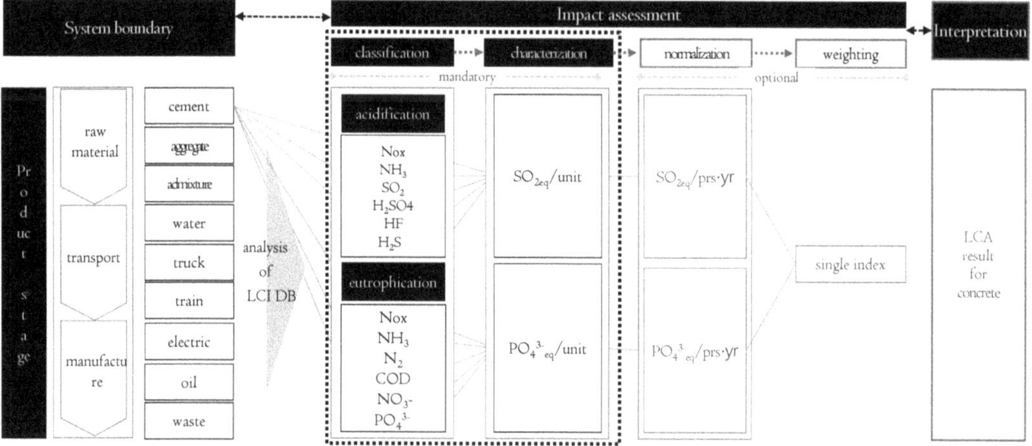

Figure 3. Process of life cycle impact assessment for concrete.

According to ISO 14044, the classification and characterization steps are mandatory assessment steps, and the normalization and weighting steps may be optionally assessed depending on the assessment purpose. In this study, assessment was performed for the classification and characterization steps because concrete-related normalization and weighting factor for Korea are yet to be developed.

Classification is done by categorizing and compiling the inventory items according to the environmental impact categories. By linking the inventory items derived from the LCI database to the environmental impact categories and integrating them by category, the environmental impact of each inventory item can be clearly identified. For example, inventory items for AP are nitrogen oxide (NO_X), ammonia (NH_3), and hydrogen fluoride (HF), sulfuric acid (H_2SO_4), with sulfur dioxide (SO_2) as the standard substance, and their respective classifications based on the Korean LCI database for OPC are 7.18×10^{-4} kg-NO_X/kg, 1.81×10^{-8} kg-NH_3/kg, 1.33×10^{-8} kg-HF/kg, 1.78×10^{-7} kg-H_2S/kg, 5.64×10^{-4} kg-SO_2/kg, 5.69×10^{-14} kg-H_2SO_4/kg [17].

Table 1 shows example classifications for OPC, coarse aggregate, diesel fuel among the LCI database classification items.

Acidification potential (AP) varies widely according to regional characteristics and atmospheric environments, and this research was applied to the AP index presented by Heijung *et al.* and Hauschild *et al.* [18] and is applicable to all regional types. A total of inventory items linked to the acidification category, including sulfur dioxide (SO_2), hydrogen sulfide (H_2S), and hydrogen fluoride (HF), are expressed in terms of their standard substance SO_2. Likewise, the index proposed by Heijung *et al.* and Hauschild *et al.* was applied for the classification of the eutrophication potential (EP), with phosphate ($PO_4{}^{3-}$) used as the standard substance for a total of 11 inventory items including phosphate ($PO_4{}^{3-}$), ammonia (NH_3), and nitrogen oxides (NO_x) [19].

Characterization is a process of quantifying the environmental loads of inventory items itemized for each category in the classification step.

In the classification step, inventory items are assigned to their respective environmental impact categories, but there is a limitation in quantifying the potential impacts of inventory items in common metrics due to different impact potentials. Category indicator results, *i.e.* characterization values, are calculated in the characterization step where the environmental load (=inventory data) of each inventory item is multiplied with the characterization factor (=impact potential) unique to the impact category concerned, and the resulting environmental loads thus converted into impacts are aggregated within each impact category to yield the overall environmental impact of that category. Equation (1) expresses this process:

Taking the acidification category of OPC as an example, which involves three inventory items in addition to SO_2 (standard substance)—H_2SO_4, HF, and NH_3—the acidification potential (AP) of H2SO4, HF, and NH_3 are 1 kg-SO_2/kg-SO_2, 0.65 kg-SO_2/kg-H_2SO_4, 1.6 kg-SO_2/kg-HF, 1.88 kg-SO_2/kg-NH_3, respectively, as calculated by multiplying their environmental loads (index data) with the characterization factor of the acidification category of OPC.

The total environmental impacts (=category indicator) on the acidification of OPC can be then obtained by adding the AP of the inventory items involved.

$$\mathrm{CIi} = \sum \mathrm{CIi,j} = \sum (\mathrm{Loadj} \times \mathrm{eqvi,j}) \quad (1)$$

Here, CI_i is the size of impact that all the list items (j) included in the impact category i have, on the impact category that they are included in. $CI_{i,j}$ is the size of impact that the list item j has on impact category i, $Load_j$ is the environmental load of the jth list item, and $eqv_{i,j}$ is the characterization coefficient value of jth list item within impact category i.

The standard substance for assessing AP is SO_2. The category indicator of AP is expressed by Equation (2):

$$\mathrm{AP} = \sum \mathrm{Load(i)} \times \mathrm{AP(i)} \quad (2)$$

where Load(i) is the experimental load of the acidification inventory item (i) and AP(i) is the characterization factor of inventory item (i) of the acidification category.

The standard substance for EP is PO_4^{3-}. The category indicator of EP is expressed by Equation (3):

$$EP = \sum Load(i) \times EP(i) \quad (3)$$

where Load(i) is the environmental load of the EP inventory item (i) and EP(i) is the characterization factor for the EP inventory item (i).

4. Deduction of Major Impact Substance

To determine which major substances in the concrete production process have impacts related to acidification and eutrophication, the LCI DB (life cycle index database) of the raw materials and energy used (e.g., normal cement, aggregate, admixture, oil, and electric power) were analyzed as shown in Tables 2 and 3.

Also, the characterized values calculated were divided into those affecting ecological systems (air and water). The major substances with impacts that corresponded to 90% or more of the cut-off of the total value were determined.

The major substances with impacts on eutrophication were found to be NO_X, NH_3, N_2, and NO_3 in the case of air, and NH_3, $NH4^+$, COD, NO_3^-, HNO_3, N_2, PO_4^{3-}, and NO_2 in the case of water systems. The major substances with impacts on acidification were found to be NO_X, NH_3, HF, H_2S, SO_2, H_2SO_4, HCL, and SO_3 in the case of air, and HF, H_2S, H_2SO_4, and HCL in the case of water systems.

The major substances in cement (the main raw material of concrete) that impact eutrophication were found to be NO_X, NH_4, COD, NO_3^-, and PO_4^{3-}, and those that impact acidification were found to be NO_X, SO_2, and H_2SO_4. The major substances in natural aggregate and recycled aggregate that have an impact on eutrophication were found to be NO_X, NH_4^+, COD, and PO_4^{3-}, and NO_X was found to have an impact on acidification. The major substances in GGBS and fly ash that impact eutrophication were found to be NO_X, NH_4^+, and PO_4^{3-}, and those that impact acidification were found to be NO_X and H_2S.

The major substances in the mixing water and admixture that impact eutrophication were found to be NO_X, NO_3^-, and PO_4^{3-}, and those that impact acidification were found to be NO_X, SO_2, and H_2S. The major substances in the energy sources related to electric power (e.g., diesel and kerosene) that impact eutrophication were found to be NO_X, NH_4^+, COD, and PO_4^{3-}, and those that have an impact on acidification were found to be NO_X and SO_2 [20].

Table 2. Major impact substance of acidification potential.

	Acidification Potential ($kgSO_2$-eq/kg, kWh)											
	Air Sector								Water Sector			
	NO_x	NH_3	HF	H_2S	SO_2	H_2SO_4	HCl	SO_3	HF	H_2S	H_2SO_4	HCl
C	7.18×10^{-4}	1.81×10^{-8}	1.33×10^{-8}	1.78×10^{-7}	5.64×10^{-4}	5.69×10^{-14}	Not included	Not included	2.94×10^{-25}	3.52×10^{-12}	3.90×10^{-9}	Not included
G	1.96×10^{-2}	4.78×10^{-5}	1.76×10^{-7}	3.08×10^{-7}	1.23×10^{-4}	1.01×10^{-13}	1.71×10^{-7}	Not included	Not included	Not included	Not included	Not included
RG	2.92×10^{-5}	7.3×10^{-8}	2.7×10^{-10}	4.7×10^{-10}	4.27×10^{-8}	1.5×10^{-16}	2.6×10^{-10}	Not included	Not included	Not included	Not included	Not included
S	1.03×10^{-2}	1.28×10^{-5}	1.04×10^{-7}	2.23×10^{-6}	6.63×10^{-4}	7.35×10^{-13}	4.99×10^{-7}	Not included	6.42×10^{-19}	Not included	Not included	9.01×10^{-12}
RS	6.47×10^{-4}	1.6×10^{-6}	5.8×10^{-9}	1.1×10^{-9}	9.52×10^{-8}	3.5×10^{-16}	3.5×10^{-9}	Not included	Not included	Not included	Not included	Not included
FA	1.04×10^{-4}	5.45×10^{-6}	2.66×10^{-8}	2.35×10^{-8}	3.50×10^{-6}	1.87×10^{-15}	1.01×10^{-7}	2.00×10^{-13}	Not included	3.38×10^{-6}	Not included	Not included
GGBS	4.25×10^{-5}	1.01×10^{-7}	4.80×10^{-9}	4.04×10^{-9}	1.20×10^{-6}	7.50×10^{-16}	4.23×10^{-8}	5.36×10^{-14}	Not included	2.27×10^{-21}	Not included	Not included
W	1.92×10^{-4}	4.83×10^{-7}	Not included	Not included	Not included	Not included	Not included	Not included	1.70×10^{-6}	Not included	Not included	9.59×10^{-10}
AE	1.92×10^{-5}	1.01×10^{-7}	6.50×10^{-9}	8.05×10^{-9}	5.32×10^{-6}	4.82×10^{-14}	9.86×10^{-8}	2.57×10^{-15}	Not included	1.13×10^{-7}	Not included	Not included
Diesel	1.76	1.05×10^{-4}	3.34×10^{-4}	1.30×10^{-2}	1.18	Not included	3.21×10^{-3}	Not included	Not included	Not included	Not included	Not included
Kerosene	1.36	0.84×10^{-4}	2.15×10^{-4}	1.11×10^{-2}	1.13	Not included	2.15×10^{-3}	Not included	Not included	Not included	Not included	Not included
Electric	8.35×10^{-4}	2.11×10^{-6}	7.42×10^{-9}	Not included	Not included	Not included	4.18×10^{-9}	Not included	Not included	Not included	Not included	Not included

C: Cement; G: Coarse aggregate; S: Fine aggregate; RG: Recycled coarse aggregate; RS: Recycled fine aggregate; FA: Fly ash; GGBS: Ground-granulated blast-furnace slag; W: Water; AE: Chemical admixture.

Table 3. Major impact substance of eutrophication potential.

	Eutrophication Potential (kgPO_4^{3}-eq/kg, kWh)											
	Air Sector			Water Sector								
	NO_x	NH_3	N_2	NO_3^-	NH_3	NH_4^+	COD	NO_3^-	HNO_3	N_2	PO_4^{3-}	NO_2
OPC	1.33×10^{-4}	3.37×10^{-9}	7.63×10^{-11}	Not included	3.37×10^{-9}	1.84×10^{-7}	1.77×10^{-7}	6.24×10^{-8}	6.30×10^{-10}	6.65×10^{-10}	6.63×10^{-8}	Not included
G	3.65×10^{-3}	8.90×10^{-6}	Not included	Not included	4.01×10^{-9}	2.26×10^{-6}	4.74×10^{-6}	3.99×10^{-7}	Not included	Not included	1.89×10^{-7}	Not included
RG	5.42×10^{-6}	1.37×10^{-8}	Not included	Not included	6.14×10^{-12}	3.47×10^{-9}	7.28×10^{-9}	6.12×10^{-10}	Not included	Not included	2.90×10^{-10}	Not included
S	1.91×10^{-3}	2.39×10^{-6}	2.53×10^{-13}	Not included	2.78×10^{-8}	6.39×10^{-7}	2.30×10^{-6}	1.11×10^{-7}	Not included	Not included	8.55×10^{-7}	Not included
RS	1.20×10^{-4}	3.04×10^{-7}	Not included	Not included	1.37×10^{-11}	7.70×10^{-8}	1.57×10^{-7}	1.36×10^{-8}	Not included	Not included	2.92×10^{-9}	Not included
FA	1.92×10^{-5}	2.72×10^{-6}	Not included	1.06×10^{-11}	Not included	3.28×10^{-12}	Not included	1.09×10^{-8}	Not included	7.98×10^{-10}	4.74×10^{-5}	6.33×10^{-12}
GGBS	7.89×10^{-6}	5.04×10^{-8}	Not included	1.32×10^{-11}	Not included	1.18×10^{-12}	Not included	2.81×10^{-9}	Not included	2.59×10^{-10}	9.22×10^{-4}	2.05×10^{-12}
W	3.56×10^{-5}	8.99×10^{-8}	Not included	Not included	Not included	2.28×10^{-8}	Not included	3.00×10^{-5}	Not included	Not included	4.56×10^{-11}	Not included
C.A	3.56×10^{-6}	1.79×10^{-7}	Not included	1.68×10^{-11}	Not included	1.10×10^{-12}	Not included	4.12×10^{-8}	Not included	5.50×10^{-10}	3.15×10^{-4}	4.36×10^{-12}
Diesel	3.27×10^{-1}	1.95×10^{-5}	Not included	Not included	1.69×10^{-4}	1.98×10^{-4}	6.22×10^{-3}	2.15×10^{-5}	Not included	Not included	4.86×10^{-3}	Not included
Kerosene	1.26×10^{-1}	0.78×10^{-5}	Not included	Not included	1.17×10^{-4}	1.32×10^{-4}	4.87×10^{-3}	1.95×10^{-5}	Not included	Not included	$3.45\times10^{-3}3$	Not included
Electric	1.55×10^{-4}	3.92×10^{-7}	Not included	Not included	Not included	9.94×10^{-8}	2.02×10^{-7}	1.75×10^{-8}	Not included	Not included	3.26×10^{-9}	Not included

C: Cement; G: Coarse aggregate; S: Fine aggregate; RG: Recycled coarse aggregate; RS: Recycled fine aggregate; FA: Fly ash; GGBS: Ground-granulated blast-furnace slag; W: Water; AE: Chemical admixture.

Through this process, the major substances that impact eutrophication during the entire process of concrete production were deduced to be NO_X and NH_3 in the case of the air and NH_4^+, COD, NO_3^-, and PO_4^{3-} in the case of water systems. The environmental emission load of each substance was 4.59×10^{-1} kg NO_X/kg, 4.23×10^{-5} kg NH_3/kg, 3.33×10^{-4} kg NH_4^+/kg, 1.11×10^{-4} kg COD/kg, 7.17×10^{-5} kg NO_3^-/kg, and 9.6×10^{-3} kg PO_4^{3-}/kg. These were found to be mostly attributable to oil, fly ash, and coarse aggregate.

The major substances impacting acidification were deduced to be NO_X and SO_2 in the case of the air, and H_2S and H_2SO_4 in the case of water systems [21]. The environmental load emission of each substance was 3.15 kg NO_X/kg, 2.31 kg SO_2/kg, 3.49×10^{-6} kg H_2S/kg, and 3.94×10^{-9} kg H_2SO_4/kg, and these were found to be mostly attributable to oil, fly ash, and cement. In particular, NO_X was found to be a major impact substance that corresponded to the 90% cutoff (or higher) in relation to acidification and eutrophication for all the items analyzed, such as raw materials and energy sources.

5. Analysis of Life Cycle Acidification and Eutrophication

5.1. Method

Evaluation of 1 m^3 of concrete with a strength of 24 MPa produced by Concrete Manufacturer A (in Korea), was carried out using the acidification and eutrophication assessment technique for the entire life cycle of concrete. The production stage (cradle to gate) was selected as the scope of the LCA and the assessment information shown in Table 4 was investigated. Also, the quantities of normal cement and natural aggregate usually mixed into the concrete were substituted with GGBS and recycled aggregate. The acidification and eutrophication impacts before and after substitution were compared in Table 5. The mixing ratios of GGBS tested were 0, 10, 20, and 30%, and those of recycled aggregate were 0%, 20%, 30% and 40%, taking into consideration the law setting a mandatory amount of recycled aggregate that might be used [22].

Table 4. Information of analysis object concrete.

Raw Material	Strength (MPa)	Mixing Design (kg/m^3)					
		C	W	G	S	GGBS	AE
	24	297	160	931	896	33	2.6
Transport	Supplier region	Chungcheongbuk-do	Water supply	Incheon	Gyounggi-do	Chungcheongnam-do	Gyounggi-do
	Distance (km)	201	-	14	66	122	90
Manufacture	Product amount (m^3/year)	Electric (kWh/year)		Diesel (L/year)		Kerosene (L/year)	
	506,739	1,895,631		1270		175	

However, the energy consumption during transportation and manufacturing stages were assumed to be the same. As shown in Table 6, the Korean and overseas LCI DBs were applied to concrete raw materials, energy, and transportation.

Table 5. Mix design of analysis object concrete.

<table>
<tr><th rowspan="2">Strength (MPa)</th><th colspan="8">Mixing Design (kg/m³)</th></tr>
<tr><th>OPC</th><th>GGBS</th><th>G</th><th>RG</th><th>S</th><th>RS</th><th>W</th><th>AE</th></tr>
<tr><td rowspan="8">24</td><td>100%</td><td>0%</td><td colspan="6" rowspan="4">equally application</td></tr>
<tr><td>90%</td><td>10%</td></tr>
<tr><td>80%</td><td>20%</td></tr>
<tr><td>70%</td><td>30%</td></tr>
<tr><td rowspan="4">90%</td><td rowspan="4">10%</td><td>100%</td><td>0</td><td>100%</td><td>0</td><td colspan="2" rowspan="4">equally application</td></tr>
<tr><td>80%</td><td>20%</td><td>80%</td><td>20%</td></tr>
<tr><td>70%</td><td>30%</td><td>70%</td><td>30%</td></tr>
<tr><td>60%</td><td>40%</td><td>60%</td><td>40%</td></tr>
</table>

* OPC: Cement; G: Coarse aggregate; S: Fine aggregate; W: Water; RG: Recycled coarse aggregate; RS: Recycled fine aggregate; GGBS: Ground-granulated blast-furnace slag; AE: Chemical admixture.

Table 6. LCI database applied for LCA of concrete.

Division	Life Cycle Index Data Base (LCI DB)	Reference
Raw material	ordinary cement	South Korea
	coarse aggregate	South Korea
	fine aggregate	South Korea
	recycled coarse aggregate	South Korea
	recycled fine aggregate	South Korea
	ground-granulated blast-furnace slag	Swiss
	fly ash	Swiss
	water	South Korea
	chemical admixture	Swiss
Energy	electric	South Korea
	diesel	South Korea
	kerosene	South Korea
Transport	truck	South Korea

5.2. Result

Acidification and eutrophication were assessed for the entire life cycle of the concrete, as shown in Tables 7 and 8. In addition, acidification and eutrophication results were compared according to the mixing ratio of admixture (%) and the mixing ratio of recycled aggregate (%).

Table 7. Acidification result of analysis object concrete.

Strength (MPa)	Raw Material								Transport	Manufacture	Acidification ($kgSO_2$-eq/m^3)		
	OPC	GGBS	G	RG	S	RS	W	AE			Air	water	total
	4.23×10^{-1}	0									$2.79\times10^{+1}$	1.98×10^{-6}	$2.79\times10^{+1}$
	3.81×10^{-1}	1.40×10^{-4}	$1.82\times10^{+1}$	0	9.23	0	3.07×10^{-2}	6.47×10^{-5}			$2.79\times10^{+1}$	1.85×10^{-6}	$2.79\times10^{+1}$
24	3.38×10^{-1}	2.80×10^{-3}							7.78×10^{-7}	1.15×10^{-2}	$2.79\times10^{+1}$	1.72×10^{-6}	$2.79\times10^{+1}$
	2.96×10^{-1}	4.21×10^{-3}									$2.78\times10^{+1}$	1.60×10^{-6}	$2.78\times10^{+1}$
			$1.82\times10^{+1}$	0	9.23	0					$2.79\times10^{+1}$	1.72×10^{-6}	$2.79\times10^{+1}$
	3.38×10^{-1}	1.40×10^{-3}	$1.46\times10^{+1}$	5.43×10^{-3}	7.38	1.16×10^{-1}	3.07×10^{-2}	6.50×10^{-5}			$2.25\times10^{+1}$	1.72×10^{-6}	$2.25\times10^{+1}$
			$1.28\times10^{+1}$	8.14×10^{-3}	6.46	1.74×10^{-1}					$1.98\times10^{+1}$	1.72×10^{-6}	$1.98\times10^{+1}$
			$1.09\times10^{+1}$	1.09×10^{-2}	5.54	2.32×10^{-1}					$1.71\times10^{+1}$	1.72×10^{-6}	$1.71\times10^{+1}$

Table 8. Eutrophication result of analysis object concrete.

Strength (MPa)	Raw Material								Transport	Manufacture	Eutrophication potential ($kgPO_4{}^{3-}$ eq/m^3)		
	OPC	GGBS	G	RG	S	RS	W	AE			Air	Water	Total
24	4.41×10^{-2}	0	3.4[illegible]	0	1.71	0	1.05×10^{-2}	8.41×10^{-4}	9.45×10^{-5}	1.55×10^{-3}	5.16	1.58×10^{-2}	5.18
	3.96×10^{-2}	3.07×10^{-2}									5.16	4.62×10^{-2}	5.20
	3.52×10^{-2}	6.14×10^{-2}									5.15	7.66×10^{-2}	5.23
	3.08×10^{-2}	9.21×10^{-2}									5.15	1.07×10^{-1}	5.26
	3.52×10^{-2}	3.07×10^{-2}	3.4[illegible]	0	1.71	0	1.05×10^{-2}	8.41×10^{-4}			5.15	4.62×10^{-2}	5.20
			2.7[illegible]	1.01×10^{-3}	1.37	2.16×10^{-2}					4.15	4.43×10^{-2}	4.19
			2.3[illegible]	1.52×10^{-3}	1.20	3.24×10^{-2}					3.65	4.33×10^{-2}	3.69
			2.0[illegible]	2.02×10^{-3}	1.03	4.32×10^{-2}					3.15	4.23×10^{-2}	3.19

* OPC: Cement; G: Coarse aggregate; S: Fine aggregate; RG: Recycled coarse aggregate; RS: Recycled fine aggregate; GGBS: Ground-granulated blast-furnace slag; AE: Chemical admixture; W: Water.

The contribution to acidification and eutrophication of the test concrete was shown to be 27.9 kg SO_2-eq/m^3 and 5.21 kg PO_4^{3-}eq/m^3, respectively. The material that had the biggest impact on acidification and eutrophication was found to be aggregate. As shown in Figure 4, coarse aggregate was found to contribute acidification of 18.2 kg SO_2-eq/m^3, which accounted for about 70% of the total. Fine aggregate acidification was 9.23 kg SO_2-eq/m^3, which accounted for about 30% of the total. As shown in Figure 5, NO_x accounted for most of the coarse aggregate acidification, and NO_x and HCl accounted for 90 and 10% of the fine aggregate acidification, respectively.

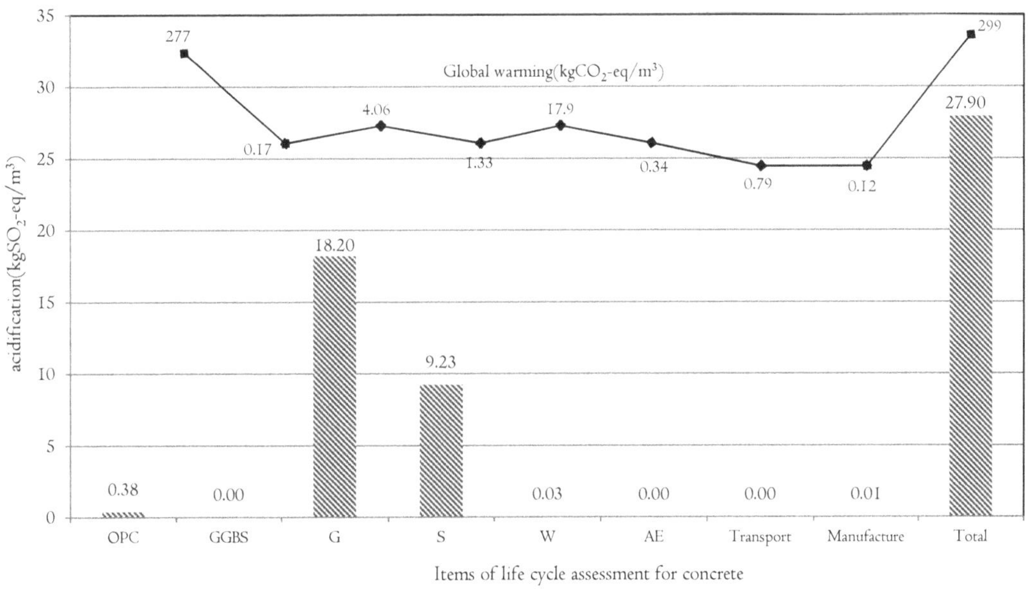

Figure 4. Acidification result of analysis object concrete.

As shown in Figure 6, coarse aggregate eutrophication was found to be 3.42 kg PO_4^{3-}-eq/m^3, which accounted for about 65% of the total. Fine aggregate eutrophication was found to be 1.71 kg SO_2-eq/m^3, which accounted for about 30% of the total eutrophication. As shown in Figure 7, NO_x and $NH4^+$ were determined to contribute 70% and 30% of coarse aggregate eutrophication, respectively, and NO_x, NH_4, and PO_4^{3-} accounted for 60%, 20%, and 10% of fine aggregate eutrophication, respectively. Figure 4 shows the contribution of concrete to global warming, acidification, and eutrophication.

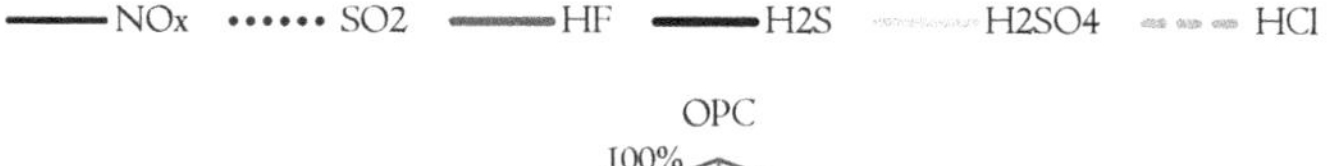

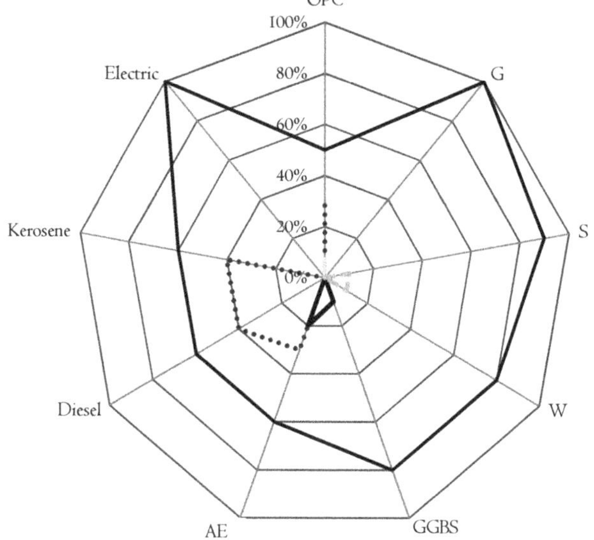

Figure 5. Acidification analysis.

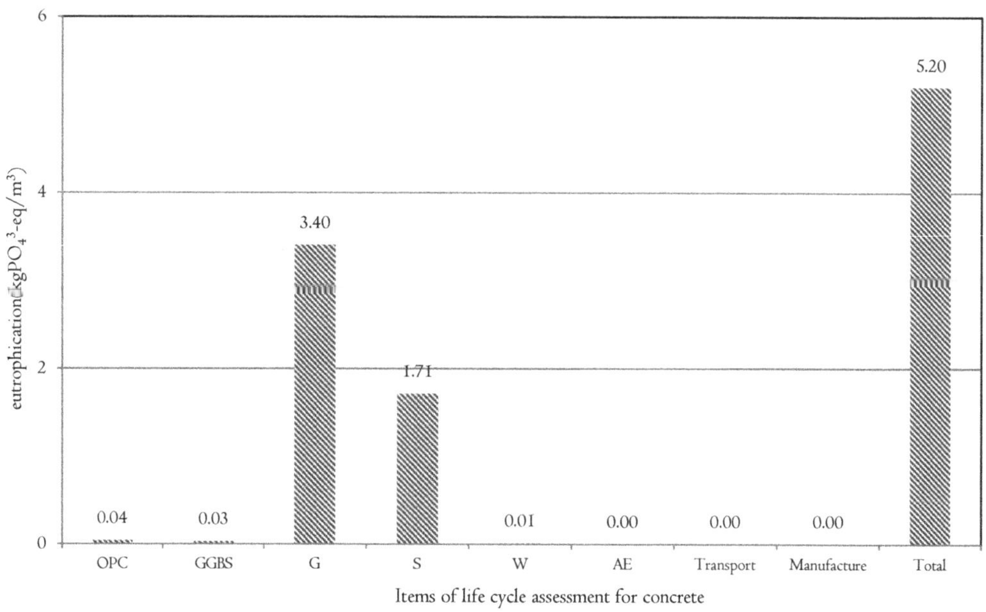

Figure 6. Eutrophication result of analysis object concrete.

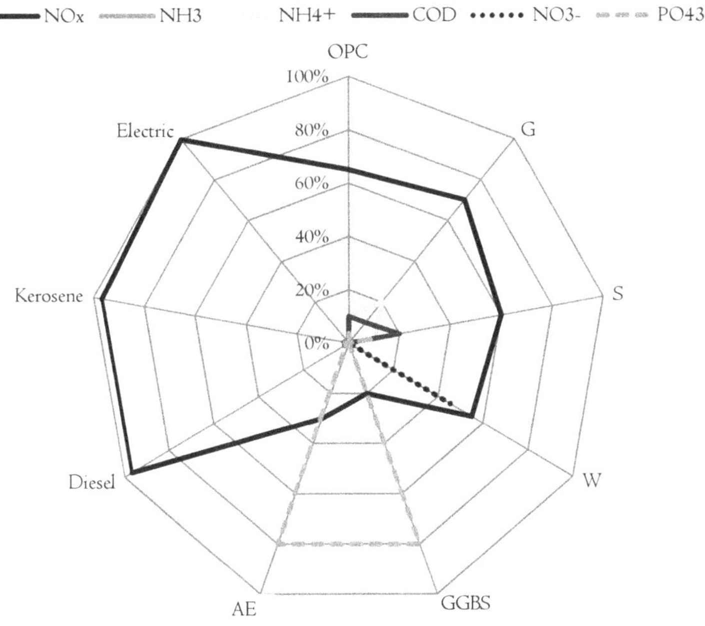

Figure 7. Eutrophication analysis.

As to global warming, the content based on a study of the author was described [23]. In the LCA of concrete, the material that had the biggest impact on global warming during the production process was found to be normal cement, and the impact of coarse aggregate and fine aggregate on global warming was very small.

The reason why coarse aggregate and fine aggregate, which have almost no impact on global warming, have big impacts on acidification and eutrophication is because of major emissions such as NO_x, SO_2, and H_2SO_4. It is because of the major emissions of aggregates that impacts on acidification and eutrophication are greater than that of normal cement, as analyzed in Tables 2 and 3. As a result of analyzing the major emissions of aggregate, it was found to have greater emissions of NO_x (to air), NH_4 (to water systems), and $PO_4{}^{3-}$ (to water systems) than the major emissions of normal cement.

The impacts of the cement (the raw material of the concrete with strength of 24 MPa) on acidification and eutrophication were very small, with values of 0.38 kg SO_2-eq/m^3 and 0.04 kg $PO_4{}^{3-}$-eq/m^3, respectively. The emissions from cement that could have an impact on acidification of ecosystems were found to be NO_x, SO_2, and H_2SO_4, which accounted for about 50, 30, and 20%, respectively. Also, the emissions that have an impact on eutrophication were NO_x, NH_4, NO_3^-, and PO_4^{3-}, which accounted for 70%, 10%, 10%, and 10%, respectively [24].

5.2.1. Analysis According to Ground-Granulated Blast-Furnace Slag (GGBS) Mixing

As shown in Figure 8, acidification and eutrophication resulting from mixing of GGBS with 24 MPa concrete were assessed. There was almost no change in acidification and eutrophication even when the mixing ratio of GGBS was increased. This is because the amount of GGBS or normal mixed cement does not have any impact because the aggregate accounts for 95% of the impacts of concrete production on acidification and eutrophication. The impacts on acidification and eutrophication were analyzed in two sectors: air (atmospheric) and water (hydrologic) systems. As the mixing ratio of GGBS increased, the acidification and eutrophication in the atmospheric sector decreased.

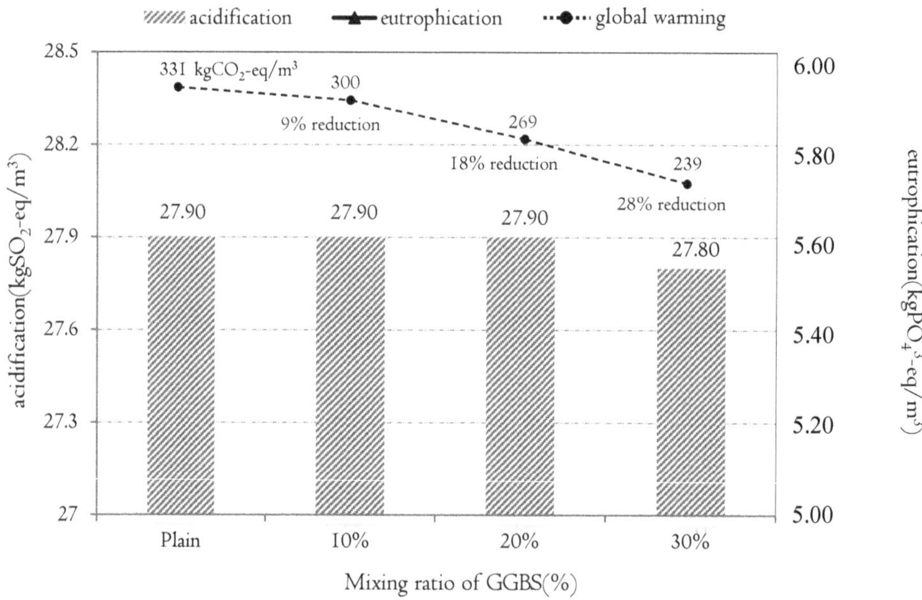

Figure 8. Analysis of acidification and eutrophication potential by GGBS mixing.

The main reason for this is because GGBS has less impact on acidification and eutrophication via the air than normal cement does. As shown in Figure 9, sulfur dioxide and sulfuric acid are discharged due to the use of dynamite (which comprises sulfuric acid, nitric acid, and sulfur) during mining of limestone and iron ore, the main raw materials of normal cement. NO_x and PO_4^{3-} are emitted by energy sources used for the electric power used in the pulverization of the mined ore and clinker. Sintering is the process into which the greatest amount of energy is input. Thus, the greatest amounts of substances with environmental impacts are emitted during production of normal cement.

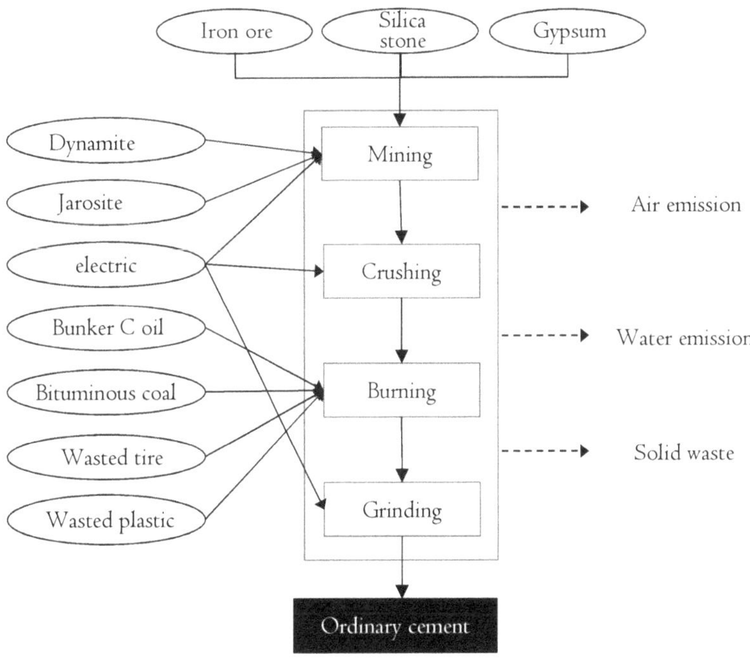

Figure 9. Analysis of acidification and eutrophication according to cement production.

To increase the temperature of the rotary kiln to the range 1000–1450 °C for production of clinker, fuels such as B-C (Bunker C) oil, bituminous coal, waste tires, and waste plastic are burned. During combustion of these fuels, substances such as ammonia, nitrate, sulfuric acid, and phosphorus are emitted in large amounts. In the case of GGBS, electric power and diesel fuel are used in the pulverizing and mixing process related to blast-furnace slag and natural gypsum. These are the main raw materials, as shown in Figure 10. Ammonia (NH_3), ammonium (NH_4^+), phosphate (PO_4^{3-}), nitrogen oxide (NO_x), *etc.*, are emitted as a result of using the diesel and electricity. The difference between the emissions of NO_x and PO_4^{3-} in the production processes of normal cement and GGBS was found to be the major cause of reduction in acidification and eutrophication.

Moreover, as the GGBS mixing ratio increased, acidification in the hydrologic sector decreased whereas eutrophication increased. The reason why acidification in water systems decreased was because the amount of H_2SO_4 (the major emission that has an impact on acidification) from normal cement was much smaller than that of H_2S (the major emission from GGBS). The reason why eutrophication in water systems increased was because the amount of PO_4^{3-} (the major emission that has an impact on eutrophication) from GGBS was bigger than the emissions of NH_4^+, NO_3^-, and PO_4^{3-} (the major emissions from normal cement).

However, such increase/decrease phenomena failed to change the total acidification and eutrophication results for the concrete with which GGBS was mixed. This is because the impact of aggregate on acidification and eutrophication is large, as explained earlier. As shown in Figure 8, although global warming (kg CO_2-eq) was shown to decrease by up to 28% as the mixing ratio of GGBS increased, the increase or decrease in acidification or eutrophication was insignificant [25,26].

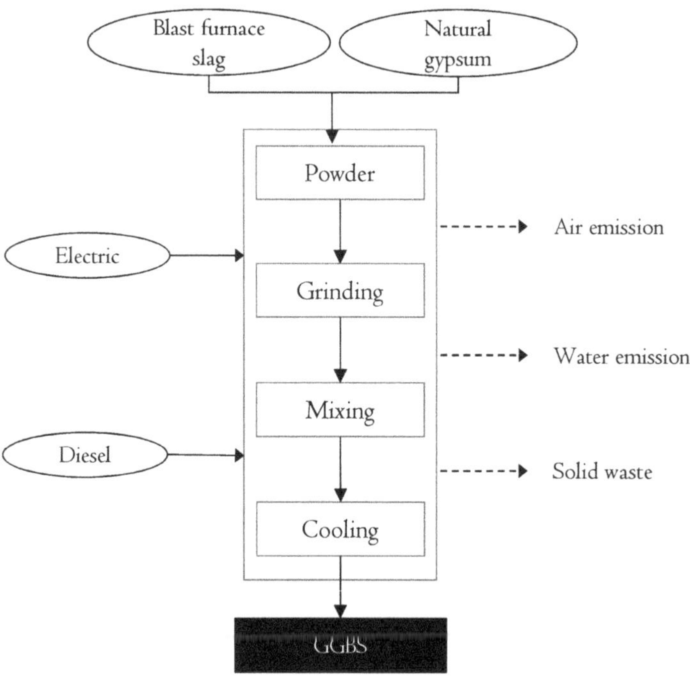

Figure 10. Analysis of acidification and eutrophication according to GGBS production.

5.2.2. Analysis According to Recycled Aggregate Mixing

As shown in Figure 11, acidification and eutrophication were assessed according to the mixing ratio of recycled aggregate with 24 MPa concrete. The results of the analysis showed that, as the mixing ratio of recycled aggregate increased, acidification and eutrophication decreased. When up to 40% of natural aggregate was replaced with recycled aggregate during concrete mixing, acidification and eutrophication were decreased by about 38%. Changes in acidification and eutrophication in the ecosystem resulting from mixing of recycled aggregate were analyzed in two sectors: atmospheric and hydrologic. When recycled aggregate was mixed in, there was no big impact on acidification and eutrophication of water systems, but the impact on the atmospheric sector was greatly reduced.

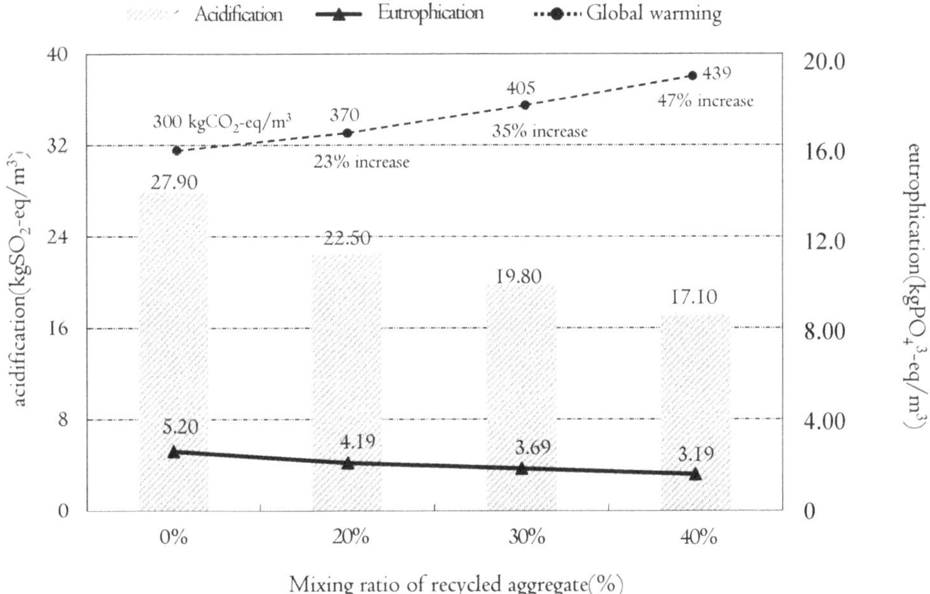

Figure 11. Analysis of acidification and eutrophication potential by recycled aggregate mixing.

The reason why acidification and eutrophication in the atmospheric sector was reduced was determined to be because NO_x emission was reduced more by mixing in recycled aggregate, than by using natural aggregate. The processes such as logging and pressure blasting are carried out first to produce natural aggregate, as shown in Figure 12. The main ingredients of the lubricant and dynamite, aside from the energy used to carry out these processes, are coal-based minerals and sulfuric acid, respectively.

For this reason, substances that have an impact on acidification and eutrophication, including SO_2 (sulfur dioxide), H_2SO_4 (sulfuric acid), and NO_3^- (nitrate), are emitted. Also, when pulverizing blasted rocks after collecting them, energy such as diesel and electricity is used. The substances emitted as a result of using diesel and electric energy are ammonia (NH_3), ammonium (NH_4^+), phosphate (PO_4^{3-}), and nitrogen oxide (NO_x). In comparison to this, the amount of energy input into the production process of recycled aggregate is very small.

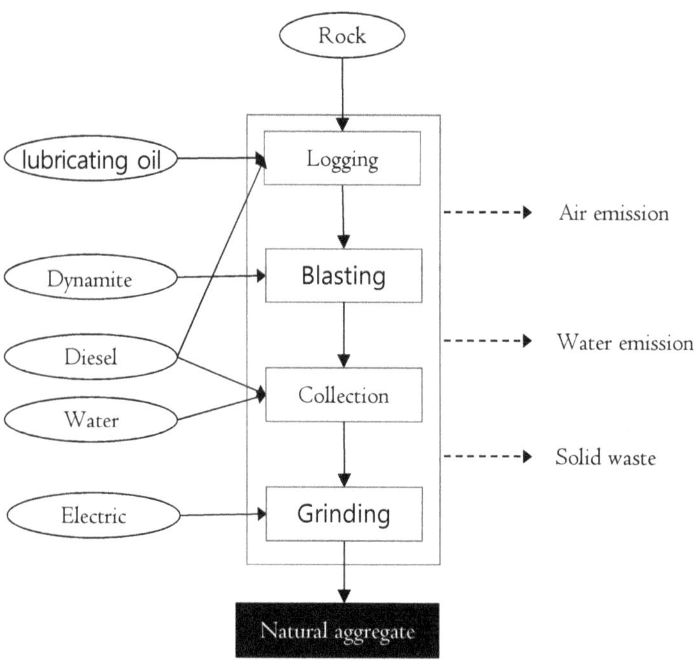

Figure 12. Analysis of acidification and eutrophication according to aggregate production.

Therefore, the emission of the substances that impact acidification and eutrophication of ecosystems via the atmosphere is smaller when producing recycled aggregate than when producing natural aggregate. As the mixing ratio of recycled aggregate is increased, no change in acidification of water systems occurred; eutrophication was reduced a little but there was no significant impact on overall eutrophication.

While global warming (kg CO_2-eq) increased to a maximum of 47% as the mixing ratio of recycled aggregate increased, acidification and eutrophication were shown to decrease [27,28].

6. Discussion and Limitation

This study attempted to analyze the effects of the matters produced during concrete production on the acidification and eutrophication of ecosystems. Then, it proposed a method to reduce concrete causing acidification and eutrophication.

However, this study has limitations in a reliability test.

There are diverse life cycle assessment (LCA) methodologies such as the conventional process LCA, EIO-LCA and hybrid LCA. This study adopted the process LCA.

Under the process LCA, this study assessed acidification (AP) and eutrophication (EP) during concrete production. However, the results were less reliable because they were not compared to the results of the other LCA method. Therefore, there should be additional studies and analyses.

In this study, the effects were assessed by concrete strength, mix rate of admixtures and mix rate of recycled aggregate, and the results found the following:

As the GGBS mix rate increased at the mix design of concrete, GWP decreased while AP and EP remained almost unchanged because the increase or decrease in the mixing amount of cement and GGBS hardly had any influence on AP and EP.

As the mix rate of recycled aggregate increased, GWP increased while AP and EP decreased because recycled aggregate is lower than natural aggregate in terms of NO_x emissions. Therefore, it would be good for the reduction of concrete AP and EP.

However, this study failed to analyze sensitivity on input and output while deriving the results.

This limitation should be overcome in future studies.

7. Conclusions

The objective of this study was to analyze the impact of the substances emitted during the concrete production process on environmental acidification and eutrophication.

In this study, a Korean model of the life cycle assessment (LCA) technique for assessing the impact of concrete on acidification and eutrophication was proposed, in accordance with the international standard. The major substances emitted during concrete production, with environmental impacts related to acidification and eutrophication, were determined.

The major substances impacting eutrophication were found to be NO_x and NH_3 in the case of air, and NH_4^+, COD, NO_3^-, and PO_4^{3-} in the case of water systems. The major substances impacting acidification were found to be NO_x, and SO_2 in the case of air, and H_2S and H_2SO_4 in the case of water systems.

As a result of carrying out an LCA of concrete, the mixing ratio of normal concrete was shown to have almost no impact on the increase or decrease of acidification and eutrophication, which was different from the result for global warming. On the other hand, the mixing ratio of aggregate, which had little impact on global warming, was found to have substantial impacts on acidification and eutrophication.

Also, when the mixing ratio of ground-granulated blast-furnace slag (GGBS) (in place of ordinary Portland cement) in concrete mixing was increased by 30%, there was almost no impact on the increase or decrease of acidification and eutrophication. This is because the amount of GGBS or normal mixed cement had little relative impact because acidification and eutrophication caused by aggregate accounted for about 85% of the total.

However, when the mixing ratio of recycled aggregate was increased in the place of natural aggregate, the acidification and eutrophication indices were found to be reduced by a maximum of about 40%. The major reason for this was found to be because the NO_X emission during the production of recycled aggregate is much less than that for the production of natural aggregate.

This paper proposed a concrete design in which recycled aggregate is mixed into the concrete as a way to reduce acidification and eutrophication.

Many studies have already reported that concrete accounts for a large portion of the greenhouse gas emitted during the construction of a structure, and technologies to reduce this have been developed. Even though many analyses have been carried out on global warming using the LCA of greenhouse gas emission resulting from concrete, this is the only study focused on acidification and eutrophication.

However, the study was limited by the assessment values calculated for acidification and eutrophication in the current analysis not clearly encompassing the impacts on all compartments of the ecosystem, such as air, soil, and water. Accordingly, it is believed that future studies will need to consider application of an end-point technique that shows the impact of environmental impacts that include not only global warming, but also acidification and eutrophication, on human health, social properties, and the extinction of living things.

Acknowledgments: This research was supported by a grant (Code 11-Technology Innovation-F04) from Construction Technology Research Program (CTIP) funded by Ministry of Land, Infrastructure and Transport.

Author Contributions: All authors contributed substantially to all aspects of this article.

Conflicts of Interest: The authors declare no conflict of interest.

References

1. Kim, T.H. Development of Eco-efficiency Evaluation Method for Concrete Using Equivalent Durability on Carbonation. Ph.D. Thesis, Hanyang University, Seoul, Korea, 2016.
2. Kim, S.H.; Kim, J.Y.; Won, H.Y. *Environmental Issues of the Restoration Project*; the National Assembly of the Republic of Korea: Seoul, Korea, 2009.
3. International Organization for Standardization. *ISO 14044: Environmental Management, Life Cycle Assessment, Requirements and Guidelines*; International Organization for Standardization: Geneva Switzerland, 2006.
4. International Organization for Standardization. *ISO 21930: Sustainability in Building Construction-Environmental Declaration of Building Products*; International Organization for Standardization: Geneva Switzerland, 2007.

5. International Organization for Standardization. *ISO/DIS 13315-2:2014: Environmental Management for Concrete and Concrete Structures–Part 2: System Boundary and Inventory Data*; International Organization for Standardization: Geneva, Switzerland, 2014.
6. Flower, D.J.M.; Sanjayan, J.G. Greenhouse gas emissions due to concrete manufacture. *Int. J. Life Cycle Assess.* **2007**, *12*, 282–288.
7. Tait, M.W.; Cheung, W.M. A comparative cradle-to-gate life cycle assessment of three concrete mix designs. *Int. J. Life Cycle Assess.* **2016**, *21*, 847–860.
8. Huntzinger, D.N.; Eatmon, T.D. A life-cycle assessment of Portland cement manufacturing: comparing the traditional process with alternative technologies. *J. Clean. Prod.* **2009**, *17*, 668–675.
9. Li, C.; Nie, Z.; Cui, S.; Gong, X.; Wang, Z.; Meng, X. The life cycle inventory study of cement manufacture in China. *J. Clean Prod.* **2014**, *72*, 204–211.
10. Chen, W.; Hong, J.; Xu, C. Pollutants generated by cement production in China, their impacts, and the potential for environmental improvement. *J. Clean Prod.* **2015**, *15*, 61–69.
11. Scott, H.S.; Stephan, A.D. A cradle to gate LCA framework for emissions and energy reduction in concrete pavement mixture design. *Int. J. of Sustain. Built Environ.* **2016**, *5*, 23–33.
12. Serres, N.; Braymand, S.; Feugeas, F. Environmental evaluation of concrete made from recycled concrete aggregate implementing life cycle assessment. *J. Build. Eng.* **2016**, *5*, 24–33.
13. Guinee, J.B. Development of a Methodology for the Environmental Life Cycle Assessment of Products: With a Case Study on Margarines. Ph.D. Thesis, Leiden University, Leiden, The Netherlands, 1995.
14. Ministry of Land, Transport and Maritime Affairs of the Korean government. *National D/B for Environmental Information of Building Products*; Ministry of Land, Transport and Maritime Affairs of the Korean government: Sejong, Korea, 2008.
15. Korea Environmental Industry and Technology Institute. National Life Cycle Index Database Information Network. Available online: http://www.edp.or.kr (accessed on 13 June 2016).
16. The ecoinvent Database. Available online: http://www.ecoinvent.org/database (accessed on 13 June 2016).
17. Roh, S.J.; Tae, S.H.; Kim, T.H.; Kim, R.H. The comparison of characterization of environmental impact of major building material for building life cycle assessment. *Architect. Inst. Korea* **2013**, *29*, 93–100.
18. Heijungs, R.; Guinée, J.B.; Huppes, G.; Lamkreijer, R.M.; Udo de Haes, H.A; Wegener Sleeswijk, A.; Ansems, A.M.M.; Eggels, P.G.; van Duin, R.; de Goede, H.P. *Environmental Life Cycle Assessment of Products. Guide (Part1) and Background (Part 2)*; Leiden University: Leiden, The Netherlands, 1992.
19. World Metrological Organization (WMO). *Scientific Assessment of Ozone Depletion: Global Ozone Research and Monitoring Project (Geneva)*; World Metrological Organization (WMO): Genève, Switzerland, 1991.

20. Van den Heede, P.; de Belie, N. Environmental impact and life cycle assessment (LCA) of traditional and green concretes: Literature review and theoretical calculations. *Cement. Concr. Compos.* **2012**, *34*, 431–442.
21. Owaid, H.M.; Hamid, R.B.; Taha, M.R. A review of sustainable supplementary cementitious materials as an alternative to all-Portland cement mortar and concrete. *Australian J. Basic Appl. Sci.* **2012**, *6*, 2887–3303.
22. Park, J.H.; Tae, S.H.; Kim, T.H. Life cycle CO_2 assessment of concrete by compressive strength on construction site in Korea. *Renew. Sustain. Energ. Rev.* **2012**, *16*, 2490–2496.
23. Kim, T.H.; Tae, S.H.; Roh, S.J. Assessment of the CO_2 emission and cost reduction performance of a low-carbon-emission concrete mix design using an optimal mix design system. *Renew. Sustain. Energ. Rev.* **2013**, *25*, 729–741.
24. Gazquez, M.J.; Bolivar, J.P.; Vaca, F.; Garcia-Tenorio, R.; Caparros, A. Evaluation of the use of TiO_2 industry red gypsum waste in cement production. *Cement Concr. Compos.* **2013**, *37*, 76–81.
25. Kim, T.H.; Tae, S.H.; Roh, S.J.; Kim, R.H. Life Cycle Assessment for Carbon Emission Impact Analysis of Concrete Mixing Ground Granulated Blast-furnace Slag(GGBS). *Architect. Inst. Korea* **2013**, *29*, 75–82.
26. Yang, K.H.; Seo, E.A.; Jeong, Y.B.; Tae, S.H. Effect of Ground Granulated Blast-Furnace Slag on Life-Cycle Environmental Impact of Concrete. *Korea Concr. Inst.* **2014**, *26*, 13–21.
27. Jung, J.S.; Lee, J.S.; An, Y.J.; Lee, K.H.; Bae, K.S. An Analysis of Emission of Carbon Dioxide from Recycling of Waste Concrete. *Architect. Inst. Korea* **2008**, *24*, 109–116.
28. Jung, C.J.; Chang, B.K.; Lee, J.H. Experimental Study on the Recycling of Concrete Sludge Produced in Ready-Mixed-Concrete Works. *Korea Soc. Waste Manag.* **1996**, *13*, 89–95.

An Insight into the Commercial Viability of Green Roofs in Australia

Nicole Tassicker, Payam Rahnamayiezekavat and Monty Sutrisna

Abstract: Construction industries around the world have, in recent history, become increasingly concerned with the sustainability of building practices. Inherently, the development of the built environment results in partial or complete destruction of the natural environment. Advanced European and North American countries have turned to green roofs as a means of sustainable development. Australia, on the other hand, has yet to fully realize the potential of green roof technology. In the first case, an extensive review of green roof literature was undertaken to establish the dominant perspectives and over-riding themes within the established body of international literature. The collection of primary data took the form of qualitative, semi-structured interviews with a range of construction practitioners and green roof experts; landscape architects, consultants and academics. The information gained from the interviews facilitated the primary aim of the paper; to critically analyse the state-of-practice in the Australian green roof industry. Green roofs, despite their proven sustainability benefits and their international success, have experienced a relatively sluggish uptake in the Australian construction industry. With this being said, the Australian green roof industry is considered to have promising potential for the future; should there be legislative changes made in its favour or greater education within the industry. To advance the local industry, it was found that government authorities are required to adapt policy settings to better encourage the use of green roofs, whilst industry bodies are required to host better, more targeted educational programs.

Reprinted from *Sustainability*. Cite as: Tassicker, N.; Rahnamayiezekavat, P.; Sutrisna, M. An Insight into the Commercial Viability of Green Roofs in Australia. *Sustainability* **2016**, *8*, 603.

1. Introduction

The construction of the built environment inherently involves some, if not total degradation of the natural environment [1]. The processes of construction, operation and ultimate demolition of man-made structures irrevocably alters the natural ecosystem [2]. In response, the global construction industry has developed an on-going commitment to rectify unsustainable trends in development [3]. As a result of this continuing moral conscience, great interest has been shown in sustainable construction practices; particularly elements of green design.

One specific aspect of green design, green roofs, has been touted as a tool for climate change mitigation given the apparent sustainability benefits, socially, economically and environmentally [4]. The numerous and wide ranging benefits of green roof technology have been widely recorded in the international body of knowledge. The fact that green roofs are extensively used in Europe and North America and even mandated in France is a testament to their efficiency as a sustainable roofing system. With this in mind, the question is begged: why is the use of green roof technology not more prevalent in Australia?

A green roof can be defined as an engineered roofing system that features multiple functional layers, including, but not limited to, a waterproofing membrane, a drainage system, a substrate layer and, finally, a vegetated surface [5–7]. Green roofs can be one of two recognized types: "extensive" or "intensive" [8]. The categorization of green roofs into either an extensive or intensive classification is determined on the depth of the supportive and vegetative layers.

The green roof industry is very much in its infancy in the Australian construction industry [9]. The difficulties that have previously hindered the uptake of Australian green roofs are widely considered to be a lack of standards, high installation costs, climatic concerns and a lack of reliable research [10]. It is noted within the current body of knowledge that a lack of national research is one of the most recognizable barriers to wider implementation of Australian green roofs [11,12]. The current state of the Australian green roof industry is underpinned by the need for greater domestic research and the ensuing development of local knowledge.

The aim of this paper is to report a recent study on general perceptions of green roof technology within the Australian construction industry with the view to promote the local green roof industry in the years to come. The specific objectives of the paper are as follows:

(1) To bring together expert views on how to advance the local industry and the realistic likelihood of doing so.
(2) To determine the critical factors affecting the commercial viability of green roof implementation; and how these factors currently impact on the local industry.
(3) To determine the most effective ways to promote green roof promotion within the construction industry at the moment.

2. Green Roof as a Sustainable Technology

The views of Kucukvar and Omer (2013) [13] regarding the inherent conflict between the natural and built environments are shared by Carter and Laurie (2008) [14], who determine that natural ecosystems are irrevocably altered through the process of development. Kibert [15] agrees that there is an obvious need for construction technologies and development processes to become more sustainable.

In considering a universal definition of "sustainable development", the Brundtland report [16] noted the concept to be "development that meets the needs of the present without compromising the ability of future generations to meet their own needs". A number of key literature pieces recognize that green roofs represent a solution, either fully or in part, to the requirement of the construction industry to become more sustainable. Bianchini and Hewage [17] and Fietosa and Wilkinson [18] all agree that green roofs outperform conventional roofing in specific sustainability realms—social, economic and environmental aspects; the triple bottom line of sustainability [19,20].

It is widely recognized throughout the literature that green roofs offer a vast array of sustainability benefits, such as reduced stormwater runoff, mitigation of noise, favourable lifecycle costing and significant amenity and aesthetic value [21,22]. The individual benefits of green roofs, however, can all be broadly categorized into the three aspects of sustainability: social, economic and environmental.

Throughout the international body of literature, economic and environmental aspects of sustainability are widely researched at the expense of the social sustainability dimension [23]. After recognizing the lack of research, Claus and Sandra (2012) [24] converted social aspects of sustainability into quantifiable measures by undertaking a cost benefit analysis to determine the personal and social aspects within green roofs. They found that "the inclusion of social costs and benefits improves their value".

Interestingly, Gatersleben and White (2010) [25] undertook a unique qualitative study whereby they assessed the validity of people's perceptions and opinions on the aesthetic value of green roofs. They noted that people's generalizations may impact on their perception of green roofs, whilst [26] determined that some people consider green roofs "visually inappropriate".

In direct contrast to these outlying views, the predominant opinion is that people generally perceive vegetation to be more favourable than traditional built forms [27–29]. Hietanen et al. [30] and Farrell et al. [31] evidenced that aspects of green design within a building will induce a more positive experience for users.

The economic benefits of green roofs are widely documented in the international body of knowledge. Bruce et al. [32] express that despite the well-documented environmental benefits of green roofs, the relatively-high initial cost of construction presents as a significant concern for building owners and developers alike. Forbes [33] specifically labelled the construction cost of green roofs as "indefensibly high", whilst evidencing the statistic [34] that construction costs of traditional roofs range between $7 and $15/square foot compared to a rate of $15–$70/square foot for green roof systems. It was widely found from comparative lifecycle studies that despite higher initial construction costs, green roofs economically outperform traditional roofing options over their lifecycles [35,36]. It is widely noted that a reduction in the maintenance costs of green roofs is the main contributory factor to favourable lifecycle costings [37,38].

The environmental benefits of green roofs are as equally recognized as the economic benefits of green roofs within the international body of knowledge. Forbes [30] recognized that despite the limitation of green roofs to single-handedly 'solve' global environmental concerns, they are a "multifunctional approach to ameliorating many different environmental problems". The findings of [39,40] suggest that effective stormwater management is the greatest environmental benefit offered by green roof systems. McIntyre and Snodgrass [41] reaffirm this dominant view, explicitly noting that stormwater management is "the green roof benefit that has been most aptly documented and validated by research".

The local body of knowledge regarding green roof implementation in Australian climatic conditions is limited [42]. Beecham et al. [43] established that significant disparity exists between the green roof body of knowledge in Australia and that of the advanced European and North American knowledge base. James and Metternich [44], however, note that despite the apparent lack of local data, there is enough information to determine that the contribution from green roofs is positive, and thus, they are worthy of promotion. Elliot [45] indicates that green roof technology has the greatest potential in the hot-dry climatic conditions that typify Australia.

3. Methodology

Semi-structured interviews were selected as the most appropriate method of collecting primary data. The structure of the interviews was not rigid in the sequencing or wording of the questions; rather, the interviewees' responses would guide the direction of the conversation and its findings. The interviews aimed to subjectively evaluate people's perception of green roofs

Bearman and Michalski [46] specify that semi-structured interviews are particularly useful where there is little previously known about the subject matter. Given the infancy of the green roof industry in Australia, the selection of semi-structured interviews to elicit qualitative, attitudinal data was deemed most appropriate for the study. Assessment of the themes is undertaken through measuring the perceptions and opinions of those who are closely involved with the subject matter [47].

A purposive collection of academics, practicing consultants and landscape architects was recruited as the target sample. The sample consisted of nine individuals, each with a practical or theoretical knowledge of green roofs in Australia.

An indicative number of questions was sent to each of the interviewees prior to the arranged meeting to ensure that the candidates had the opportunity to familiarize themselves with the intended content. It was made clear to each interviewee that the indicative agenda of questions was flexible and able to be adapted in any way to suit their experiences. The interview questions, as provided in Appendix A, were categorized into 4 main sections, which were specifically designed to closely relate to each of the study's objectives.

The interview participants were selected from a range of backgrounds (i.e., landscape architects, academics and consultants) with varying levels of experience in the construction industry and with green roofs. The profiles of the interviewees are included in Appendix B.

4. Results

Figure 1 depicts the outcomes from the interview. It suggests that the key factors that affect the commercial viability of green roofs include the cost implications of constructing green roofs, education/awareness among practitioners, as well as the role of government bodies in facilitating its adoption, for example through the setting of national policies or the development of incentive schemes. The three main actors identified to have a critical role to play in the promotion of green roofs are private clients, government bodies and industry bodies. However, some concerns were raised about the current effectiveness of industry bodies. Detailed findings from the interview are further discussed in this section based on the three objectives highlighted.

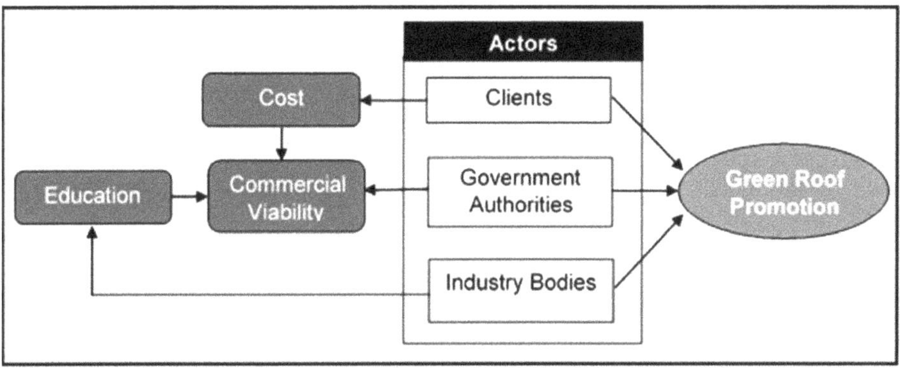

Figure 1. Network of interactions among key factors affecting the promotion of green roof.

4.1. Collective View Regarding the State of the Local Industry

The holistic construction industry and its sub-industries, such as architecture, project management and sub-contracting, are generally aware of green roof technology and its potential in the Australian marketplace. A general observation from the interviews was that the construction sub-industries were typically hyper-aware of the potential for negative outcomes to occur in green roof projects; such as roof leakages and cost blowouts.

4.1.1. Industry Perception

When asked how positively or negatively the construction industry perceives green roofs as an alternative to traditional roofing, the answers were typically dominated by negative bias. The interviewees regularly reiterated the industry's trepidation in accepting a relatively unproven technology.

> Interviewee C: "Anything new scares people ... unless you have people that are really motivated by something else, change usually scares them away".

> Interviewee F: "If I am honest, from my experience they [construction contractors] thought that green roofs were a bit painful ... the contractors thought 'well, what are we doing this for?'".

> Interviewee B: "Green roofs are not actively encouraged at the moment. I mean I think there is an awareness but I think planning won't drive this".

> Interviewee D: "Sceptical. The immediate thing would be complexity—architects would think 'it is just another thing I have to detail'".

Heller et al. [48] interestingly noted that designers are perceived to have the greatest ability (of construction professionals) to encourage greater uptake of green roofs. This raises an important point that, whilst landscape architects are typically in full support of green roof inclusions on a project [10], the construction professionals have a tendency to 'value engineer' the green roof out of the building design [48].

> Interviewee E: "Landscape architects view green roofs very positively ... "

> Interviewee A: "Typically, the key driver for those innovations on projects comes from landscape architects ... The landscape architects, you will find, are pretty good advocates for green roofs".

The view of [48] about green roofs typically being 'value engineered' out of construction projects was reaffirmed by the interview data.

> Interviewee I: "It was how they structured the green roof into the renovation build so that there was no option of the green roof getting dropped off by budget blowouts, it couldn't even be engineered out".

4.1.2. Project Champions

Heller et al. [48] go on to mention that key stakeholders are required to champion the inclusion of green roofs in building designs to ensure that they are not "value engineered" out of a project. The interviewees, however, highlighted the impact of a client suggesting that builders or construction contractors do no tend to push for green roofs; instead, the onus is on the client side.

> Interviewee I: "You have to have a champion for the project who drives it. So I don't think a builder will push it. I think you have to have a client that really wants it, and then it will happen".

> Interviewee C: "Yeah, it is certainly client driven to a large extent".

> Interviewee D: "You really need to dedicate a client that says 'I am going to do everything I can".

> Interviewee E: "I'd say it is clients—people who have seen the concept and want to go on and try it".

This finding is in opposition to the view of Heller et al. (2014) [48], who consider landscape architects to be the typical champion.

4.1.3. Potential of the Industry

Despite the negative bias surrounding green roof technology at the moment, it was agreed with a great deal of recurrence that the local green roof industry has a realistic potential to evolve in some way in the years to come. Eight of the nine interviewees answered "yes" to the question "do you believe that there is a realistic potential for the local green roof industry to evolve in the future?" Answers ranged from "absolutely, I don't think we're anywhere close to the potential of green roofs in Australia" to "Yes, but it will require policy settings." One interviewee presented an opposing view and was particularly sceptical about the future of the Australian green roof industry, noting:

> Interviewee D: "I don't think there is technical reasons that there isn't many green roofs. I think there isn't enough demand or there isn't enough push".

4.1.4. What Needs to Change

All of the interviewees who agreed that the local industry has potential to evolve went on to mention aspects of the industry that would need to change in order for progression to be possible. It was clear from these suggestions that there was one distinctive theme in advancing the industry, namely the role the government has to play: "government authority". At the federal level, cities' portfolios used to fall under the authority of the Department of Environment until the establishment of the federal government role of the Minister for Cities and the Built Environment in the third quarter of 2015, which was soon replaced in the first quarter of 2016 by another newly-created role of Assistant Minister to the Prime Minister for Cities and Digital Transformation. As far as it relates to the scope of this study, such a change highlights the enhanced notion in regards to greening cities because now, policy setting associated with the sustainability and resilience of Australian cities has entered the Prime Ministerial agenda. However, a closer look at the recent federal activities around green cities reveals that the Commonwealth government intends to transfer the responsibility in regards to green infrastructure to the local government. Accordingly, tangible actions, such as The Sydney Green Grid project and The City of Melbourne's Urban Forest Strategy, are initiated by local councils. The interviewees would tend to clarify the term "government authority" into a specific level: local, state or national. Most often, local or state governments were nominated to have the greatest potential in encouraging change. One interviewee went so far as to clarify that green roofs are not in the jurisdiction of the national government:

> Interviewee C: "It is not a Commonwealth issue".

The primary theme of "government authority" appeared to be categorized into two distinctive sub-themes: "policy settings" and "government incentives".

Government Authority: Policy Settings

The existing body of Australian literature does not extensively discuss the role and the power that governments have to play in promoting the local green roof industry. Unlike many other advanced countries, like Singapore and France [49], Australia has very little, to no national-based policies or regulations regarding green roof implementation. The suggestion of [10], who note that "policy incentives developed to increase uptake", is supported with the data collected through interviews. Five of the nine interviewees flagged a change in policy settings to be the way forward for the local industry.

> Interviewee C: "There may need to be a policy setting that supports it. I think that the policy environment has to change, around all of these things. If we don't [change] we are heading towards a train wreck".

Rayner et al. [10] further clarify and provide weight to this theme by explicitly stating that policy conditions have previously been restrictive for the local green roof industry; "While the absence of policy incentives is a barrier to the widespread uptake of green roofs ... "

The interviewees predominantly considered the main drivers for policy change to be the encouragement of green roof uptake and the subsequent reduction in cost.

> Interviewee H: "Government policy setting would encourage more people to put green roofs up and therefore more people enter the market and they become cheaper".

> Interviewee E: "Green roofs would become more regular, I suppose if the government made them conform to a certain level of environmental excellence".

These two identified drivers are directly supported in the statement of Rayner et al. (2010) [10], "recommendations and policies have not [previously] translated into financial incentives likely to stimulate the green roof industry or take the form of regulations that encourage the construction of green roofs".

Cost is widely considered to be a very significant barrier for the implementation of green roofs across the local construction industry. Given the infancy of green roofs locally, the costs associated with such projects typically present as being too overwhelming for the private industry alone. Together, the established body of literature and interviews clearly indicate the need for government assistance in the form of policy settings. Once implemented, policy settings are able to indirectly reduce the cost of green roof technology as it becomes more "common practice".

> Interviewee A: "I think that's important to understand, and without policy you are really relying on the market to drive it so, if people are more aware of it or they, I guess, see great examples of what that space [roof space] can be then the market will start to demand it".

> Interviewee B: "So that will drive it and then the next step is for the local government to accommodate that. It is not a commonwealth or state issue, it is a local government issue. The local government needs to have appropriate policies which will facilitate that [driving force] to happen".

> Interviewee H: "I keep coming back to how do we drive green roofs ... that is to construct more of them. How do we do that? We change our policy settings and there is a number of examples around the world where they can do that ... "

Importantly, a couple of interviewees flagged that the Green Building Council's Green Star rating system does not appropriately provide concessions for green roofs. The Green Star rating system, an initiative of the Green Building Council of Australia, is legislatively supported in Section J of the National Construction Code [50]. Two of the nine interviewees explicitly noted their concern over the "inequitable" point's concession for green roofs.

> Interviewee H: "Green roofs are not really integrated into the Green Star systems so they [the key stakeholders of building that won an award for the most sustainable building in the world] wanted the green roof because it got the ecology bonus ... that was how they got the points".

> Interviewee I: "You can get more points through having a bike rack".

> Interviewee H: "I'd argue that those building stars don't acknowledge the multiple benefits of green roofs enough".

In direct comparison to the views presented by the above interviewees reflecting on the Australian Green Star rating system, a project manager who worked on a green roof project in London specifically linked the inclusion of a green roof to the achievement of a Building Research Establishment Environmental Assessment Method (BREEAM) Excellence rating.

> Interviewee F: "I know in London, the only reason unfortunately, it's the reality though, the only reason why we did have a green roof on the project and spent the extra money was that it gave us a BREEAM 'Excellent' rating. It was a commercial incentive of leasing the building. The only reason we included a green roof was because it actually ticked a box and allowed us to get a BREEAM 'Excellent' rating. That additional rating was a huge driver for us to spend the money".

BREEAM is the United Kingdom's Environmental rating system for buildings. It is evident that in London, the BREEAM rating system played a significant role in the incorporation of a green roof in that specific project.

Legislative conditions and their influence on the construction of green roofs are distinctly different between the United Kingdom and Australia. The Green Star rating system, unlike the U.K.'s BREEAM system, provides very little commercial incentive to include green roofs within a building's design.

Enhancing this finding, Rayner et al. (2010) [10] note that "policy makers are likely to be reluctant to include green roofs in building codes and planning guidelines until there is quality data assessing their costs and benefits in an Australian context". As a solution to this, two of the nine interviewees offered an "evidence-based approach" to be the way forward in allowing policy changes to be made.

> Interviewee H: "I mean it's also going to be about an evidenced based approach looking at the benefits of green roofs".

> Interviewee I: " ... Melbourne Water [organization] .... they need the evidence base that enables them to argue for a policy change and the green roof industry".

Government Authority: Government Incentives

The central government's incentives toward green urban environment is associated with its ultimate impact on attracting highly skilled people who follow the improved lifestyle, rather than a vision that prioritises environmental programs. Envisaging New York as the model for further development of Sydney and Melbourne indeed demonstrates the loose position of greening in the national strategic plans [51]. In contrast, the central tenet of Green Plan of Singapore [52] is evolving into a city within a garden rather than revamping green canopies across a city to moderate the heat island effect or improve inhabitability. Data collected through this study also emphasizes the vision on perceiving the green roof as an integral component of the modern landscape.

> Interviewee A: "More amenity that people can actually take advantage of is going to be the focus within developments".

A comparison between long-term green schemes envisaged by the Singaporean Government and their Australian counterpart highlights the sizable impact of governmental support on the uptake of green roof technology at a national scale. The Department of the Environment in Australia shares comparable accountability in regards to the urban environment as the Singaporean Ministry of the Environment and Water Resources does. However, contradictory to Singapore, the Australian framework does not target urban development directly. While the Sustainable Singapore Blueprint [53] envisions "a liveable and endearing home [embraced with] a vibrant and sustainable city", Australia reinforces "clean up and revegetate urban environments" as a secondary matter under the "clean land" section of "a cleaner environment" plan [54]. In other words, a lack of individual attention to "skyrise greenery" in the Clean Environment Plan compared to the Blueprint surfaces as one fundamental reason behind the significant difference in the level of green roof adoption between the two nations. Yet, deploying green technologies for retrofitting heritage buildings as highlighted by Sibley and Sibley (2013) [55] certainly is aligned with two out of four core priority areas, clean land and national heritage, targeted by the Australian Government. The challenges of adopting green roofs as a retrofitting technique leverage both the design and implementation of the project. However, interview results indicate that the difficulties in design outbalance construction problems in the case of heritage buildings, while the proportion is opposite for new developments.

> Interviewee F: "I've worked with two green roofs on the project [in the capacity of project manager]. A green roof was on an existing building ... an existing buildings roof, it was a 150 year old building. Basically we had to change the structure of the roof component and

obviously also the waterproofing and everything else that goes with it; so that was quite interesting. As I say I was involved in the design and implementation of that green roof; the implementation part of these projects is the easy bit. The second component of the project involved a large section of green roof to a new building. This was a lot easier in the sense that the design. The structural design was built in and it (green roof) wasn't retrofitted. That was really good."

Financial incentives were additionally alluded to as being a potential way to evolve the local green roof industry. Whilst incentives were raised on a couple of occasions within the transcripts, the interviewees rarely went on to explain the role that incentives had to play.

> Interviewee I: "Some sort of policy and being incentivized".

> Interviewee F: "Green roofs would become more regular if the government incentivized it".

> Interviewee E: "I think government incentives would help".

> Interviewee G: "You need a series of incentives that encourage green roofs to happen".

An observation from within the data corpus was that whilst incentives do have some value in advancing the local green roof industry, that value is unspecified and appears to generally be less influential than other forms of government-driven change, such as the creation of new policy settings.

> Interviewee D: "I don't think it is subsidies, I think that if there were requirements ... "

> Interviewee A: "Well, I don't know if subsidies is the right approach. Policy can equate to subsidy in some ways ... "

4.2. Commercial Viability of Local Green Roof Industry

Three distinctive critical factors, namely cost, education and government authority, all featured high rates of recurrence across the data corpus. When asked what the greatest barrier has been for local green roof implementation, six of the nine interviewees nominated cost as the most critical factor. Conversely to this, when asked how to overcome the barriers to local green roof technology, five of the nine interviewees suggested education to be a viable solution, and three interviewees noted the role that government authorities have to play in promoting the industry. The other, less significant critical factors were considered to be "value proposition" and "industry conservatism".

4.2.1. Cost Implications

It is widely known, in the literature and amongst practitioners, that the cost of green roof construction is more expensive than the cost of constructing a traditional roofing system [48]. What is not widely agreed upon, however, is why the cost of constructing green roofs is "indefensibly high", as noted by Forbes (2010) [33]. Without a doubt, the structural and non-structural components that go together to form the make-up of the green roof are more expensive than the construction materials of a typical roof. However, many prominent pieces of international literature agree that the greater initial investment of constructing a green roof as opposed to a traditional roof will be more than compensated by the savings on maintenance and the subsequent favourable lifecycle costings. Therefore, why then is the Australian construction industry, an industry inherently focused on the bottom line, not embracing green roofs in the way that international construction industries are? It predominantly comes down to risk. Of the six interviewees who noted cost as the greatest barrier to local green roof implementation, three of these interviewees inextricably linked the cost factor to risk.

The study of Rayner et al. (2010) [10] briefly addresses the effect of risk in green roof projects in Australia. Their study raises the concern that reliance on Northern Hemisphere research is problematic and highly risky, noting that such practices "May introduce unacceptable levels of risk and unnecessary expense to development projects considering green roofs". Contributing further to this finding, risk is not only incurred through reliance on international practices, but also inherently exists in the uncertainties of the local industry in its relative beginnings. Risk, as defined by the International Standards Organization [56], is "the effect of uncertainty on objectives". During the interview process, two interviewees labelled the construction industry as being "incredibly risk averse", whilst another one of the interviewees described the industry as being "conservative and loathed to change". This view is specifically supported by Heller et al. (2014) [48], who also note "builders are quite risk averse". More broadly than this though, the interviewee explained from personal experience;

> Interviewee C: "the Australian construction industry is incredibly risk adverse ... when you're talking about doing something different, because there is uncertainty that something could go bad, people do really get nervous because the risk is quite high ... the costs are quite high".

The interviewee additionally recognized that the cost of risk is not only monetary, but also transpires into other facets of a person's life, such as job security and reputation. The monetary impact of risk on green roof construction projects is borne by each project stakeholder—the main contractor, sub-contractors, suppliers and last, but not least, the client. In an interview with a project manager, it was noted that, for contractors, there was a cost associated with owning that commercial risk and uncertainty;

> Interviewee F: "Owning the risk, the contractor is going to obviously cost that risk of an unproven technology".

Two of the three interviewees who linked cost to risk raised a discussion about contractual arrangements and noted that a project's contract has significant weight in determining the value of the risk and where it lies. One of the two above-mentioned interviewees, an academic, disclosed information on a green roof, public-private partnership project where the contractual arrangement focused on the effect of cost sharing.

> Interviewee C: "Cost sharing between the design and construction components; there was risk sharing. It was a tender contract where if things went well and ran smoothly then the contractors got paid more, whereas if it didn't then the contractors would lose money. It gave the contractors the opportunity to deal a little bit more with uncertainty, otherwise these contracts they just get so tightly restricted ... "

The interviewee focused on the partial freedom afforded to a specialist green roof, or construction contractor in the case of a contractual public-private partnership. Conversely to this, the project manager specifically highlighted the limitations of a lump sum contract in relation to green roof projects;

> Interviewee F: "Especially in lump sum contracts, it comes back to the developer and to the client .... If you take on a lump sum contract you are taking on some risk and you price for that. Both at the contractor and subcontractor levels. If you are smart enough as a contractor to pass the risk on cheaply to the subcontractor then you would. But you also have to be there when things go wrong, so you have to tread carefully and negotiate".

The project manager suggested that contractors are inherently going to try and pass on as much commercial risk as possible; lump sum contracts were additionally labelled, by another interviewee, as being "so tightly restricted".

4.2.2. Construction and Utilization

The Hanging Gardens of Babylon are probably the most well-known adoption of the concept of green roofing throughout history. According to United Nations Educational, Scientific and Cultural Organization (UNESCO), the Persian Garden evolved in ancient Persia to honour sky, earth, water and plants. Interestingly, these core elements not only are entirely consistent with the modern environmental policy spotlight on air, land and water as advertised, for example, by the Australian Government, but also exclusively address greenery. Like its contemporary equivalent, a sophisticated irrigation system, as well as a consistent combination of vegetation facilitated the adoption of the nowadays called green roof a couple of thousand years ago [57]. Similarly, engineered solutions aligned with today's advanced construction procedures are required to allow widespread implementation of the green roof. The mounting mechanism, overlooking concerns, maintenance, constructability, flora selection and, more importantly, understanding the different performance aspects of the green roof are a few examples articulated by the participants.

> Interviewee A: "There is a lot of engineers that are excited by the challenge of integrating things into their structures".
>
> Interviewee B: "Whereas green walls [in contrast to green roofs] are straight forward, they are not visually intrusive and they are quite aesthetic".
>
> Interviewee E: "An understanding of both the design but also the maintenance aspects of having plants on roofs (is essential to promote the technology)".
>
> Interviewee F: "Risk is involved with the buildability of these things (green roof)".
>
> Interviewee G: "Long, hot and dry summers all with little to no rain (...) is pretty challenging for any kind of plant."
>
> Interviewee I: "The industry need to be more realistic about these things [performance of green roof]. We have evidence around the energy savings, we have evidence for biodiversity benefits. However, that far that the coolness (i.e., the reduction in the urban heat island effect) extends we don't know".

4.2.3. Engineered Functions

As the green roof becomes more prevalent, studies on the long-term performance of this technology emerge. While the positive impact of a green roof on urban storm water quality and quantity in the short term is well evidenced [58], Speak et al. [59] highlight the two-fold potential of the technology in dispositioning urban atmospheric pollutants, as well as becoming a source of legacy metal pollution in the long term. Advocating the positive side of the results reported by Speak et al. (2014) [59], one of the interviewees pointed out that customized green roof soil is able to accumulate nutrients available in the storm water to support vegetation growth.

> Interviewee E: "It [the advanced green roof design] is a different system that essentially strips nutrients from storm water basically through soil".

Concerns regarding the lack of profound knowledge about the multifaceted performance criteria of the green roofs, specifically the shortage of longitudinal studies, have been echoed by the interviewees.

> Interviewee A: "There hasn't necessarily been the data there before to really support the more humanistic elements of design (for green roof)".
>
> Interviewee B: "The unknown of what comes with green roofs".
>
> Interviewee D: "I don't think green roofs have been tested enough".
>
> Interviewee E: "Landscape architects view green roofs very positively but maybe with a slight sense of trepidation regarding the number of unknown factors".

Therefore, it is interpreted that filling the knowledge gap in the area of green roofs enhances the competitive advantage of the technology in the construction market and eventually contributes to extended application. The result of studies on competing technologies, such as cool roofing [60], is a proper benchmarking tool to evaluate the competitiveness of the green roof against other advanced building technologies, particularly in regards to thermal insulation and energy performance. Pisello et al. [61] suggest that a combination of the two aforementioned technologies, the cool-green roof, is a promising solution where other invasive mitigation strategies are not practical to address the urban heat island phenomenon.

> Interviewee B: "The way that people are dealing with the heat island effect in infill cities is to increase the amount of greenery in the place and again that feeds into the green roof".
>
> Interviewee C: "You have the heat island effect also." The interviewee explained that his source is common knowledge and not actual measurement "Aside from my government stuff around sustainability projects and that sort of thing".
>
> Interviewee E: "Green roofs (. . .) could alleviate the Heat Island Effect. Based on the data I collected, it looked like the green roof would actually reduce the energy bills of the shopping centre significantly, almost as much that it would pay for the construction, and then also double the retail area as well. So, yeah there is really good possibilities for green roofs in that sense".

Interviewee G: "In my role as mayor, we've been looking at the role that green roofs and green walls can play in terms of urban design in the CBD primarily and around looking at the Heat Island Effect but also around urban biodiversity".

Li and Norford [62] demonstrated that with the city-scale implementation, the two technologies complement each other, as a cool roof is more effective during the daytime and a green roof adapts to the cooling cycle at night-time.

Interviewee H: "If you want to reduce the urban heat island effect you need an irrigated green roof on the hottest days. Our substrates here store a lot of heat and they'll re-radiate that heat at night but they won't store a lot of heat if you irrigate it because that energy will be evaporated off".

4.2.4. Education

When asked whether or not there was a viable way to mitigate the challenges presented by green roof technology in Australia, five of the nine interviewees suggested that education has a significant scope and leverage for improving the current state of the local green roof industry. Through the process of conducting interviews, the term "education" was found to be synonymous with "research" and "raising awareness".

It was found that every stakeholder group, from project managers to subcontractors to clients, generally requires some form of further education about green roofs if the local industry is to progress in the future. The existing body of local literature supports this finding, in recognizing that the research gap stands as a barrier to more widespread use of green roofs in Australia [63,64].

Interviewee E: "Well really it is education and greater (more) information through the construction industry, builders and also the government at all levels. I think that is quite important".

Of particular note was an outlying, but informed response, which noted the education programs currently existing within the industry relating to "green roofs". Australasia seminars and workshops are not "hitting" those with the power to make the required changes in the industry. That is, decision makers rarely have enough information or education regarding the logistics of how green roof projects work and thereby do not opt to undertake such work given the uncertainty and associated risks.

Interviewee E: "Education isn't perhaps available to the people in decision making positions. If people have been in their industries for 10–15 years and are at the director level—they are the people that really drive a lot of change. Something that I have found with green roof infrastructure projects is . . . it is not that these people do not want to implement these things, it is that they weren't taught how to do them".

In recognizing that people in upper management positions require greater education and information about green roofs, the need for education for the other constituents in a supply chain is not to be ignored. Of the nine interviewees, five recognized that education is required to increase awareness and subsequently decrease the inherent uncertainty of current green roof projects.

> Interviewee C: "It comes down to research. People who are doing work around bedding down that uncertainty would then share the info and take some risk out of them (green roofs)".

4.2.5. Government Authority

A third of the interviewees recognized that government authorities, whether they be local, state or national, have the ability to mitigate some of the challenges identified for the local green roof industry. These participants raised a small range of government-driven-specific solutions; "the right research and the right incentives", "the legislative side of things" and "demonstration projects".

Governments at all levels are empowered, in some way, to make a positive change to the local green roof industry. As evidenced in more than one interview, the government was identified as the stakeholder group that is needed to drive the local green roof industry. Private companies alone are typically unwilling to take on green roof projects given the current commercial risk; herein lies the need for government assistance.

> Interviewee G: "There is an important role for both government at the national level but also at the local level to provide an example of how we design our cities and I think it needs that level of support because this is not going to be something that many companies will do on their own".

> Interviewee E: "Certainly developers aren't so keen to do demonstration projects because of the risk and monetary expenses".

The government is a unique stakeholder in the construction industry and is perhaps the best equipped at dealing with commercial risk.

> Interviewee G: "I think this is where governments have a role to play in terms of having the right incentives and the right research happening to enable these things to cross from 'risky' to 'mainstream'".

Public-private partnerships present as an opportunity for the inclusion of green roofs within building designs. Of course, the approach is not "one size fits all";

not every building constructed under public-private partnerships will be able to incorporate a green roof practically. However, some success has been shown in projects with these contractual arrangements due to cost-sharing and risk sharing between the government and private industry.

> Interviewee C: "Well the reason why this project was even able to go into that territory is because they did a public-private partnership type thing so there was cost sharing between the design and construction components; there was risk sharing. It was a tender contract where if things went well and ran smoothly then the contractors got paid more, whereas if it didn't then the contractors would lose money. It gave the contractors the opportunity to deal a little bit more with uncertainty ... "

Additionally, government-led demonstration projects or "pilot projects" were regularly raised as a solution in removing some of the risk factors and associated expenses of current green roof projects.

> Interviewee C: "Yes, that's where pilot projects can be really good. Pilot projects would allow policy makers to get comfortable with 'ok, maybe we can change the rules'".

> Interviewee B: "People are incredibly risk averse so yes, pilot projects are good at creating a bit of comfort".

> Interviewee F: "In time it will become more common place and to do that you have to go back to the legislative side of things, making green roofs something that people want to do".

The proactive strategy applied in Singapore removes the need for further reactive efforts to rectify issues associated with the micro-climate formed by density living. Gliedt and Hoicka [65] share a similar notion categorizing green roofs as an "environmental sustainability initiative", rather than a "financial investment". Actually, from a risk management perspective, Singaporean policy makers have targeted the root causes of excessive heat amassed in building materials by looking at natural greenery as a piece of infrastructure, rather than shaping policy around responsive actions to improve urban residents' wellbeing compromised by overdevelopment. Therefore, a vertically-inclusive policy mechanism is required to facilitate long-term cooperation in all tiers of government in order to transform the development of green cities in Australia. For example, land value capture models, which intend to partially shift the cost of infrastructure development to surrounding land owners that benefit from the development potentially, can be adopted to promote the diffusion of green roof technology within the local construction industry subject to the fundamental change in regards to accepting urban canopies as infrastructure. Such an interpretation implied in guidelines and policies issued by local governments needs to be disseminated from the bottom toward to the top of the regulatory system in Australia. Both the Green Roofs Policy and the Walls Policy issued by the City of Sydney [66] in 2014 and Victoria's Guide to Green Roofs, Walls and Facades [67], issued by four inner Melbourne local governments in 2013, approach the green roof as a component of "green infrastructure". Victoria's Guide highlights the fundamental need for this type infrastructure to aid urban cooling, reduce energy consumption, manage storm water, improve air quality and enhance wellbeing. Sydney's policy further discusses the benefit of the green roof as a facility for food production and absorption of carbon dioxide, a catalyst of solar panels' performance, as well as a solution to address biodiversity.

4.3. Green Roof Promotion

Through the interviewing process, green roofs were shown to be familiar to a range of stakeholder groups within the construction industry. A number of construction sub-industries are "aware" of the concept of green roofs, but know little about the technicalities of the roofing system.

> Interviewee A: "Inherently it requires a lot more technical input . . . That is why these things can be viewed by people who have never done green roofs as being really difficult things to do".

Whilst many industry stakeholders are "aware" of the green roof concept, private clients, governments and, to a lesser degree, industry bodies (e.g., Green Roofs Australasia or the Australian Institute of Landscape Architects) were noted as being the biggest drivers for local green roof technology. When asked who the biggest driver for green roofs in Australia was, four interviewees noted "private clients", three interviewees noted "government authorities" and the remaining two interviewees suggested 'industry bodies' to be the biggest driver.

4.3.1. Private Clients

Much of the current green roof demand in Australia has seemingly come from proactive private clients.

> Interviewee B: "It is going to be the individuals and we are seeing it now. They will drive that, they've been driving it anyway".

> Interviewee F: "Obviously demand will drive these things".

4.3.2. Government Authorities

Once there is sufficient demand from the private clients within the industry, many interviewees agreed that the responsibility then becomes that of the local government and policy makers. The onus turns to the policy makers who, in order to promote the industry, must create policy settings that accommodate and encourage the demand from private clients. The government authority's role is one of facilitation, rather than promotion.

> Interviewee A: "I guess through other mechanisms and procedures the government is able to make the case for green roofs be important".

The importance of an evidence-based approach to policy making was highlighted by two interviewees.

> Interviewee H: "They [authorities] need the evidence base that enables them to argue for a policy change ... "

> Interviewee G: "I mean it's also going to be about an evidenced based approach looking at the benefits of green roofs ... "

4.3.3. Industry Bodies

Interviewees generally perceived one particular industry body, which cannot be named for anonymity reasons, as having a positive influence on the development of the local green roof industry. As one interviewee divulged, the industry body is assumed to undertake media releases and training workshops as a means of promoting the local industry.

> Interviewee E: "I guess they do media releases. That would be a big one and also training workshops".

It was the general consensus amongst interviewees that the industry body is tasked with providing the promotion of the industry in its 'early-days', that is until the private industry more widely adopts the technology.

> Interviewee D: "They [the industry body] push it until green roofs are taken up by a wide part of the construction industry ... "

> Interviewee G: "The industry body. They are certainly pushing them".

Heller et al.'s [48] focus group study found that "professional bodies have a role to play with the provision of best practice guidelines and notes for members so that they are able to learn about the technical issues and factors to take into account in decision making".

Despite some interviewee's support for the industry body, concerns have been raised about its efficiency in promoting the local use of green roof technology. Two interviewees, who have previously had working relationships with the organization, explicitly labelled the unnamed industry body as being "totally defunct".

> Interviewee I: "We don't take (the industry body) with much weight, they are defunct."
>
> Interviewee H: "It is not actually a functional organization at all. It is pretty much, not an organization, they don't do anything. I wouldn't use their membership as being something that indicates "involvement" because I know a number of people who do green roof stuff that are not necessarily members of (the industry body), because it is totally defunct".
>
> Interviewee I: "People do not see the benefit in being a member".

Given the specificity of these comments, further investigation into the role of industry bodies in the Australian green roof industry would be of significance.

5. Verifying the Consistency of the Findings and the Collected Data

As discussed, the green technology adoption in Australia is significantly new, and consequently, the market is comparatively small. Therefore, the data collected through the interviews can confidently be used to draw a picture of the market. According to the nature of the interviews, some of the participants may have not commented on some particular topics, due to a lack of knowledge, for example. The authors have tried to reflect where the majority of participants consistently agree/disagree with something. However, where opposing comments have been noticed, existing literature has been cited to demonstrate that inconsistency is inherent to that particular topic and does not compromise the statistical soundness of the results. However, in order to verify that the results are reflecting participants' input, free of investigator bias, this study applies a novel statistical approach. The intention is to ensure that the articulated findings are consistently supported by the participants' input. To do so, according to the structure of the paper, we divide the current manuscript into three categories corresponding to Sections 4.1–4.3. Then, we extract a library of keywords from each category and use them to develop three histograms of common vocabularies for each individual interview. This means that each interview is analysed against the libraries representing the three categories. We will then test the consistency of the histogram representing each category and the ones obtained from each distinct interview. The rationale behind using this manuscript to develop the library is that it is extracted from all interviews inclusively. Therefore, the procedure of extracting the findings is verified to be statistically sound, if the comparison of the final product and each interview manuscript results in the conclusion that both are comprised of the same building elements. This is, of course, shown by testing the hypothesis that the two histograms are drawn from the same distribution. It must be highlighted that this procedure only verifies that the results emerged from the interview data, but it does not measure the level of agreement/disagreement on a topic. For example, the outcome of this statistical analysis is an indicator of the significance of the impact of one particular interview on the overall conclusions presented in each category.

An online text analyser called VOYANT [68] has been used to screen the Results Section of this manuscript. The components consisting of literature surveys were deliberately included in the input as they are mainly used as the grounds for triangulation to shape different classifications emerging from interview data. In other words, mining the reference library from the selected pieces of literature together with interview results enhances the reliability of the discussed findings in terms of highlighting any potential controversy between the accepted trend in the body of knowledge and what was outlined in the interviews. Given N_i as the total number of words and n_i as the number of unique words in each category, the average occurrence of one word in each category, $\overline{n}_i$, can be calculated.

$$\overline{n}_i = \frac{N_i}{n_i} \; for \; i = 1,2,3, \tag{1}$$

$\overline{n}_i$ is then used to assign an "impact" index, $I_{i,j}$, to each unique word, $u_{i,j}$, in different categories. Note that $j = 1, ..., n'_i$ for each i; with $j = 1$ representing the highest frequency. The "impact" of each word is assumed to be directly correlated to its frequency as long as the word passes the first filtration process to confirm its relevance to provide a discussion in the context of green roofs. The modified number of words in each category is represented by n'_i. To remain on the conservative side, words such as green, roof(s), etc., that inevitably are used when discussing green roofs are excluded from the libraries. Including these words will increase the likelihood of verifying the consistency, while it does not exist, therefore, decaying the power of the test.

$$I_{i,j} = \frac{n_{i,j}}{\overline{n}_i}, \tag{2}$$

where $n_{i,j}$ is the word count associated with word $u_{i,j}$. The impact index enables us to classify unique words used in the paper into three tiers:

a. Uniquely important: referring to words that exclusively represent one category. For example, the word "risk" with $n_{2,1} = 31$, resulting in an impact index of 9.6, is strongly correlated with the content of Category 2 only.

b. Equally important: covering words with a high impact index in all four categories. For example, the word "local" with $I_{1,4} = 5.2$, $I_{2,5} = 5.0$, $I_{3,3} = 3.3$ falls in this tier.

c. Mutually important: including words with a high impact index in more than one category, but not all three. For example, the word "client(s)" with $I_{1,13} = 1.5$ and $I_{3,5} = 2.5$ contributes to both Categories 1 and 3.

The four reference libraries providing the basis for the consistency test are listed in Table 1.

Table 1. Keyword distributions in the three reference categories.

Reference Library	Category 1			Category 2			Category 3		
	Rank	I	Tier	Rank	I	Tier	Rank	I	Tier
Australia(n)/national	3	6.8	c						
Authority(s)	11	1.8	c				9	1.2	c
Benefit(s)									
Body(is)							2	5.7	c
Change(s)	6	4	a						
Client(s)	13	1.5	c				5	2.5	c
Commercial				19	1.6	a			
Construction	7	4	c	4	5.3	c			
Contract(ual)				12	2.8	a			
Contractor(s)				7	4.3	a			
Cost/Money				3	6.2	c			
Demand/Need	12	1.8	c	18	1.9	c	7	2	c
Design	10	2.2	c	15	2.2	c			
Development(s)									
Education/Knowledge				8	4	c			
Energy				20	1.6	a			
Environment(al)/Sustainable(ity)									
Government(s)	2	7.1	c	6	4.7	c	6	2.5	c
Heat				9	3.4	a			
Incentive(s)	8	3.7	a						
Industry(s)	4	6.2	b	2	6.5	b	1	9.4	b
Island				16	2.2	a			
Local	5	5.2	b	5	5	b	3	3.3	b
People				10	3.1	a			
Policy(s)	1	7.7	a						
Potential	9	2.8	a						
Private				17	2.2	a	4	2.9	c
Risk(s)				1	9.6	a			
Technology(s)				13	2.8	a			
Uncertainty(s)				14	2.5	a			
Urban				11	3.1	a			
Value	14	1.5	a						

Determining the number of words in each bin, a histogram can be created to visualize the distribution of the key words in each interview. In our case, we are only interested in comparing the shape of each interview's histogram with the one obtained from the reference library. Accordingly, we test the null hypothesis that the densities of the two histograms are bin-by-bin equal. The null hypothesis is tested against the hypothesis indicating that the densities of the two histograms are not bin-by-bin equal. Attention is required that we are interested in comparing the bin-by-bin means of the two histograms. For this reason, we normalize histogram bin contents before calculating the statistic for the chi-square test. The normalization is carried out using the ratio of the total number of words included in the histogram being tested over the word density of the reference library. For example, when testing the consistency of Section 4.1 of this manuscript and Interview A with a total of 183 and 81 words in their histograms, correspondingly, the number of words in each bin of the reference histogram is normalized by multiplying it by 0.44, which is $\frac{81}{183}$. This way, a normalized histogram is available to be tested against each interview's histogram for the purpose of consistency verification. The test statistic [69] is calculated using Equation (3).

$$T = \sum_{t=1}^{n'} \frac{(s_t - v_t)^2}{s_t^2 + v_t^2}, \tag{3}$$

where s_t and v_t are the number of words in each bin of the histogram being tested and the normalized reference histogram, accordingly. Using the same hypothesis as mentioned, the chi-square test will be run 27 times to compare each interview against the three reference libraries. However, it is very unlikely to get the null hypothesis accepted without eliminating the "deficiency" in histogram being tested. "Deficiency" refers to bias in the word distribution induced due to the personal preference of the interviewees that, to some extent, is inevitable without the intervention of the interviewer to trigger the use of particular words. Of course, because intervention clearly compromises the reliability of the data collection, deficiency has to be allowed, but filtered during the consistency test. Deficiency surfaces in terms of a relatively more significant contribution to the test statistic. Referring to Table 2, the words "value" and "design" collectively contribute 13 points to the calculated 20.3 statistic. Hypothesis testing using this statistic will result in the rejection of the null hypothesis while a closer look at the data contradicts this outcome. The impact index of all words in the reference histogram add to 56.3. Excluding "value" and "design" still contains 93.4% (cumulative sum of 52.6) of the total impact compared to the initial setup. Interestingly, excluding these deficiencies results in enough evidence to accept the null hypothesis and to confirm the consistency of Interview A and Section 4.1. In other words, the fact that an interviewee has not used specific keywords does not necessarily mean that s/he has not sufficiently used the remaining keywords to assure consistency. As explained, the overall impact can be used to assess links to the remaining keywords. However, in the case that removing deficiency significantly reduces the impact, as well, then the null hypothesis must be rejected.

Knowing that the data have been collected through interviews, a 50% confidence level seems reasonable to verify the consistency of findings and the interview data. Setting a higher confidence level is not favourable, because the objective of the data analysis is not to copy and paste the interview data, but is to extract evidence from the data corpus. A confidence level below 50%, on the other hand, increases the probability of incorrectly verifying consistency. Referring to the results of the consistency tests summarized in Table 3, it can be concluded that discussions and conclusions provided in this paper are all consistent with the content of the interviews with at least 50% confidence, except one instance. Of course, Table 3 can be consulted to determine the significance of different interviews' inputs on various conclusions made in the paper. For example, Interview H has the highest impact on determining the state of the industry; Interview G is the champion in predicting the commercial viability of the green roof; and Interview A has provided the most insight into the promotion strategies of the technology.

Table 2. Details of hypothesis testing and the filtration of deficiency (Interview A against Section 4.1).

Keyword	Interview Histogram	Reference Histogram	Modified Reference	Contribution to Statistic	Impact Index	Deficiency Excluded Impact
Australia(n)/national	10	22	9.7	0.003	6.8	6.8
Authority(s)	2	6	2.7	0.092	1.8	1.8
Change(s)	3	13	5.8	0.866	4	4
Client(s)	2	5	2.2	0.011	1.5	1.5
Construction	2	13	5.8	1.818	4	4
Demand/Need	1	6	2.7	0.750	1.8	1.8
Design	11	7	3.1	4.429	2.2	-
Government(s)	10	23	10.2	0.002	7.1	7.1
Incentive(s)	2	12	5.3	1.500	3.7	3.7
Industry(s)	7	20	8.9	0.216	6.2	6.2
Local	3	17	7.5	1.945	5.2	5.2
Policy(s)	11	25	11.1	0.000	7.7	7.7
Potential	3	9	4.0	0.139	2.8	2.8
Value	14	5	2.2	8.569	1.5	-
Sum	183	81	81	20.34	56.3	52.6
				Impact Covered	**100%**	**93.4%**

Table 3. Outcome of the consistency test of the interviews and their associated impact.

		Insight: Impact (Confidence Level)		
		State of the Industry	**Commercial Viability**	**Promotion**
	A	0.93 (0.77)	0.64 (0.56)	0.92 (0.62)
	B	0.48 (0.53)	0.69 (0.56)	Tie [1]
	C	0.88 (0.51)	0.92 (0.50)	0.50 (0.56)
	D	0.53 (0.51)	0.72 (0.50)	0.69 (0.77)
Interview	E	0.80 (0.51)	0.66 (0.61)	0.64 (0.58)
	F	0.58 (0.59)	0.61 (0.53)	0.61 (0.61)
	G	0.86 (0.86)	1.00 (0.72)	0.72 (0.73)
	H	0.97 (0.72)	0.73 (0.59)	0.61 (0.61)
	I	0.81 (0.59)	0.72 (0.65)	0.74 (0.52)

[1] A tie that happens when eliminating deficiencies does not lead to any confidence level higher than 0.50. Therefore, Interviewee B's input inn Section 4.3.1 must be used with more care.

6. Conclusions

The Australian construction industry, not dissimilar to the global construction community, is faced with the ever-present concern over the sustainability of development practices. Established literature and examples of functioning green roof systems prove that the technology has the potential, when used correctly, to be a sustainable means of development. Despite their benefits, the potential of green roofs in the Australian construction industry is currently unrealized. A vast proportion of Australian-based green roof literature concerns the horticultural or scientific aspects of green roof technology. The construction perspective of this study therefore stands to add value to the existing literature.

Through the research process, it was decisively found that the Australian construction industry (holistically) approaches green roof technology with scepticism and trepidation. The interviewee's responses demonstrated that the local construction industry's perception of green roofs is, in one way or another, dominated by negative stigmas. Representative of these generally-negative perceptions, terms such as "painful" and "complex" were used to explain people's first-hand experience or perception of local green roof projects.

However negative the current perceptions of green roofs are in the Australian construction industry, it was found, with overwhelming support, that green roof technology does have the potential to evolve at some point (unspecified) in the future. A dominant majority of the interviewees (eight of nine) claimed that the "local green roof industry has the potential to evolve". From these responses, "government authority" was repeatedly suggested as the key stakeholder capable of making the required changes in order to improve the current state of the industry. Governments were noted to have the power to change policy settings, which was suggested to be the first and significant step in improving the local green roof industry. The importance of policy settings was additionally addressed and supported in key Australian literature.

Two interviewees clarified the 'policy' aspect of governmental power to be, for example, energy efficiency legislation, such as the Green Building Council of Australia's Green Star Ratings. The Green Star rating system was found to inappropriately recognize the sustainability benefits of including green roofs in building designs. To heighten this apparent legislative shortfall, one interviewee reflected on the efficiency of the United Kingdom's BREEAM system in promoting the use of green roofs. It was found that the BREEAM system directly encouraged developers to spend additional funds to include a green roof within a building's design in return for a favourable BREEAM rating, such as BREEAM "Excellent".

Cost was found to be, by far, the greatest barrier to more widespread implementation of green roofs in the Australian construction industry. Six of the nine interviewees nominated cost as being the most apparent barrier, and this finding is extensively supported in the Australian-based literature. Of the six interviewees who noted cost to be the most critical factor affecting green roof implementation in the local industry, a further three noted an inextricable link to commercial risk. The construction industry was labelled, by two of the interviewees, as being "conservative and loathed to change", as well as 'incredibly risk averse', a view shared by the established literature.

The monetary risk of green roof projects is typically born by each of the project stakeholders; though, it was found that contractual arrangements, such as public-private partnerships (featuring cost sharing) are a way of "managing" risk levels. Conversely, lump sum contracts were labelled as being "so tightly restrictive" when considered in the context of green roof projects.

Education, or more appropriately, a relative lack-thereof, was found to be another critical factor affecting the implementation of green roofs in the Australian construction industry. Education was suggested by five of the nine interviewees as having significant scope to improve the uptake of local green roof projects. Each stakeholder group, be it the project managers, subcontractors or clients, would gain from further education in the field of Australian green roofs. Insightfully, one interviewee noted education to be ill-directed, whereby upper-management, those with the power to drive change, are not taught how to successfully undertake a green roof project. It would therefore appear that green roofs are, in some cases, 'overlooked' due to uncertainty and ill-directed educational programs.

The third factor considered as being critical by the interviewees was the role of the government authority, not only promoting the industry through policy settings, but also conducting 'pilot projects'. A third of interviewees noted the government to be a critical factor within the industry; especially in conducting demonstration projects within the public realm to reduce the uncertainty associated with green roofs currently.

It was found that private clients tended to be the stakeholder group that is currently driving the green roof industry here in Australia. Environmentally-conscious and proactive private clients are interested in the concept of green roof technology given the commercial incentives or marketability aspects. It was noted that despite the private industry's interest in green roof technology, greater assistance is needed. Assistance was specifically noted to be adapted to policy settings (government action) or greater promotion from the industry bodies.

Once people start to demand green roofs, through whatever motives or reasons, the onus turns to the policy makers who have the power to facilitate their uptake. Through legislative mechanisms, the government is able to make the case for green roofs important. Highly regarded literature agrees and further finds that changes to Australian legislation, be it building codes or planning guidelines, must be based on comprehensive research; this is the realm of industry bodies.

It was found that the role of professional bodies is to provide best practice guidelines and members' support/education. While this, in theory, is the function of industry bodies, specific concerns were raised about their efficiency. Whilst there was some support for industry bodies, two interviewees specifically challenged the role that one industry body is currently playing in the industry. Those interviewees who supported the work of the industry body, to the knowledge of the researcher team, have not had close dealings with the unnamed organization. Conversely, the two interviewees who expressed negative comments have previously had working relationships with the organization. Opinions on the industry body are divided, with further investigation into the role of industry bodies within the Australian green roof industry being a topic with further research potential.

Further research into the effect of policy changes on the cost of Australian green roofs, an analysis of the role of industry bodies in the promotion of green roofs and a lifecycle cost benefit analysis of green roofs in an Australian context are some studies that would all be of significant value for the local green roof industry. Studies that comparatively assess the cost of green roofs in Australia to the cost of traditional roofing systems would be invaluable in allowing the industry to see the potential cost savings of green roofs, which have clearly been demonstrated in the international arena. Additionally, alternative frameworks for sustainable building funds will need to be explored in further depth.

Acknowledgments: The open access processing fee has been covered through the supportive funding provided by the School of Built Environment at Curtin University.

Author Contributions: Nicole Tassicker and Payam Rahnamayiezekavat have designed the research and the interviews. The interviews have been carried out and initially analysed by Nicole Tassicker. The results have been finalized through discussion between Payam

Rahnamayiezekavat, Monty Sutrisna and Nicole Tassicker. The statistical analysis has been designed and implemented by Payam Rahnamayiezekavat. The manuscript has been structured and reviewed by Payam Rahnamayiezekavat and Monty Sutrisna.

Conflicts of Interest: The authors declare no conflict of interest.

Appendix A. Interview Questions

INTERVIEW QUESTIONS

Section 1: The Future of the Australian Green Roof Industry	
1a.	Do you believe that the Australian green roof industry has a realistic potential to evolve and adapt in the years to come?
1b.	What would be required to happen to bring about this positive change? (i.e., technological change, government subsidies/incentives etc.)?
Section 2: Barriers to Green Roof Implementation in Australia	
2a.	What would you consider to be the greatest barrier to widespread implementation of green roofs? Is it a stubbornness to change within the industry, cost etc.?
2b.	Is there a viable way to mitigate the challenges presented by this barrier, or is it too great of an issue to overcome?
Section 3: Promotion of the Australian Green Roof Industry	
3a.	In your experience, who would be the biggest driver for green roof technology in Australia? Is it the government, private industry, professional bodies etc.?
3b.	What do 'they' do to promote the industry?
3c.	Do you think there is more that 'they' can do to promote the industry?
Section 4: Demand for Green Roofs in Australia	
4a.	How would you describe the construction/design industry's perception of green roofs? Are green roofs actively encouraged throughout the industry?

Design Rationale: The first section; "The Future of the Australian Green Roof Industry", aims to assess the likelihood of advancing the local green roof industry and thereby relates directly to the first objective of this study. Questions 1a and 1b purposely assess the interviewee's opinion of the industry's potential, so that the ensuing questions could be posed in a way to extract greater information on the matter. For example, if an interviewee considered there to be minimal potential for the green roof industry in Australia, then the following sections of questions would be used to assess how or why they have formed that opinion, and vice versa.

"Section 2: Barriers to Green Roof Implementation in Australia" focusses on the critical factors affecting green roof implementation and how these are impacting the local industry; the second objective of the dissertation. Questions 2a and 2b feature an amount of inherent bias, however, focussing on discovering what have been the constraining factors in the past and how these factors can be mitigated going forward.

"Section 3: The Promotion of the Australian Green Roof Industry" inherently focusses on the positive side of the local green roof industry; a question to directly offset the negative bias of Section 2. Questions 3a, 3b and 3c critically analyse the sources of green roof promotion in the local industry and determine what each of the nominated stakeholders are currently doing within the industry; the third objective of the dissertation.

"Section 4: Demand for Green Roofs in Australia" directly addresses the level of green roof awareness and willingness that people perceive within the construction-related industries; a topic relevant to all of the objectives of the dissertation. Question 4a enabled significant insight to be gained into the current state of the local green roof industry, thereby allowing the researcher to objectively gauge the progression of the industry.

After the "structured" components of the interview had been discussed, the researcher then asked the lead in question, "is there anything else that you would like to add, or feel that I have missed?" Participants were afforded the opportunity to further explain aspects of the local green roof industry and to make recommendations or comments about the study.

Appendix B.

PROFILE OF INTERVIEWEES

Pseudonym	Gender	Background with Green Roofs
Interviewee A	**Male**	• Practicing landscape architect; • Practiced internationally for a number of years; • Working knowledge of Australian green roofs, particularly in Perth; and • Worked as a green roof consultant on the Fiona Stanley Hospital and 140 William Street projects in Perth.
Interviewee B	**Male**	• Experience as a consultant in the private industry in regards to sustainability and town planning.
Interviewee C	**Female**	• 10 years' experience; • Studied a green roof engineering firm's performance.
Interviewee D	**Female**	• Academic background in the field of architecture; and • Involved in construction of a small-scale green roof in the 1970s.

Interviewee E	**Male**	• Practicing landscape architect; • Royal Melbourne Institute of Technology (RMIT) Qualified; • Research projects into historic green roofs in Perth; and • Green roof experience in Melbourne.
Interviewee F	**Male**	• Practicing consultant; • Curtin University Construction Management and Economics graduate; • Project manager on a large-scale green roof project in London; and • Experience in green roofs and design, legislation and compliance.
Interviewee G	**Male**	• Practicing consultant; • PhD in sustainability studies; and • Research background in green infrastructure and its benefits.
Interviewee H	**Male**	• Highly regarded Australian green roof researcher; and • Authored a number of published green roof journal articles and other scholarly articles for an Australian context.
Interviewee I	**Female**	• Previous board member of Green Roofs Australasia; • Authored a number of published green roof journal articles and other scholarly articles for an Australian context;

References

1. Carter, T.; Andrew, K. Life-cycle cost–benefit analysis of extensive vegetated roof systems. *J. Environ. Manag.* **2008**, *87*, 350–363.
2. Siew, R.Y.J. A review of sustainability reporting tools (SRTs) for communities. *Int. J. Sustain. Constr. Eng. Technol.* **2014**, *5*, 39–52.
3. Halliday, S. *Sustainable Construction*; Taylor and Francis: Southport, UK, 2008.
4. Bianchini, F.; Kasun, H. Probabilistic social cost-benefit analysis for green roofs: A lifecycle approach. *Build. Environ.* **2012**, *58*, 152–162.
5. Castleton, H.F.; Stovin, V.; Beck, S.B.M.; Davison, J.B. Green roofs; building energy savings and the potential for retrofit. *Energy Build.* **2010**, *42*, 1582–1591.
6. Farrell, C.; Mitchell, R.E.; Szota, C.; Rayner, J.P.; Williams, N.S.G. Green roofs for hot and dry climates: interacting effects of plant water use, succulence and substrate. *Ecol. Eng.* **2012**, *49*, 270–276.
7. Kosareo, L.; Ries, R. Comparative environmental life cycle assessment of green roofs. *Build. Environ.* **2007**, *42*, 2606–2613.
8. Berardi, U.; GhaffarianHoseini, A. State-of-the-art analysis of the environmental benefits of green roofs. *Appl. Energy* **2014**, *115*, 411–428.

9. Butler, C.; Timothy, C. Ecological impacts of replacing traditional roofs with green roofs in two urban areas. *Cities Environ.* **2008**, *1*, 1–17.
10. Rayner, J.; Raynor, K.; Williams, N. Green roofs for a wide brown land: Opportunities and barriers for rooftop greening in Australia. *Urban For. Urban Green.* **2010**, *9*, 241–251.
11. Ely, M.; Sheryn, P. Life Support for Human Habitats: The Compelling Evidence for Incorporating Nature into Urban Environments. 2014. Available online: file:///C:/Users/263830g/Downloads/Green_Infrastructure_Evidence_Base_2014.pdf (accessed on 25 November 2015).
12. Wilkinson, S.J.; Reed, R. Green roof retrofit potential in the central business district. *Prop. Manag.* **2009**, *27*, 284–301.
13. Kucukvar, M.; Omer, T. Towards a triple bottom-line sustainability assessment of the U.S. construction industry. *Int. J. Life Cycle Assess.* **2013**, *18*, 958–972.
14. Carter, T.; Laurie, F. Establishing green roof infrastructure through environmental policy instruments. *Environ. Manag.* **2008**, *42*, 151–164.
15. Kibert, C.J. *Sustainable Construction: Green Building Design and Delivery*; John Wiley and Sons: Hoboken, NJ, USA, 2012.
16. Brundtland, G.H. Report of the World Commission on Environment and Development: Our Common Future. 1987. Available online: http://www.un-documents.net/our-common-future.pdf (accessed on 20 November 2015).
17. Fabricio, B.; Hewage, K. How "Green" are the green roofs? Lifecycle analysis of green roof materials. *Build. Environ.* **2012**, *48*, 57–65.
18. Feitosa, R.; Sara, W. Retrofitting housing with lightweight green roof technology in Sydney, Australia, and Rio de Janeiro, Brazil. *J. Sustain.* **2015**, *7*, 1081–1098.
19. Beecham, S.; Brien, C.J.; Razzaghmanesh, M. Developing resilient green roofs in a dry climate. *Sci. Total Environ.* **2014**, *490*, 579–589.
20. Dillion, M. Sky High but Down to Earth. Available online: http://www.airah.org.au/imis15_prod/Content_Files/EcoLibrium/2008/November2008/2008-11-F04.pdf (accessed on 20 November 2015).
21. Aneli, S.; Gagliano, A.; Detommaso, M.; Nocera, F. The retrofit of existing buildings through the exploitation of the green roofs—A simulation study. *Energy Procedia* **2014**, *62*, 52–61.
22. Elley, T.B.; Max, G.; David, J.S. Exploring the building energy impacts of green roof design decisions—A modeling study of buildings in four distinct climates. *J. Build. Phys.* **2011**, *35*, 372–391.
23. Rahman, S.R.A.; Ahmad, H.; Mohammad, S.; Muhamad, S.L.R. Perception of green roof as a tool for urban regeneration in a commercial environment: The secret garden, Malaysia. *Procedia Soc. Behav. Sci.* **2013**, *170*, 128–136.
24. Claus, K.; Sandra, R. Public versus private incentives to invest in green roofs: A cost benefit analysis for Flanders. *Urban For. Urban Green.* **2012**, *11*, 417–425.
25. Gatersleben, B.; White, E. Greenery on residential buildings: Does it affect preferences and perceptions of beauty? *J. Environ. Psychol.* **2010**, *31*, 89–98.
26. Smith, C.; Boyer, M. Who wants to live with a living roof? *Green Places* **2007**, *39*, 24–27.

27. Hien, W.N.; Yuen, B. Resident perceptions and expectations of rooftop gardens in Singapore. *Landsc. Urban Plan.* **2004**, *73*, 263–276.
28. Koole, S.; Van Den Berg, A.; Wulp, N. Environmental preference and restoration: (How) are they related? *J. Environ. Psychol.* **2003**, *23*, 135–146.
29. Ulrich, R.S. Aesthetic and affective response to natural environment. *Hum. Behav. Environ.* **1983**, *6*, 88–125.
30. Hietanen, J.; Korpela, K.; Klemettila, T. Evidence for rapid affective evaluation of environmental scenes. *Environ. Behav.* **2002**, *34*, 634–650.
31. Farrell, C.; Kate, L.; Kathryn, W.; Leisa, S.; Nicholas, W. Living roof preference is influenced by plant characteristics and diversity. *Landsc. Urban Plan.* **2013**, *122*, 152–159.
32. Bruce, T.; Jian, Z.; Raufdeen, R.; Stephen, P. Factors influencing the retrofitting of existing office buildings using Adelaide, South Australia as a case study. *Struct. Surv.* **2015**, *33*, 150–166.
33. Forbes, D.J. An Analysis of Municipal Tools for Promoting Green Roof Technology into Dense Urban Development. Masters' Thesis, Tufts University, Boston, MA, USA, Auguest 2010.
34. Weiler, S.K.; Katrin, S.B. *Green Roof Systems: A Guide to the Planning, Design and Construction of Landscapes over Structure*; John Wiley & Sons: Hoboken, NJ, USA, 2009.
35. Perini, K.; Paolo, R. Cost-benefit analysis for green façades and living wall systems. *Build. Environ.* **2013**, *70*, 110–121.
36. Ong, C.L.; Angelia, S.; Taya, S.F.; Wonga, N.H.; Wonga, R. Life cycle cost analysis of rooftop gardens in Singapore. *Build. Environ.* **2003**, *38*, 499–509.
37. Adriaens, P.; Corrie, C.; Brian, T. Green roof valuation: A probabilistic economic analysis of environmental benefits. *Environ. Sci. Technol.* **2008**, *42*, 2155–2161.
38. Finkbeiner, M.; Peri, G.; Rizzo, G.; Traverso, M. The cost of green roofs disposal in a life cycle perspective: Covering the gap. *Energy* **2012**, *48*, 406–414.
39. Andresen, J.; Bradley, R.; Clayton, L.R.; Lan, X.; Van Woert, N.D. Watering regime and green roof substrate design affect sedum plant growth. *HortScience* **2005**, *40*, 659–664.
40. Lundholm, J.; Scott, M. Performance evaluation of native plants suited to extensive green roof conditions in a maritime climate. *Ecol. Eng.* **2011**, *37*, 407–417.
41. McIntyre, L.; Snodgrass, E. *The Green Roof Manual: A Professional Guide to Design, Installation, and Maintenance*; Timber Press Incorporated: Portland, OR, USA, 2010.
42. Bradbury, D.; Hughes, R.; Jones, N.; Rayner, J.; Williams, N. The performance of native and exotic species for extensive green roofs. In Proceedings of the International Conference on Landscape and Urban Horticulture, Bologna, Italy, 9–13 June 2009.
43. Beecham, S.; Kazemi, F.; Razzaghmanesh, M. The growth and survival of plants in urban green roofs in a dry climate. *Sci. Total Environ.* **2014**, *476*, 228–297.
44. Nina, J.; Metternich, G. How to grow a green roof industry. In Proceedings of the 11th Annual Green Roof and Wall Conference, San Francisco, CA, USA, 23–26 October 2013.
45. Elliott, T. Green roofs growing in popularity. *Sydney Morning Herald*, 9 September 2008.
46. Bearman, C.; Michalski, D. Factors affecting the decision making of pilot who fly in outback Australia. *Saf. Sci.* **2014**, *68*, 288–293.

47. Bryman, A. *Quantity and Quality in Social Research*; Routledge: New York, NY, USA, 1988.
48. Heller, A.; Lammond, J.; Manion, J.; Proverbs, D.; Sharman, L.; Wilkinson, S. Technical considerations in green roof retrofit for stormwater attenuation in the Central Business District. *Struct. Surv.* **2014**, *33*, 36–51.
49. International Monetary Fund. Classifications of Countries Based on Their Level of Development: How It Is Done and How It Could Be Done. 2011. Available online: https://www.imf.org/external/pubs/ft/wp/2011/wp1131.pdf (accessed on 10 October 2015).
50. Australian Building Codes Board. National Construction Code Series-Volume One. 2016. Available online: http://www.abcb.gov.au/Resources/Publications/NCC/NCC-2016-Volume-One (accessed on 2 May 2016).
51. Government News. Available online: http://www.governmentnews.com.au/2015/10/new-cities-and-built-environment-taskforce/ (accessed on 18 May 2016).
52. Ministry of Environment and Water Resources. Available online: http://www.mewr.gov.sg/grab-our-research/singapore-green-plan-2012 (accessed on 16 May 2016).
53. Ministry of Environment and Water Resources. Available online: http://www.mewr.gov.sg/ssb/ (accessed on 16 May 2016).
54. Department of the Environment. Available online: https://www.environment.gov.au/cleaner-environment (accessed on 16 May 2016).
55. Sibley, M.; Sibley, M. Hybrid green technologies for retrofitting heritage buildings in North African medinas: Combining vernacular and high-tech solutions for an innovative solar powered lighting system for hammam buildings. *Energy Procedia* **2013**, *42*, 718–725.
56. International Standards Organisation. ISO 31000: 2009 Risk Management—Principles and Guidelines Sydney: Standards Australia & Standards New Zealand. 2009. Available online: http://www.iso.org/iso/catalogue_detail?csnumber=43170 (accessed on 18 May 2016).
57. UNESCO. The Persian Garden. Available online: http://whc.unesco.org/en/list/1372 (accessed on 25 May 2016).
58. Kok, K.H.; Mohd, S.L.; Chow, M.F.; Zainal Abidin, M.R.; Basri, H.; Hayder, G. Evaluation of green roof performances for urban stormwater quantity and quality controls. *Int. J. River Basin Manag.* **2016**, *14*, 1–7.
59. Speak, A.F.; Rothwell, J.J.; Lindley, S.J.; Smith, C.L. Metal and nutrient dynamics on an aged intensive green roof. *Environ. Pollut.* **2014**, *184*, 33–43.
60. Pisello, A.L.; Pignatta, G.; Castaldo, V.L.; Cotana, F. Experimental analysis of natural gravel covering as cool roofing and cool pavement. *Sustainability* **2014**, *6*, 4706–4722.
61. Pisello, A.L.; Piselli, C.; Cotana, F. Thermal-physics and energy performance of an innovative green roof system: The cool-green roof. *Solar Energy* **2015**, *116*, 337–356.
62. Li, X.X.; Norford, L.K. Evaluation of cool roof and vegetations in mitigating urban heat island in a tropical city, Singapore. *Urban Clim.* **2016**, *16*, 59–74.
63. Fexas, M. Green roofs and vegetated systems for a sustainable future. *Aust. Gard. Hist.* **2010**, *21*, 8–12.

64. Beecham, S.; Kazemi, F.; Razzaghmanesh, M. The role of green roofs in water sensitive urban design in South Australia. In Proceedings of the 7th International Conference on Water Sensitive Urban Design, Melbourne, Australia, 21–23 February 2012.
65. Gliedt, T.; Hoicka, C.E. Energy upgrades as financial or strategic investment? Energy Star property owners and managers improving building energy performance. *Appl. Energy* **2015**, *147*, 430–443.
66. City of Sydney. Available online: http://www.cityofsydney.nsw.gov.au/vision/towards-2030/sustainability/greening-the-city/green-roofs-and-walls#page-element-dload (accessed on 27 May 2016).
67. Growing Green Guide. Available online: http://www.growinggreenguide.org/ (accessed on 27 May 2016).
68. VOYANT See through Your Text. Available online: http://voyant-tools.org/ (accessed on 2 June 2016).
69. Porter, F.C. *Testing Consistency of Two Histograms*; Cornell University Libarary: Ithaca, NY, USA, 2008; Available online: http://www.hep.caltech.edu/~fcp/statistics/hypothesisTest/PoissonConsistency/PoissonConsistency.pdf (accessed on 5 June 2016).

Evaluation Analysis of the CO_2 Emission and Absorption Life Cycle for Precast Concrete in Korea

Taehyoung Kim and Chang U. Chae

Abstract: To comply with recent international trends and initiatives, and in order to help achieve sustainable development, Korea has established a greenhouse gas (GHG) emission reduction target of 37% (851 million tons) of the business as usual (BAU) rate by 2030. Regarding environmentally-oriented standards such as the IGCC (International Green Construction Code), there are also rising demands for the assessment on CO_2 emissions during the life cycle in accordance with ISO (International Standardization Organization's Standard) 14040. At present, precast concrete (PC) engineering-related studies primarily cover structural and construction aspects, including improvement of structural performance in the joint, introduction of pre-stressed concrete and development of half PC. In the manufacture of PC, steam curing is mostly used for the early-strength development of concrete. In steam curing, a large amount of CO_2 is produced, causing an environmental problem. Therefore, this study proposes a method to assess CO_2 emissions (including absorption) throughout the PC life cycle by using a life cycle assessment (LCA) method. Using the proposed assessment method, CO_2 emissions during the life cycle of a precast concrete girder (PCG) were assessed. In addition, CO_2 absorption was assessed against a PCG using conventional carbonation and CO_2 absorption-related models. As a result, the CO_2 emissions throughout the life cycle of the PCG were 1365.6 (kg-CO_2/1 PCG). The CO_2 emissions during the production of raw materials among the CO_2 emissions throughout the life cycle of the PCG were 1390 (kg-CO_2/1 PCG), accounting for a high portion to total CO_2 emissions (nearly 90%). In contrast, the transportation and manufacture stages were 1% and 10%, respectively, having little effect on total CO_2 emissions. Among the use of the PCG, CO_2 absorption was mostly decided by the CO_2 diffusion coefficient and the amount of CO_2 absorption by cement paste. The CO_2 absorption by carbonation throughout the service life of the PC was about 11% of the total CO_2 emissions, which is about 16% of CO_2 emissions from ordinary Portland cement (OPC) concrete.

Reprinted from *Sustainability*. Cite as: Kim, T.; Chae, C.U. Evaluation Analysis of the CO_2 Emission and Absorption Life Cycle for Precast Concrete in Korea. *Sustainability* **2016**, *8*, 663.

1. Introduction

Internationally, greenhouse gases (GHGs) are arguably the most prevalent global environmental problem. In an effort to curb the release of GHGs, there has been an assortment of international movements that aspire to cap and/or reduce GHG emissions. To comply with this international trend, and to achieve sustainable development, Korea has established the GHG emission reduction target of 37% (851 million tons) of business as usual (BAU) rates by 2030 [1].

The building construction industry has played a role in impoverishing the environment; developments have occurred for the sake of improving our quality of life, but at a great cost of impact to the environment [2]. It is therefore incumbent upon the industry to endeavor to mitigate the effects from building construction projects on our environment [3,4].

The major construction materials accounting for about 65% of building greenhouse gas emissions include concrete and reinforcement steel. Among the CO_2 emissions generated by these major construction materials, concrete accounts for 40% [5,6].

Therefore, an assessment of the CO_2 emissions throughout the concrete life cycle has been conducted. Regarding environmentally-oriented standards such as the IGCC (International Green Construction Code) [7], there are also rising demands for the assessment on CO_2 emissions during the life cycle in accordance with ISO (International Standardization Organization's Standard) 14040 [8]. Recently, concrete research institutes in Northern Europe, such as the Swedish Cement and Concrete Institute (CBI) [9], insist that CO_2 emissions by concrete are overestimated if only processes of production are considered, and thus propose that CO_2 absorption by carbonation during the use of a structure should also be considered.

The system boundaries designed to assess CO_2 emissions throughout the concrete life cycle were drawn between the following stages: raw material, transportation, manufacture, use. However, it has been very hard to find a study on the assessment method and analysis of CO_2 emissions of precast concrete (PC). PC engineering is defined as a process of transporting the manufactured concrete member to a construction and civil-engineering site and assembling it properly. It is widely used in construction sites, such as in the underground spaces of apartments and stadiums due to its easy process management and great constructability. Now, PC is perceived as the future of construction engineering because of the shortening of the construction period, quality improvement, decrease in accidents, and eco-friendly concrete option that it provides to the construction industry. PC engineering was first introduced to the Republic of Korea in the early 1970s with the goal of supplying houses in large quantities. After reaching a peak in the late 1980s, it has lost its competitiveness due to poor technology and quality. Entering the new millennium, it is widely used again in various fields such as in stadium, underground parking lot, discount store, warehouse, and factory construction [10].

At present, PC engineering-related studies primarily cover structural and construction-related aspects, including the improvement of structural performance in the joints, the introduction of prestressed concrete, and the development of half PC. In the manufacture of PC, steam curing is mostly used for the early-strength development of concrete. At least 10 h of steam curing is used every day. In steam curing, a large amount of CO_2 is produced, thereby causing an environmental problem [11].

Hence, this study proposes a method to assess CO_2 emissions (including absorption) throughout the PC life cycle, using a life cycle assessment (LCA) method. Using the proposed assessment method, CO_2 emissions during the life cycle of PC girders (PCGs) were assessed. In addition, CO_2 absorption was assessed for PCGs, using conventional carbonation and CO_2 absorption-related models.

2. Literature Review

2.1. CO_2 Emission of Precast Concrete

Victor et al. described a methodology to optimize cost and CO_2 emissions when designing precast-prestressed concrete road bridges with a double U-shape cross-section: To this end, a hybrid glowworm swarm optimization algorithm (SAGSO) was used to combine the synergy effect of the local search with simulated annealing (SA) and the global search with glowworm swarm optimization (GSO) [12].

Duo et al. developed the precast concrete panel by substituting blast-furnace slag for part of the unit weight of cement in the precast concrete mix. A life cycle assessment technique was used to estimate the carbon dioxide reduction. Carbon reduction in the materials, as well as during the production phase, was considered [13].

Carlo et al. focused on the analysis of the entire main input inventory data used for assessing the environmental impacts linked to the life cycle of a precast concrete shed: great importance was given to the use of on-site collected specific data which was carefully verified in order to assure its quality and reliability [14].

Ya et al. compared the carbon emissions of precast and traditional cast-in-situ construction methods based on a case study of a private residential building in Hong Kong [15]. The objective of this study was to develop energy-efficient algorithms of the steam curing for the in situ production of PC members. The results of this study will provide basic information for subsequent efforts to implement an energy-efficient in situ PC production system [16].

Cassgnabere et al. discussed the results of a hydration study performed in order to explain the significant increase in compressive strength at one day of age observed on steam cured mortars when 25% by mass of cement was replaced with a metakaolin [17].

2.2. Carbonation and Absorption of Precast Concrete

Pade et al. [18] stated that CO_2 produced during cement sintering is mostly absorbed through concrete carbonation if the structure's life cycle (100 years) is considered. Liwu et al. established an understanding of the effectiveness of accelerating the carbonation process. Pressurized CO_2 (up to 1.0 MPa) was employed to enhance the carbonation of mortar blends consisting of Portland cement, fly ash, and reactive MgO [19].

Lee et al. described a numerical procedure to quantitatively evaluate carbon dioxide emissions and the absorption of ground granulated blast furnace slag (GGBFS) blended concrete structures. Based on building scales and drawings, the total volume and surface area of concrete were calculated [20].

Elke et al. performed accelerated carbonation tests on concrete specimens containing different amounts of blast-furnace slag (BFS) after different curing times. The tests revealed that, although BFS concrete has a lower carbonation resistance than ordinary Portland cement (OPC) concrete, the depth of carbonation at the end of the concrete's life (50 years) can still be acceptable in normal environments [21].

Gajda stated that among the service life of concrete structures, carbonation-based CO_2 retention capacity accounted for 3%–4% of the CO_2 emissions from the production of cement [22]. Lee et al. suggested that in concrete structures CO_2 retention capacity did not exceed 5% of the CO_2 emissions from the production of concrete [23]. In contrast, Pade et al. noted that if a recycling stage is considered along with the use stage, CO_2 emitted by chemical response during the sintering process designed to produce clinker can be collected through carbonation [18]. Yang et al. described a mathematical procedure which can estimate CO_2 retention capacity through the carbonation of concrete in each stage suggested for the assessment of the CO_2 emissions during the concrete life cycle in a reasonable fashion [24].

3. Assessment of CO_2 Emission in the Precast Concrete

The PC manufacturing process can be divided into three stages: raw material, transportation, and manufacture. In the raw material stage, the mixed ingredients of concrete (e.g., cement, aggregate, admixture, etc.) and reinforcing bar (rebar) used to produce PC are individually manufactured and transported to the PC manufacturer. These materials are weighed in a certain ratio and cast into a cement mold to produce PC. The PC production process is shown in Figure 1:

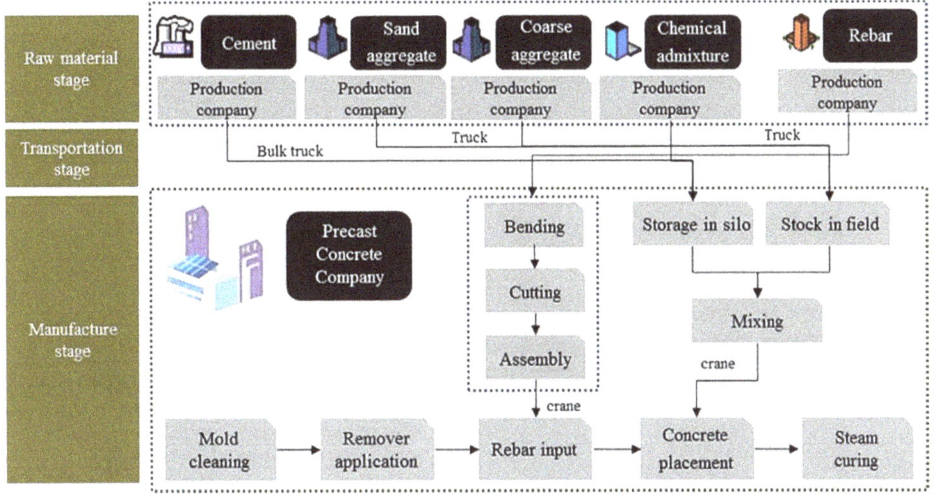

Figure 1. Production process of precast concrete.

3.1. *Raw Material Stage*

(1) Concrete

Using the CO_2 emission factors emitted during the production of concrete mix ingredients (m^3) which are used to manufacture PC, CO_2 emission for concrete is estimated by the accumulation of multiplication between the mixing volume of each ingredient per 1 m^3 and a greenhouse gas emission factor for the production of concrete (kg) [25].

For the CO_2 emission factors of concrete ingredients (ordinary Portland cement, aggregate, admixture, and water), the Korean life cycle inventory (LCI) database was adopted [26]. For the chemical admixtures without CO_2 emission factors in Korea, a foreign LCI database (Table 1) was applied [27].

(2) Reinforcing Bar

CO_2 emission for reinforcing bar is estimated by the accumulation of multiplication between the input (kg/rebar) and CO_2 emission factor for the production of PC.

$$CO_2M = \sum(M(i) \times CO_2 \text{ emission factor M}) \quad (1)$$
$$(i = 1\text{: cement, 2: aggregate, 3: admixture, 4: water, 5: reinforcing bar})$$

Here, CO_2M is the CO_2 emission quantity at the raw material stage of the production of a unit of concrete (kg-CO_2/m^3); M(i) is the amount of material used (kg/m^3) in the concrete; and the CO_2 emission factor M is the CO_2 emission factor for each material (kg-CO_2/kg).

Table 1. Life cycle inventory (LCI) Database (DB) reference.

Material	Unit	Reference Basis [26,27]
Ordinary Portland Cement	kg	National LCI DB (South Korea)
Coarse aggregate	kg	National LCI DB (South Korea)
Fine aggregate	kg	National LCI DB (South Korea)
Blast-furnace slag powder	kg	Overseas LCI DB (ecoinvent)
Fly ash	kg	Overseas LCI DB (ecoinvent)
Water	kg	National LCI DB (South Korea)
Chemical Admixture compound	kg	Overseas LCI DB (ecoinvent)
Reinforcing bar	kg	National LCI DB (South Korea)
Truck	km	National LCI DB (South Korea)
Train	km	National LCI DB (South Korea)
Diesel	L	National LCI DB (South Korea)
Kerosene	L	National LCI DB (South Korea)
LPG	m^3	National LCI DB (South Korea)
Electricity	kwh	National LCI DB (South Korea)

3.2. Transportation Stage

Among the concrete mix ingredients, ordinary Portland cement (OPC) concrete is transported from the production plant to a transit point through a bulk train. The cement at the transit point is then delivered to a ready-mix concrete manufacturing plant through a bulk cement truck (BCT). If the plant was situated in a coastal area, railroad transport accounted for about 25% of total cement transportation. When it was located in an inland region, railroad transport was as high as 60%–70%. For the assessment of CO_2 emissions during the transportation stage, the number of transportation-related vehicles/pieces of equipment is estimated with the input of concrete ingredients and reinforcing bars and the load of transportation-related vehicles/pieces of equipment. Considering distance and fuel efficiency in addition to the number of the estimated transportation-related vehicles/pieces of equipment, the consumption of diesel oil and CO_2 emissions are assessed. The calculation formula for CO_2 emissions during the transportation stage is shown in Equation (2):

$$CO_2T = \sum[(M(i)/Lt) \times (d/e) \times CO_2 \text{ emission factor T}]$$
$$(i = 1\text{: cement, } 2\text{: aggregate, } 3\text{: admixture } 4\text{: reinforcing bar}) \quad (2)$$

Here, CO_2T is the quantity of CO_2 emitted during the transportation of a unit of produced concrete (kg-CO_2/m^3); M(i) is the amount of material used (kg/m^3) in the concrete; Lt is the transportation load (tons); d is the transportation distance (km);

e is the fuel efficiency (km/L); and CO_2 emission factor T is the CO_2 emission factor of the energy resource (kg-CO_2/kg).

3.3. Manufacturing Stage

The CO_2 emissions during the manufacturing stage are estimated after measuring the amount of energy consumed during the unloading of raw materials and the manufacturing of reinforcing bars (Figure 2). For this, the amount of energy consumption should be estimated first. After investigating the type and specification of the facilities which consume electricity, diesel oil, LNG (Liquefied Natural Gas), and water, annual energy consumption and output are analyzed. Then, energy consumption and CO_2 emissions can be estimated for the manufacture of the PC product 1. The calculation formula for CO_2 emissions during the manufacturing stage is shown in Equation (3):

$$\begin{gathered} CO_2F = \sum[(E(i)/R) \times CO_2\,\text{emission factor F}] \\ (i = 1\text{: electricity usage, 2: oil usage, 3: water usage}) \end{gathered} \tag{3}$$

Here, CO_2F is the amount of CO_2 emitted during the concrete manufacturing stage for producing a unit of concrete (kg-CO_2/m^3); R denotes the annual RMC (Ready-mixed Concrete) production (m^3/year); E(i) denotes the annual energy usage (unit/year), and CO_2 emission factor F is the CO_2 emission factor of an energy resource(kg-CO_2/kg).

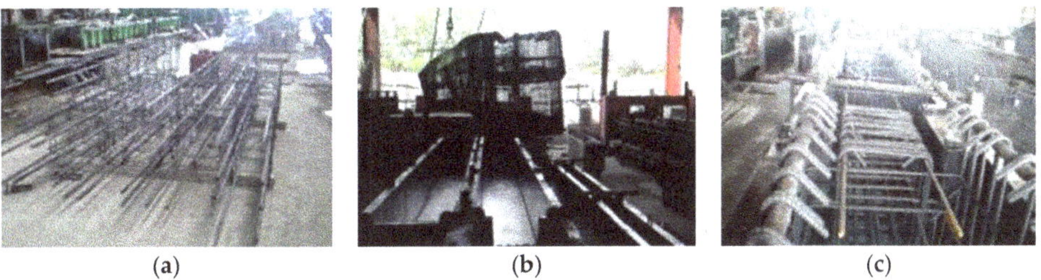

(a) (b) (c)

Figure 2. Assembly (**a**); insertion of steel (**b**); embedment installation (**c**) of reinforcing bars for precast concrete.

4. Assessment of CO_2 Absorption in Precast Concrete

The CO_2 absorption by concrete carbonation during the use of PC only was considered. In terms of a service life, 40 years are set according to the standard useful life of buildings under the Enforcement Rules of Corporate Tax Act [28]. The CO_2 absorption during the use of PC is determined by the depth of carbonation. To assess CO_2 absorption by concrete carbonation, it is required to predict the exact depth of carbonation by age. Since CO_2 absorption is assessed during the use of a building, environmental factors and concrete exposure conditions should also be considered with significance in predicting the depth of carbonation.

The CO_2 absorption (CO_2 U(t)) by carbonation for t(days) is stated below:

$$CO_2\,U(t) = A_{CO2}(t) \times W_a \times Kc(t)\ (g) \tag{4}$$

Here, $A_{CO2}(t)$ refers to the amount of absorbable CO_2 (g/cm^3) by carbonation at age t days while W_a represents the surface area (cm^2) of the concrete member exposed to CO_2. In addition, Kc(t) denotes the depth of carbonation at age t days.

4.1. *Absorbable CO_2*

Among the minerals which constitute cement paste and hydrates, $A_{CO2}(t)$ which is decided by the water concentration of carbonation-enabled factors can be estimated as follows [29]:

$$A_{CO2(t)} = a_h(t) \times M_d(t) \times M_{CO2} \times 10^{-6}\,(g/cm^3) \tag{5}$$

Here, ah(t) refers to the degree of hydration of cement paste at age t days while $M_d(t)$ represents the water concentration (mol/cm^3) of carbonation-enabled factors in cement paste per concrete unit volume at age t days. In addition, M_{CO2} (=44 g/mol) means molar mass of CO_2.

Papadakis et al. [29] suggested a mathematical model to set $M_d(t)$ in a fully hydrated state based on the chemical response of hydrates. Because the molar concentration of carbonation-enabled elements are estimated based on the molar mass of major cement ingredients, $M_d(t)$ is greatly influenced by the amount of unit cement. $M_d(t)$ gradually decreases over time. After one year, it converged to an almost constant value.

Therefore, $M_d(t)$ can be estimated in a simple fashion:

$$M_d(t) = 8.06\,C(\times 10^{-6}\,mol/cm^3) \tag{6}$$

The cement with general fineness is not fully hydrated even for 100 years. The degree of cement hydration by age reveals a parabola which converges to the extreme degree of hydration (∞). The pores of cement paste decline over time due to the progress of hydration and carbonation. After 100 days, however, the slope of decrease is close to zero. Yang et al. [24] modeled ah(t) with a water-cement ratio (W/C) based on experimental results [30].

$$Ah(t) = t/(2.0 + t) \times A(\infty) \tag{7}$$

$$A(\infty) = 1.03(W/C)/(0.19 + W/C) \quad (8)$$

In the service life (four decades) of the concrete structure from Equations (7) and (8) above, the difference between Ah(t) and $A(\infty)$ is small enough to be ignored.

4.2. Carbonation Depth and CO_2 Diffusion Coefficient

In general, the carbonation depth of concrete (Kc(t)) is generalized as a carbonation velocity-time function as follows:

$$Kc\,(t) = \sqrt{\frac{2D_{CO2}\,(t)}{A_{CO2}\,(t)} \times S_{CO2}} \quad (9)$$

Here, D_{CO2} refers to the diffusion coefficient (cm^2/day) at age t days while S_{CO2} represents CO_2 mass concentration (g/cm^3) on the concrete surface. The volume concentration of CO_2 (ppm) is converted into mass concentration using the ideal gas theory. In concrete, the diffusion velocity of CO_2 is influenced by its exposure conditions (relative humidity, temperature, concrete surface conditions) as well as by material properties (water-cement ratio, degree of hydration, pore size distribution, degree of saturation). As Pommer et al. [31] pointed out, in addition, supplementary cementing materials (SCMs) also have a significant effect on the diffusion velocity of CO_2. The concrete surface's finish blocks the penetration of carbonic acid gas and slows down carbonation [32]. Table 2 reveals calibration coefficients for the admixture replacement [31] while Table 3 represents calibration coefficients [32] for the finished material.

Table 2. Correction factor for the substitution of supplementary cementitious materials (SCMs).

Type	Substitution Level of SCMs (%)					
	0–10	10–20	20–30	30–40	40–50	60–80
FA	1.05	1.05	1.10	1.10	-	-
GGBS	1.05	1.10	1.15	1.20	1.25	1.30
SF	1.05	1.10	-	-	-	-

FA: fly ash; GGBS: ground granulated blast-furnace slag, SF: silica fume.

Table 3. Correction factor for the finishing materials on concrete surface.

Finishing Condition	Indoor Area							Outdoor Area			
	No Finishing	Plaster	Mortar + Plaster	Mortar	Mortar + Paint	Tile	Paint	No Finishing	Mortar	Paint	Tile
Value	1.0	0.79	0.41	0.29	0.15	0.21	0.57	1.0	0.28	0.8	0.7

5. Case Study: CO_2 Emission Assessment of Precast Concrete Girders

The CO_2 emissions and CO_2 absorption capacity by carbonation were assessed for precast concrete girders (PCGs) manufactured by a PC concrete manufacturer (Figure 3) in the Republic of Korea [33–35].

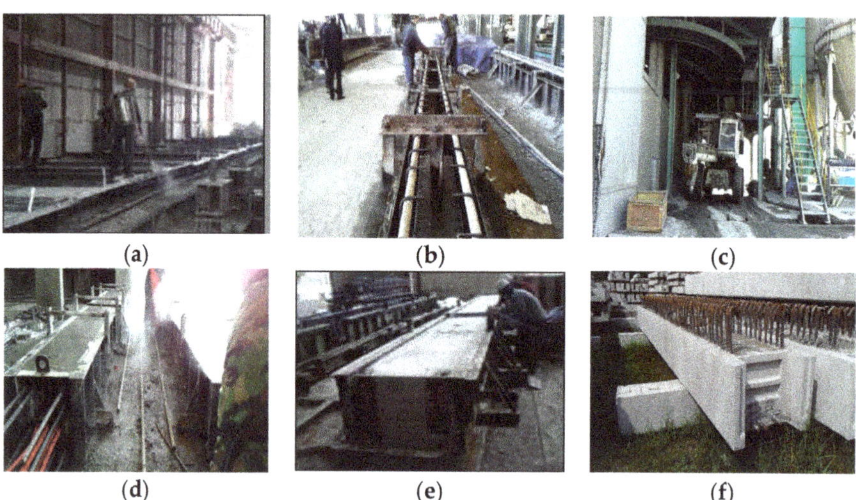

Figure 3. Production process of precast concrete girders (PCGs): (**a**) Spreading form oil; (**b**) rebar insert; (**c**) pouring concrete; (**d**) curing; (**e**) removal; (**f**) precast concrete girder.

The environmental conditions (average temperature: 15 °C, average relative humidity: 66%, CO_2 concentration) were estimated based on Korean data from the year 2012 [36]. The system boundary on the CO_2 emissions at the concrete production stage is drawn after the pre-construction stage (Figure 4). The production stage thus includes all of the following: (1) purchase of precast concrete materials from the cradle to gate; (2) transport of components to the precast concrete factory; (3) manufacturing at the precast concrete factory; (4) transport of the concrete to the construction site.

In terms of CO_2 concentration, 380 ppm and 2000 ppm were assumed for outdoor and indoor environments, respectively [37]. In terms of the expected life of concrete structure, 40 years were set.

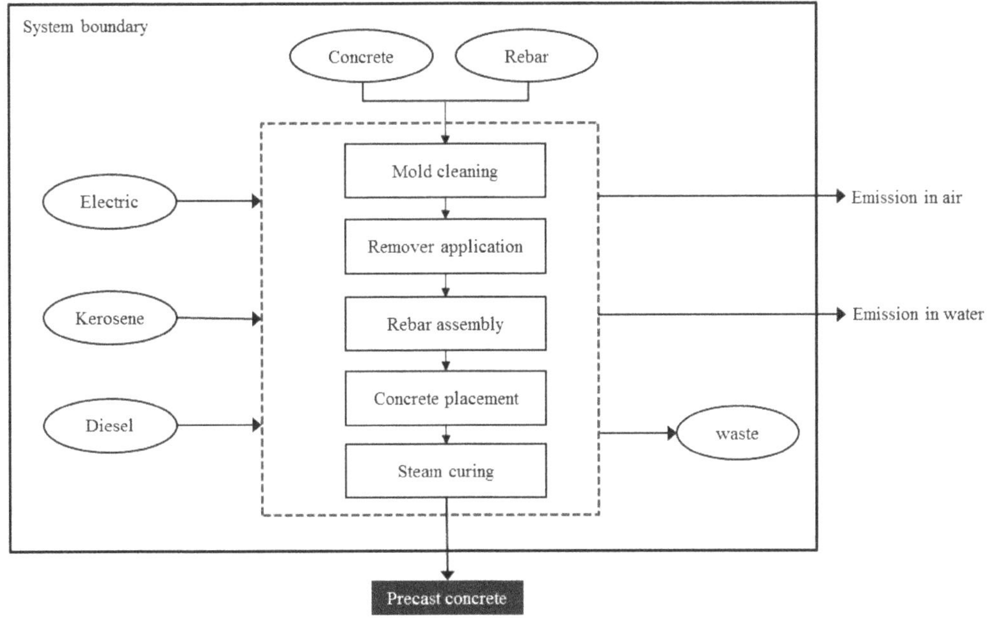

Figure 4. System boundaries for the life cycle assessment (LCA) of precast concrete.

5.1. CO_2 Emissions throughout the PC Life Cycle

The CO_2 emissions throughout the life cycle of PCGs were 1365.6 (kg-CO_2/1 PCG), as shown Table 4.

In particular, the CO_2 emissions during the raw material stage were 1390 (kg-CO_2/1 PCG), accounting for the highest portion in total CO_2 emissions. As the concrete used to produce one PCG is about 2.2 m^3, concrete mixing ingredients were considered as follows: ordinary Portland cement 1100 (kg/1 PCG), coarse aggregates 2030 (kg/1 PCG), fine (sand) aggregates 1751 (kg/1 PCG), admixtures 11 (kg/1 PCG), and mixed water 352 (kg/1 PCG). The percentage of the CO_2 emissions from the ordinary Portland cement in the raw material stage was about 67%.

Deformed reinforcing bars were factored into the production of PCG. The amount of deformed bars consumed for the production of each PCG was 425 (kg/1 PCG). The CO_2 emissions from the production of reinforcing bars were 163.6 (kg-CO_2/rebar), accounting for 11% of the CO_2 emissions form the raw material stage.

Table 4. Life cycle CO_2 emission and absorption of precast concrete girders (PCGs).

	Raw Material Stage			**Transportation Stage (from Gate to Ready-Mixed Concrete Plant)**		
Unit: 1 PCG (Concrete 2.2 m^3)		**Concrete Component**				
Unit Item	A	B	C = A × B	D	E	F = A × D × E
	kg/unit	kg-CO/kg	Kg-CO_2/unit	Distance (km)	kg-CO_2/kg·km	kg-CO_2/kg
Ordinary Portland cement (OPC)	1100	9.48×10^{-1}	1.04×10^{3}	106	6.06×10^{-5}	7.07
Water	352	1.31×10^{-4}	4.61×10^{-1}	-	-	-
Sand aggregate	1751	1.52×10^{-4}	2.66	32	1.16×10^{-5}	6.50×10^{-1}
Coarse aggregate	2300	7.74×10^{-3}	1.78×10^{2}	32	1.16×10^{-5}	8.54×10^{-1}
Chemical admixture	11	2.05×10^{-3}	7.22×10^{-1}	77	1.16×10^{-5}	3.14×10^{-1}
Reinforcing rebar	425	3.85×10^{-1}	1.64×10^{2}	161	4.29×10^{-5}	2.94
	Sum		1.39×10^{3}	Sum		1.18×10^{1}
	Manufacturing Stage			**Transportation Stage (from Batch Plant to Construction Site)**		
Unit Item	A	B	C = A × B	D	E	F = D × E
	Input/unit	kg-CO_2/input	kg-CO_2/input	Distance (km)	kg-CO_2/kg·km	kg-CO_2/kg
Electric (kwh)	57.4	4.88×10^{-1}	2.80×10^{1}			
Kerosene (L)	29.3	3.17	9.29×10^{1}			
Diesel (L)	4.7	3.19	1.50×10^{1}	50	2.81×10^{-2}	1.41
Remover (L)	1.3	1.45×10^{-3}	1.89×10^{-3}			
	Sum		1.36×10^{2}			
			Use stage of structure			
A	B	C	D	E	F	G = D × E × F
Service life	Type	Finishing material	Exposed surface area (m^2)	aCO_2 (g/cm^3)	xc (cm)	CO_2 absorption (kg/m^3)
40 years	Outdoor area	Paint	9.27	0.32	5.79	−171.8
Total = 1365.6 kg-CO_2/1 PCG (=1537.4 (CO_2 emission due to concrete) − 171.8 (absorption due to carbonation))						

In addition, the CO_2 emissions from the transport of the concrete mixing materials and reinforcing bars to the PCG factory were about 12 (kg-CO_2/1 PCG). The departure distances for the transport of ordinary Portland cement, fine and coarse aggregates, and admixtures to the PCG factory (Chungbuk) were 106 km (Gangwon-do), 32 km (Chungcheongbuk-do), and 77 km (Gyeonggi-do), respectively. Among the mixing ingredients, however, mixing water was excluded from the assessment because it was supplied through the waterworks of the PCG factory. Among the concrete mixing ingredients, the CO_2 emissions of ordinary Portland cement which was transported from the most distant region were about 7.11 (kg-CO_2/1 PCG), accounting for about 60% of CO_2 emissions from the transportation stage. Furthermore, in the case of reinforcing bars, the distance from the deformed bar factory in Incheon to the PCG manufacturing plant is 161 km, and the CO_2 emissions were 2.9 (kg-CO_2/1 PCG), accounting for 25% of those from the transportation stage.

The CO_2 emissions from the consumption of electricity and diesel oil during the manufacturing stage were 136.2 (kg-CO_2/1 PCG). To manufacture PCG, the following were used: 57.4 kwh (electricity), 29.3 L (kerosene), 4.7 L (diesel oil), and 1.3 L (remover). For the assessment of CO_2 emissions from oil, direct and indirect (production and combustion) emissions were considered.

Regarding wastewater, it is relevant that it is discharged after being filtered in the factory. Therefore, the influence of wastewater was not applied. The CO_2 emissions from the consumption of kerosene among the CO_2 emissions from the manufacture stage were 92.9 (kg-CO_2/1 PCG), accounting for the highest portion in the manufacturing stage. The CO_2 emissions (including production and combustion) from electricity and diesel oil were 28.2 (kg-CO_2/1 PCG) and 15.1 (kg-CO_2/1 PCG), respectively (Figure 5).

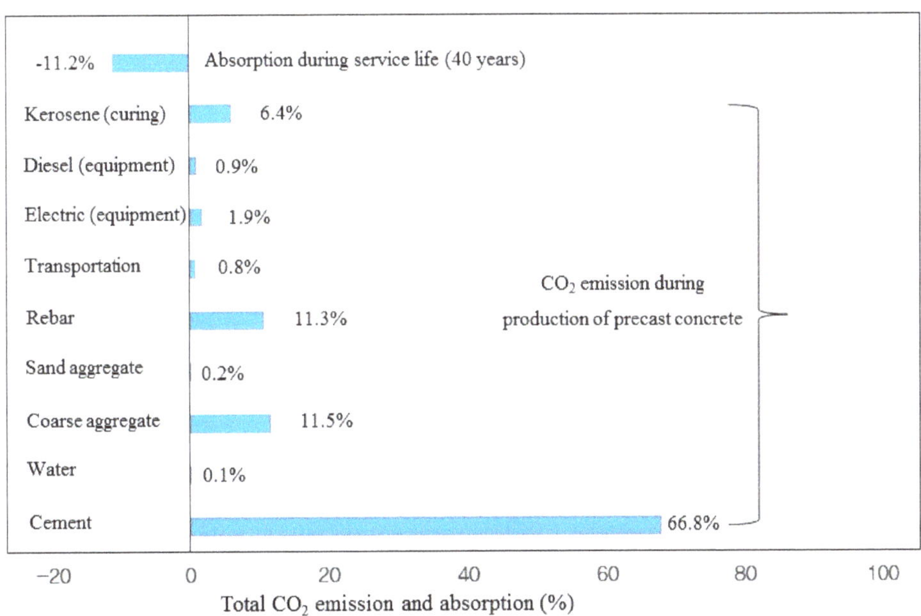

Figure 5. Detailed result of the total CO_2 emission-absorption for PCGs.

5.2. CO_2 Absorption during Service Life (Four Decades)

This study assessed CO_2 absorption assuming that the PCG was exposed to the external environment for four decades. It was also presumed that the external surface of the PCG was painted and that the external surface area was 9.27 m^2. $A_{CO2}(t)$ and xc(t) of the PC concrete from having a 40-year age were 0.32 g/cm^3 and 5.79 cm in an outdoor surface (Table 4).

Therefore, it was concluded that the CO_2 absorption throughout the service life of the PC was 172 (kg-CO_2/1 PCG), which is about 11.2% (16.5% of the CO_2 emissions by cement) of the CO_2 emissions from the production stage, as shown in Figure 6.

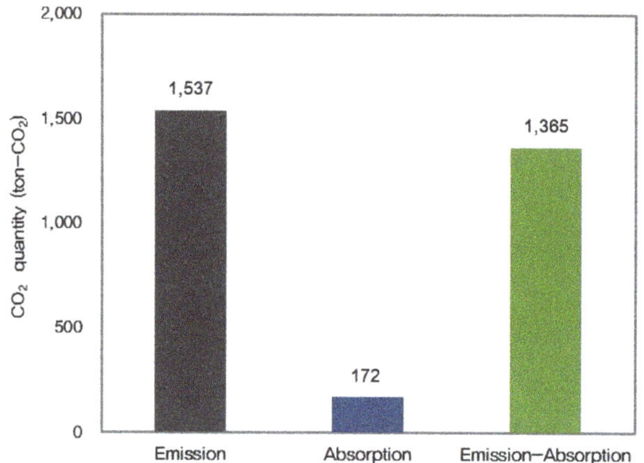

Figure 6. Total CO_2 emission-absorption of a precast concrete girder (PCG).

6. Conclusions

This study proposed the assessment of CO_2 emissions throughout the life cycle of PC and assessed CO_2 absorption by carbonation during its service life (40 years). For the assessment of CO_2 emissions during the life cycle of the PC, this study covered raw material, transportation, manufacturing, and use stages in accordance with the ISO 14044 (LCA). In case study of CO_2 emissions and absorption throughout the life cycle of a PCG, this study found that CO_2 emissions throughout the life cycle of the PCG were 1365.6 (kg-CO_2/1 PCG).

The CO_2 emissions during the production of raw materials were 1390 (kg-CO_2/ 1 PCG), thus accounting for a high portion to total CO_2 emissions with approximately 90% of the total. In contrast, the transportation and manufacturing stages accounted for 1% and 10%, respectively, having little effect on total CO_2 emissions.

Among the use of the PCGs, CO_2 absorption was mostly decided by the CO_2 diffusion coefficient and the amount of CO_2 absorption by cement paste. The CO_2 absorption by carbonation throughout the service life of the PC was about 11% of the total CO_2 emissions, which is about 16% of CO_2 emissions from the ordinary Portland cement.

However, this study has the following limitations: First, it has poor reliability for the assessment of CO_2 emissions from the proposed PC because it handled only one case. Therefore, there should be further verifications through diverse case studies. Second, the results for the case study on the PC were obtained from the firms in the Republic of Korea only. Therefore, they would not necessarily be applicable to the PC abroad. Hence, this study needs to improve reliability through assessment of CO_2 emissions throughout the life cycle of the PC in foreign countries as well.

Acknowledgments: This research was supported by the Korean Institute of Civil Engineering and Building Technology (KICT); No. 16-4-Green remodeling 3, 20160124-001.

Author Contributions: The paper was developed, written and revised by Tae Hyoung Kim. Chang U. Chae conducted the experimental and analytical work. Both authors contributed to the analysis and conclusion and have read and approved the final manuscript.

Conflicts of Interest: The authors declare no conflict of interest.

References

1. Ministry of Environment. *The UNFCCC COP21 for a New Climate Regime to Take Place in Paris*; Ministry of Environment: Sejong City, Korea, 2015.
2. Cabeza, L.F.; Rincón, L.; Vilariño, V.; Pérez, G.; Castell, A. Life cycle assessment (LCA) and life cycle energy 32 analysis (LCEA) of buildings and the building sector: A review. *Renew. Sustain. Energy Rev.* **2014**, *29*, 394–416.
3. Lee, K.H.; Tae, S.H.; Shin, S.W. Development of a Life Cycle Assessment Program for building (SUSB-LCA) in South Korea. *Renew. Sustain. Energy Rev.* **2009**, *13*, 1994–2002.
4. Ortiz, O.; Castells, F.; Sonnemann, G. Sustainability in the construction industry: A review of recent developments based on LCA. *Constr. Build. Mater.* **2009**, *23*, 28–39.
5. Dimoudi, A.; Tompa, C. Energy and environmental indicators related to construction of office buildings. *Resour. Conserv. Recycl.* **2008**, *53*, 86–95.
6. Pacheco-Torgal, F.; Cabeza, L.; Labrincha, J.; De Magalhaes, A. *Eco-Efficient Construction and Building Materials: Life Cycle Assessment (LCA), Eco-Labelling and Case Studies*; Elsevier: Cambridge, UK, 2014; Volume 1, pp. 624–630.
7. International Code Council (ICC). *International Green Construction Code (IgCC)*; International Code Council (ICC): Washington, DC, USA, 2010; pp. 1–95.
8. International Standardization Organization (ISO). *ISO 14044(Environmental Management-Life Cycle Assessment Principles and Framework)*; International Standardization Organization (ISO): Geneva, Switzerland, 2006.
9. Cement och Betong Institutet. *Carbon Dioxide Uptake during Concrete Life Cycle State of the Art*; Swedish Cement and Concrete Research Institute: Stockholm, Sweden, 2005.
10. Korea Society of Architectural Hybrid System. Available online: http://www.ahs.or.kr/ (accessed on 23 March 2016).
11. Min, T.B.; Cho, I.S.; Lee, H.S. Fundamental Study on the Development of Precast Concrete without Steam Curing. *J. Archit. Inst. Korea Struct. Constr.* **2012**, *28*, 61–68.

12. Víctor, Y.; José, V.M.; Tatiana, G.-S. Cost and CO_2 emission optimization of precast-prestressed concrete U-beam road bridges by a hybrid glowworm swarm algorithm. *Autom. Constr.* **2015**, *49*, 123–134.
13. Zhang, D.; Shao, Y. Early age carbonation curing for precast reinforced concretes. *Constr. Build. Mater.* **2016**, *113*, 134–143.
14. Ingrao, C.; Lo Giudice, A.; Mbohwa, C.; Clasadonte, M.T. Life Cycle Inventory analysis of a precast reinforced concrete shed for goods storage. *J. Clean. Prod.* **2014**, *79*, 152–167.
15. Dong, Y.H.; Jaillon, L.; Chu, P.; Poon, C.S. Comparing carbon emissions of precast and cast-in-situ construction methods—A case study of high-rise private building. *Constr. Build. Mater.* **2015**, *99*, 39–53.
16. Ilwoo, W.; Youngju, N.; JeongTai, K.; Sunkuk, K. Energy-efficient algorithms of the steam curing for the in situ production of precast concrete members. *Energy Build.* **2013**, *64*, 275–284.
17. Cassagnabère, F.; Escadeillas, G.; Mouret, M. Study of the reactivity of cement/metakaolin binders at early age for specific use in steam cured precast concrete. *Constr. Build. Mater.* **2009**, *23*, 775–784.
18. Pade, C.; GuimaraesPade, C.; Guimaraes, M. The CO_2 Uptake of Concrete in a 100 Year Perspective. *Cem. Concr. Res.* **2007**, *37*, 1348–1356.
19. Liwu, M.; Feng, Z.; Min, D.; Daman, K.P. Effectiveness of using CO_2 pressure to enhance the carbonation of Portland cement-fly ash-MgO mortars. *Cem. Concr. Compos.* **2016**, *70*, 78–85.
20. Lee, H.S.; Wang, X.Y. Evaluation of the Carbon Dioxide Uptake of Slag-Blended Concrete Structures, Considering the Effect of Carbonation. *Sustainability* **2016**, *8*, 312.
21. Elke, G.; van den Philip, H.; de Nele, B. Carbonation of slag concrete: Effect of the cement replacement level and curing on the carbonation coefficient—Effect of carbonation on the pore structure. *Cem. Concr. Compos.* **2013**, *35*, 39–48.
22. Gajda, J. *Absorption of Atmospheric Carbon Dioxide by Portland Cement*; Portland Cement Association (PCA) R&D: Skokie, IL, USA, 2001.
23. Lee, S.H.; Park, W.J.; Lee, H.S. Life cycle CO_2 Assessment Method for Concrete Using CO_2 Balance and Suggestion to Decrease $LCCO_2$ of Concrete in South-Korean Apartment. *Energy Build.* **2013**, *58*, 93–102.
24. Yang, K.H.; Seo, E.A.; Tae, S.H. Carbonation and CO_2 Uptake of Concrete. *Environ. Impact Assess. Rev.* **2014**, *46*, 43–52.
25. Taehyoung, K.; Sungho, T.; Seongjun, R. Life Cycle Assessment for Carbon Emission Impact Analysis of Concrete Mixing Ground Granulated Blast-furnace Slag (GGBS). *Archit. Inst. Korea* **2013**, *29*, 75–82.
26. National Life Cycle Index Database Information Network. Available online: http://www.edp.or.kr (accessed on 11 August 2015).
27. The Ecoinvent Database. Available online: http://www.ecoinvent.org/database (accessed on 11 August 2015).
28. Ministry of Strategy and Finance. Implementing Regulations in Corporate Tax Act, 2013.

29. Papadakis, V.G.; Vayenas, C.G.; Fardis, M.N. Physical and Chemical Characteristics Affecting the Durability of Concrete. *ACI Mater. J.* **1991**, *88*, 186–196.
30. Cha, S.W. Modeling of Hydration Process and Analysis of Thermal and Hygral Stresses in Hardening Concrete. Ph.D. Thesis, Seoul National University, Seoul, Korea, 1999. pp. 30–36.
31. Pommer, K.; Pade, C. *Guidelines. Uptake of Carbon Dioxide in the Life Cycle Inventory of Concrete*; Danish Technological Institute: Taastrup, Denmark, 2005; p. 45.
32. Oshida, F.; Izumi, I.; Kasami, H. *Effects of Cement Type, Mixture Proportion and Curing Condition on Carbonation of Concrete*; Architectural Institute of Japan: Tokyo, Japan, 1985; pp. 111–114.
33. Taehyoung, K.; Sungho, T.; Seongjun, R. Assessment of the CO_2 emission and cost reduction performance of a low-carbon-emission concrete mix design using an optimal mix design system. *Renew. Sustain. Energy Rev.* **2013**, *25*, 729–741.
34. Junghoon, P.; Sungho, T.; Taehyoung, K. Life Cycle CO_2 Assessment of Concrete by compressive strength on Construction site in Korea. *Renew. Sustain. Energy Rev.* **2012**, *16*, 2940–2946.
35. Taehyoung, K.; Sungho, T.; Changu, C. Analysis of Environmental Impact for Concrete Using LCA by Varying the Recycling Components, the Compressive Strength and the Admixture Material Mixing. *Sustainability* **2016**, *8*, 361.
36. Kwon, S.J.; Park, S.S.; Nam, S.H. A Suggestion for Carbonation Prediction Using Domestic Field Survey Data of Carbonation. *J. Korea Inst. Struct. Maint. Insp.* **2007**, *11*, 81–87.
37. Jung, S.H. Diffusivity of Carbon Dioxide and Carbonation in Concrete through Development of Gas Diffusion Measuring System. Ph.D. Thesis, Seoul National University, Seoul, Korea, 2003. pp. 7–19.

Proposal for the Evaluation of Eco-Efficient Concrete

Taehyoung Kim, Sungho Tae, Chang U. Chae and Kanghee Lee

Abstract: The importance of environmental consequences due to diverse substances that are emitted during the production of concrete is recognized, but environmental performance tends to be evaluated separately from the economic performance and durability performance of concrete. In order to evaluate concrete from the perspective of sustainable development, evaluation technologies are required for comprehensive assessment of environmental performance, economic performance, and durability performance based on a concept of sustainable development called the triple bottom line (TBL). Herein, an assessment method for concrete eco-efficiency is developed as a technique to ensure the manufacture of highly durable and eco-friendly concrete, while minimizing both the load on the ecological environment and manufacturing costs. The assessment method is based on environmental impact, manufacturing costs, and the service life of concrete. According to our findings, eco-efficiency increased as the compressive strength of concrete increased from 21 MPa to 40 MPa. The eco-efficiency of 40 MPa concrete was about 50% higher than the eco-efficiency of 24 MPa concrete. Thus eco-efficiency is found to increase with an increasing compressive strength of concrete because the rate of increase in the service life of concrete is larger than the rate of increase in the costs. In addition, eco-efficiency (KRW/year) was shown to increase for all concrete strengths as mixing rates of admixtures (Ground Granulated Blast furnace Slag) increased to 30% during concrete mix design. However, when the mixing rate of admixtures increased to 40% and 60%, the eco-efficiency dropped due to rapid reduction in the service life values of concrete to 74 (year/m^3) and 44 (year/m^3), respectively.

Reprinted from *Sustainability*. Cite as: Kim, T.; Tae, S.; Chae, C.U.; Lee, K. Proposal for the Evaluation of Eco-Efficient Concrete. *Sustainability* **2016**, *8*, 705.

1. Introduction

As argued by the "Declaration of Concrete Environment (2010)" of Korea and the "Declaration of Asian Concrete Environment (2011)" of six Asian countries, concrete has been shown to have an extremely large impact on environmental issues, including climate change [1]. Assessment of the environmental impacts of concrete materials and production has considerable importance. The concrete industry is shifting its paradigm from a simple construction industry that exploits resources and has environmental consequences to a sustainable development industry that preserves and maintains a robust global environment.

In addition, with the global topicality of energy conservation and greenhouse gas reduction, it is essential for the green growth of Korea to improve upon the environmental impacts of conventional cement and concrete industries. Based on the results of a recent evaluation of greenhouse gas emissions from materials used during the construction of buildings, concrete was found to be one of the major construction materials (along with reinforcement, steel frames, glass, paint, and insulation materials) that contribute to approximately 95% of overall greenhouse gas emissions. Because concrete is responsible for about 70% of this figure, there is a demand for techniques to evaluate and reduce greenhouse gas emissions from the production and use of concrete in construction [2].

However, there are a variety of significant pollution problems related to the environmental impact of concrete (in addition to global warming due to greenhouse gas emission), including depletion of the ozone layer, which predates global warming as an important environmental issue. In fact, the "Montreal Protocol on Substances that Deplete the Ozone Layer" came into effect in 1989 to stipulate regulations on the generation and use of chlorofluorocarbons (CFCs) that deplete the ozone layer. Sulfur dioxide (SO_2) and nitrogen oxides (NO_x), factors known to affect acidification, are specified as factors that affect human health, and are regulated as targets in the management of clean air environments.

Accordingly, diverse substances with an environmental load must be considered in addition to greenhouse gas emissions. To date, however, most of the research on concrete focuses almost exclusively on ways to improve the physical performance of concrete, such as crack control and increased strength and durability.

Currently, the development of eco-friendly concrete is receiving greater attention due to a growing focus on the effects of concrete production and use on the global, regional, and surrounding environments. Without comparing environmental emissions from the unit production of concrete with other construction materials, and considering the fact that large amounts of concrete and concrete products are required to construct commercial buildings and civil structures, efforts to quantitatively evaluate and reduce the effects of concrete production and use on the environment are essential to move the concrete industry toward sustainability. For this reason, studies are currently underway to evaluate and reduce the environmental impact of concrete. Nonetheless, existing domestic studies have mostly concentrated on technologies that reduce and evaluate greenhouse gas emissions among substances with an environmental impact.

There is an extreme scarcity of studies on technologies that comprehensively consider components of concrete, energy costs, and the durability of concrete as a structural material.

High-strength concrete has been evaluated to determine whether it is responsible for greater environmental emissions in comparison to normal-strength concrete in the context of environmental impact. It turns out, however, that high-strength concrete is in fact responsible for a lower proportion of emissions in terms of environmental impact and compressive strength (Figure 1) [3].

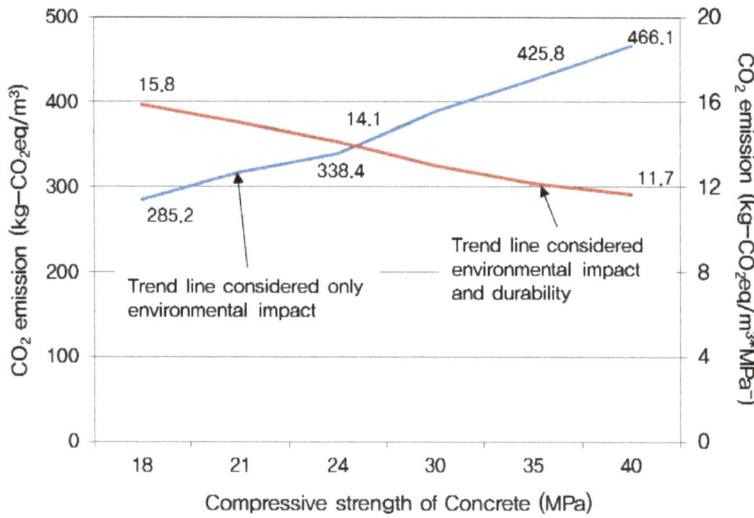

Figure 1. Environmental load considering environmental impact and durability aspect.

The importance of ecosystem effects of diverse substances with environmental impacts emitted during the production of concrete is recognized, but environmental performance has been evaluated separately from (and prioritized less than) the economic performance and durability performance of concrete. Eco-efficiency is an advanced concept used to evaluate the eco-friendliness of concrete [4]. The purpose of this study is to develop a method of evaluating the eco-efficiency of concrete based on environmental load emissions, manufacturing costs, and durability in the concrete production process. The technique proposed in this study intends to produce environment-friendly and highly durable concrete while minimizing both the environmental load on the ecosystem and manufacturing costs associated with the production and use of concrete. Our metrics are based on the results of service life assessments of various concrete samples. The technique can be utilized to evaluate the eco-friendly concrete.

Furthermore, the vision of this research is to contribute to broad implementation of environment-friendly concrete and construction industries, as shown in Figure 2.

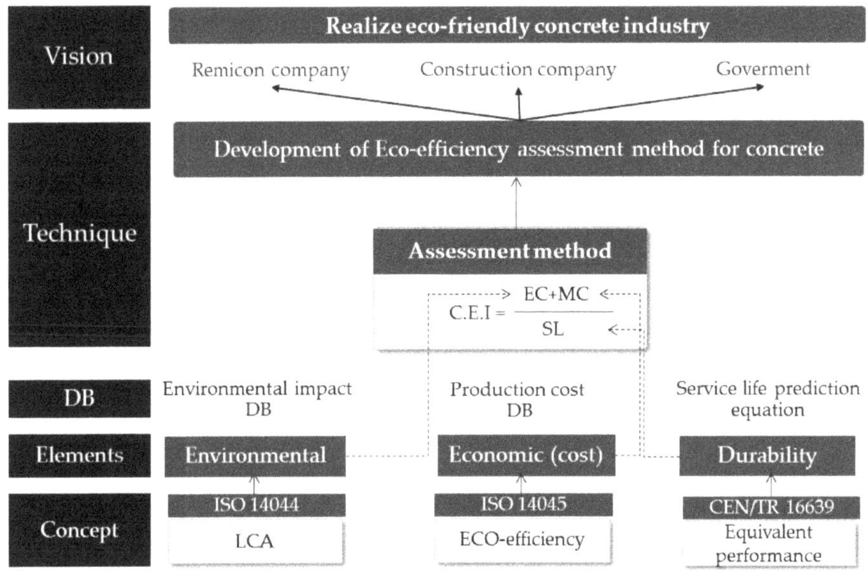

Figure 2. Research vision.

2. Literature Review

2.1. Concept of Eco-Efficiency

Through world summits on sustainable development such as the Rio Summit (1992) and the Johannesburg Summit (2002), sustainability has become the most important issue of the twenty-first century.

Eco-efficiency can be defined as a specific methodology to accomplish sustainability in the behaviors of governments, companies, or individuals. Accordingly, the World Business Council for Sustainable Development (WBCSD) was established to inspire companies to recognize the potential importance of eco-efficiency. There are some differences in the conceptual definitions of eco-efficiency and indicator development methodologies among nations, but the fundamental concept is identical to the original definition established by the WBCSD [5]. Eco-efficiency is based on the logic that companies with better eco-efficiency are more environmentally sustainable and show better environmental performance [6].

Eco-efficiency was initially utilized as an indicator to measure environmental efficiency as signified by environmental damage in comparison to the input of resources. However, eco-efficiency now includes environmental and economic implications, as it intends to create greater value using fewer resources. Eco-efficiency is a general concept that combines notions of economic efficiency and technical efficiency according to the conservation of resources, together with the concept of environmental efficiency, which aims to minimize environmental impacts from pollution (Equation (1)). As such, eco-efficiency is a concept for the sustainable development of products with competitive prices that simultaneously satisfies both environmental and social purposes.

$$Eco\text{-}efficiency = Economic\ \ Value/Environmental\ \ Impact \tag{1}$$

2.2. Eco-Efficiency Evaluation Methodology

Important details of the proposed eco-efficiency evaluation methodology can be summarized in four points.

First is the measurement of environmental impact and economic value, which constitute the denominator and numerator of the eco-efficiency index. As shown in Table 1, this can also be measured as the economic value of each unit element, or we can use environmental impact as the denominator and conversion factor. In addition, environmental emissions and economic value can be applied as cradle-to-grave or cradle-to-gate during the evaluation of product eco-efficiency [7].

Table 1. Evaluation factors for eco-efficiency.

Category	Environmental Impact	Economic Value
Element type	Unit elements (resources, water, energy, waste)	Output
	General elements (comprehensive environmental impact)	Sales
		Productivity
		ROE, ROI
Evaluation boundary	Gate-to-gate	
	Cradle-to-gate	
	Cradle-to-grave	

Second, there is a demand for an alternative to weakness of eco-efficiency as a relative index. If the economic value of a product increases by four times and the environmental impact increases by two times, then the eco-efficiency of the product is increased by two times. However, this leaves some doubt as to whether it is appropriate to increase environmental emissions by two times.

Third, eco-efficiency evaluation is done in the form of an index.

An index is a dimensionless value that requires the denominator and numerator to have the same dimension. It is difficult to convert eco-efficiency into an index because its denominator and numerator have different dimensions.

The fourth point relates to the method of including the effects of recycling in the evaluation of eco-efficiency from a lifecycle perspective. An increase in eco-efficiency due to recycling is not included in the scope of evaluation.

For example, industries and companies that use large amounts of recycled resources pursue environmental management to reduce their environmental impact. Regardless, such efforts are not reflected in the evaluation of eco-efficiency according to the methodology herein.

2.3. Research Trends

The WBCSD has been developing eco-efficiency indicators that can be used to measure economic development and environmental sustainability in corporate management. It defines eco-efficiency as a means to minimize environmental impact from the use of resources, waste, and pollution, and to maximize the creation of corporate value.

To use indicators more effectively based on the same concept, the WBCSD classifies indicators into two types. The first type includes generally applicable indicators, which can be effectively applied to all companies, although the indicators do not have the same value or importance across varying companies. The second type includes business-specific indicators, which reflect specific characteristics of each business. Examples of generally applicable indicators include financial profitability and value-added benefits of companies in terms of product and service value, and emissions of acidification substances to the atmosphere and the total amount of waste in terms of the environmental impact of products and services.

Pacheco has introduced important problems related to crude oil, the main feedstock of most polymer-based materials. These problems include interstate wars and environmental disasters. The latter is the most worrisome, such as the recent Deep Water Horizon oil spill that released approximately 780 million liters of crude oil on the Gulf of Mexico. Some historical examples on the use of bio-admixtures in construction materials are presented [8].

Zuoren et al. has focused on the following major issues: (1) introduction to eco-materials; (2) life-cycle thinking and LCA methodology; and (3) LCA practice for materials products and development of eco-materials [9].

Agustin et al. has presented the environmental value proposed by the Multilayer Structural Panels technology. This is an overall approach that considered the structural aspects and the environmental problems involved [10].

Jiaying et al. has analyzed the eco-footprint and eco-efficiency for the life cycle assessment of an exhibition hall in China. In addition, the measures were presented to reduce the eco-footprint and improve the eco-efficiency [11].

Dezhi et al. has presented an analysis methodology for energy evaluation of building construction. In addition, the amount of energy consumed is selected as the indicator of product value and the environmental impact [12].

3. Proposed Assessment Method for the Eco-Efficiency of Concrete

3.1. Overview

The proposed assessment method for the eco-efficiency of concrete was developed as a technique to ensure the high durability and eco-friendliness of concrete, while minimizing both the environmental load and manufacturing costs of concrete, as shown in Figure 3 [13].

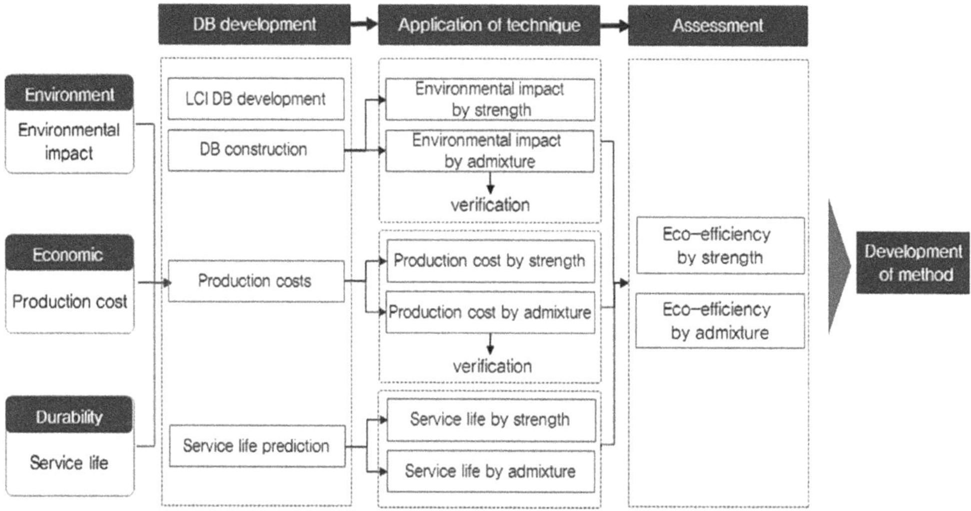

Figure 3. Process for the assessment of the eco-efficiency of concrete

3.2. Definition of Concrete Eco-Efficiency

There are limitations to assessing the eco-friendliness of concrete according to environmental emissions alone, without consideration of functionality (durability) and economic performance of concrete as an essential construction material.

Therefore, the concept of eco-efficiency was applied to assess the eco-friendliness of concrete, as shown in Figure 4.

The proposed concrete eco-efficiency assessment method is a technique to ensure the production of highly durable and environment-friendly concrete, while minimizing both the environmental load and manufacturing costs on the ecological environment based on the results of the assessment of environmental load, manufacturing costs, and the service life of concrete. Concrete eco-efficiency is improved by increasing the service life (years) of concrete and reducing the costs (KRW, Korea Won) of production. Thus, concrete with a low eco-efficiency value (KRW/year) among other concrete samples of the same strength is regarded as concrete with improved eco-efficiency.

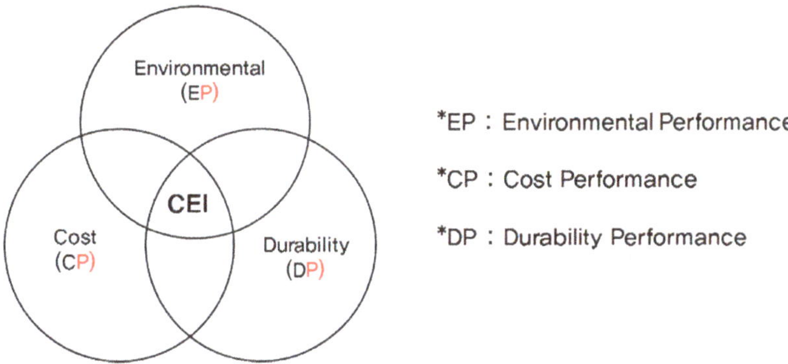

Figure 4. Eco-efficiency concept for concrete.

Equation (2) shows the calculation method for concrete eco-efficiency.

$$CEI = (EC + MC)/SL \tag{2}$$

Here, CEI is the concrete eco-efficiency index (KRW/year), EC is the environmental cost (KRW/m^3), MC is the manufacturing cost (KRW/m^3), and SL is the service life of concrete (year/m^3).

3.3. Eco-Efficiency Assessment Process

Concrete eco-efficiency assesses the environmental performance, cost performance, and durability performance of concrete according to input materials and energy sources during the production of concrete (cradle-to-gate). The environmental aspect assesses the environmental load of raw materials and energy sources used during the production of concrete based on a lifecycle assessment (LCA) technique.

The lifecycle impact assessment of concrete was performed separately for raw materials, transportation, and manufacturing stages. For the impact assessment, a damage-oriented (endpoint) method was applied to identify the degree of effects of environmental emissions from concrete production on human beings and the ecosystem [14].

Six types of environmental impacts, global warming, acidification, eutrophication, resource consumption, effects on the ozone layer, and creation of photochemical oxides, were selected as subjects of assessment. We analyzed the effects of these six environmental impact categories on safeguard subjects, including human health, social assets, fauna, flora, and primary manufacturing.

Database presented in the previous study [15] was applied to convert the effects into cost values.

The previous study is the first research of its kind in Korea to create lifecycle environmental impact indicators in the current unit. Because there are no other theories or DB specialized for construction materials and concrete in Korea, the DB of our previous work is appropriate for calculating the environmental costs of concrete herein.

Economic performance is deduced by combining different lifecycle costs, such as raw materials and energy sources (oil and electric). This study used DB on standard unit prices of materials for the production of concrete and energy sources provided by the Korean Price Information System [16].

Durability evaluates the expected service life of concrete based on carbonation among diverse deterioration phenomena. Among existing carbonation prediction equations, a formula presented by the Japanese Society of Civil Engineers was used herein to derive this study's equations for the service life of concrete.

3.4. *Eco-Efficiency Assessment Method of Concrete*

3.4.1. Environmental Performance (EP)

For the assessment of concrete eco-efficiency, environmental performance was assessed using a lifecycle assessment (LCA) technique.

(1) Definition of purpose and scope

General concrete was selected as the product subject to LCA, and the formation of concrete structure and concrete product was selected as the primary function of concrete. The functional unit was defined as 1 m^3 of concrete based on the primary function.

(2) Life Cycle Impact Assessment (LCIA)

Life Cycle Impact Assessment (LCIA) is a process that assesses the potential effects of indicators (input and output) created during lifecycle impact analysis on the environment.

Concrete is a mixture of cement, aggregates, and admixtures. Large amounts of energy are consumed during the collection of limestone or clay and the manufacture of clinker. Soil erosion or destruction of the ecosystem may occur from the collection of aggregates. In addition, because energy is also used during the transport of materials such as cement and aggregate to concrete manufacturers and during the production of concrete, diverse substances that incur environmental loads on air, water systems, and soil are emitted.

Because different environmental impacts occur during the lifecycle of concrete, multiple environmental impact assessments must be conducted on various aspects from production to use and disposal of functional units of concrete. Indeed, the assessment of a single environmental impact, such as global warming, fails to comprehensively assess the environmental performance of concrete (Figure 5). Accordingly, Life Cycle Impact Assessment (LCIA) on concrete was conducted based on standard substances and impact indicators for each of the six environmental impact categories, global warming, resource consumption, acidification, eutrophication, effects on the ozone layer, and creation of photochemical oxides [17].

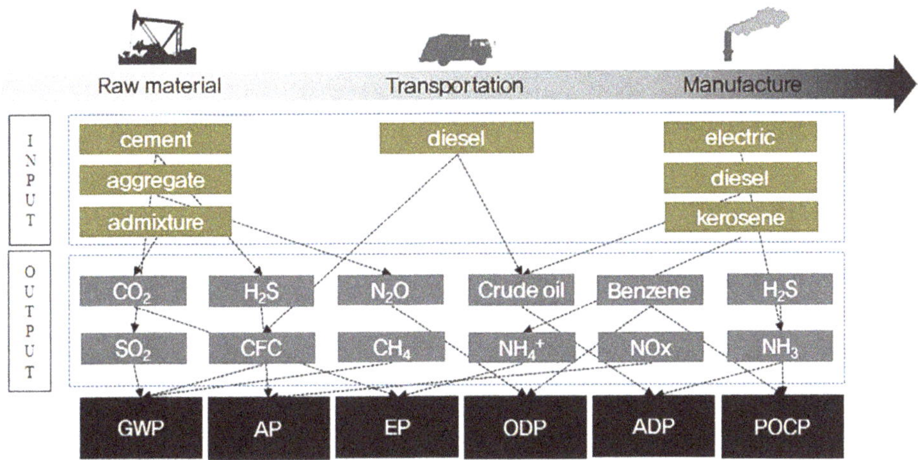

Figure 5. Environmental impacts that occur during the product lifecycle of concrete.

Life Cycle Impact Assessment (LCIA) is divided into four steps: (1) classification in which the inventory items extracted from the inventory analysis are assigned to the corresponding impact categories; (2) characterization in which the impact of each item classified into its impact category on each category is quantified; (3) normalization in which the environmental impact exerted on the environmental categories are divided into local or global environmental impacts; and (4) weighting in which relative importance among the impact categories is determined. According to ISO 14044 [18], the classification and characterization steps are mandatory assessment steps, and the normalization and weighting steps may be optionally assessed depending on the assessment purpose. In this study, assessment was performed for the classification and characterization steps because factors for concrete-related normalization and weighting suitable for Korean situations are yet to be developed.

The standard substances and impact potentials for environmental impact categories were applied in accordance with the respective databases used in the Ministry of Environment for the eco-labeling of the Environmental Declaration of Products [19,20]. The classification and characterization steps of assessment were performed on the basis of the previously selected LCI (Life Cycle Inventory) database [21].

Classification is done by categorizing and compiling the inventory items according to the environmental impact categories. By linking the inventory items derived from the LCI database to the pertinent environmental impact categories and integrating them by category, the environmental impact of each inventory item can be clearly identified.

CO_2, CH_4, and N_2O belong to the 23 GHGs specified in the Intergovernmental Panel on Climate Change (IPCC) guidelines [22], of which the standard substance is CO_2.

The classification of abiotic depletion potential (ADP) based on the standards provided by Guinee (1995) [23], takes into account a total of 89 resource items including crude oil, natural gas, and uranium (U). Acidification potential (AP) varies widely according to regional characteristics and atmospheric environments, and we applied the AP index presented by Heijung et al. and Hauschild and Wenzel [24] applicable to all regional types. A total of 23 inventory items linked to acidification category, including sulfur dioxide (SO_2), hydrogen sulfide (H_2S), and hydrogen fluoride (HF), are expressed in terms of their standard substance SO_2. Likewise, the index proposed by Heijung et al. and Hauschild and Wenzel was applied for the classification of the eutrophication potential (EP), with phosphate (PO_4^{3-}) used as the standard substance for a total of 11 inventory items including phosphate (PO_4^{3-}), ammonia (NH_3), and nitrogen oxides (NO_x). For the ozone depletion potential (ODP), we applied the ODP index specified in the World Meteorological Organization (WMO) [25] for a total of 23 inventory items, including CFC-11, Halon-1301, and CFC-114, with trichloro-fluoro-methane (CFC-11) as the standard substance. For the photochemical oxidant creation potential (POCP), a total of 128 inventory items were considered, including ethylene, NMVOC, and ethanol, with ethylene being the standard substance, thereby applying the POCP index proposed by Derwent et al. [26] and Jenkin and Hayman [27].

Characterization is a process of quantifying the environmental impact of inventory items itemized for each category in the classification step.

In the classification step, inventory items are assigned to their respective environmental impact categories, but there is a limitation in quantifying the potential impacts of inventory items in common metrics due to different impact potentials. Category indicator results, i.e., characterization values, are calculated in the characterization step where the environmental impact (=inventory data) of each inventory item is multiplied with the characterization factor (=impact potential) unique to the impact category concerned, and the resulting environmental impact thus converted into impacts are aggregated within each impact category to yield the overall environmental impact of that category (Equations (3)–(8)).

$$\textit{Global Warming Potential (GWP)} = \text{Environmental impact} \times \text{Characterization factor for GWP} \quad (3)$$

$$\textit{Ozone Depletion Potential (ODP)} = \text{Environmental impact} \times \text{Characterization factor for ODP} \quad (4)$$

$$\textit{Acidification Potential (AP)} = \text{Environmental impact} \times \text{Characterization factor for AP} \quad (5)$$

$$\textit{Abiotic Depletion Potential (ADP)} = \text{Environmental impact} \times \text{Characterization factor for ADP} \quad (6)$$

$$\textit{Photochemical Oxidant Creation Potential (POCP)} = \text{Environmental impact} \times \text{Characterization factor for POCP} \quad (7)$$

$$\textit{Eutrophication Potential (EP)} = \text{Environmental impact} \times \text{Characterization factor for EP} \quad (8)$$

(3) Damage assessment

Environmental cost, shown by Equation (9), converts the damage done to human beings and the ecosystem by the production of concrete into an economic value. Environmental cost is calculated by multiplying the damage factor of each safeguard subject and the economic value of each damage indicator for influential substances that belong to an environmental impact category (Figure 6).

$$D_k = \sum (Load_{i,j} \times DF_{k,i,j} \times K_c) \quad (9)$$

Here, $Load_{i,j}$ is the environmental load of influential substance j that belongs to impact category i, $DF_{k,i,j}$ is the damage factor for the damage caused by influential substance j that belongs to impact category i on safeguard subject k, and K_c is the economic value (KRW) of safeguard subject k.

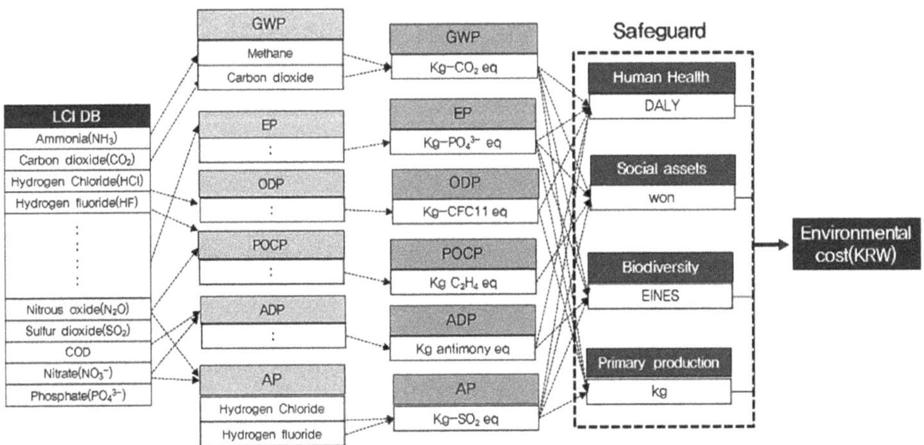

Figure 6. Assessment process for the environmental costs of concrete.

(a) Safeguard subject

From the viewpoint of environmental ethics, the environment is mainly divided into the human environment and the ecosystem. The human environment can be subdivided into human health and social assets that sustain human society, such as crops and resources.

Accordingly, as shown in Table 2, human health, biodiversity, social assets, and primary production were selected in this study as four safeguard subjects. Rather than the Eco-indicator 99 [28] (which reflects the environmental situation of Europe), the Korean lifecycle impact assessment index based on damage-oriented modeling (KOLID) [29] benchmarks the lifecycle impact assessment method based on endpoint modeling (LIME) of Japan [30], which clearly describes the philosophical grounds for the selection of safeguard subjects. Therefore, it is appropriate for the current analysis to use the safeguard subjects of KOLID for the calculation of the environmental costs of concrete.

Table 2. Safeguard and damage indicators.

Safeguard			Damage Indicators
Safeguard	Human	Human Health	DALY *
		Social assets	Monetary value (KRW)
	Ecosystem	Biodiversity	EINES **
		Primary production	NPP ***
Final evaluation result			Monetary value (KRW)

* DALY: disability-adjusted life year; ** EINES: expected increase in the number of extinct species; *** NPP: net primary production.

(b) Damage indicators

Table 2 shows damage indicators for each safeguard subject, unit of each damage indicator, and scope of assessment necessary for quantification of damage received by the four safeguard subjects according to environmental load. For human health, the degree of damage on human health caused by specific disease or physical factor was quantified as Disability Adjusted Life Year (DALY), a value that expresses the degree of damage on human health using the concept of life year. DALY can be used to assess cases of early death from disease or accident and cases of unhealthy life due to disability. For example, one DALY means that life year during which death from disease or accident and unhealthy life occurs is 1 year. As shown by Equation (10), DALY can be expressed as the sum of "Years of Life Lost (YLL)" from disease or accident and "Years Lost due to Disability (YLD)." YLD can be derived by reflecting importance of disability (D) and social value of age upon death. However, only importance of disability was accounted without consideration on importance of age and time discount originally considered in DALY. DALY was used as an indicator to show combined years lost from death and disability [31,32].

$$DALY = YLL + YLD \tag{10}$$

Social assets [33] can be classified into "production capacity of the ecosystem" that corresponds to the biological world and "non-biological resources" related to non-biological elements. In this study, forest resources, marine resources, and agricultural crops correspond to production capacity of the ecosystem, and fossil fuels and mineral resources correspond to non-biological elements. Current value assessment method, which uses actual market price, was applied as an indicator to measure the amount of damage for each item. When assessing biodiversity, general method involves classification of species according to the Endangered Species Act of the United States, the Convention on international angered Species of Wild Flora and Fauna (CITES), and Red Data Book (RDB) published by the International Union for Conservation of Nature (IUCN) [34]. The conventional damage oriented methods of impact assessment reflect such reality and conduct assessment based on quantifiable species [35,36]. In this study, "Expected Increase in Number of Extinct Species (EINES)" based on preservation ecology was adopted as damage indicator for ease of interpretation and agreement with data from Red Data Book. EINES, obtained from the increase in number of extinct species caused by environmental load, can be expressed as Equation (11).

$$\text{EINES} = \sum R_s = \sum [(1/T_{a,s}) - (1/T_{b,s})] \tag{11}$$

Here, R_s is the extinction risk increase by environmental impact; $T_{a,s}$ is the extinction hours after caused environmental impact; and $T_{b,s}$ is the extinction hours ago caused environmental impact.

Damage indicators to assess the effects of primary production include gross primary production (GPP), net primary production (NPP), and plant biomass. From the perspective of plant supply to consumers, plant biomass seems to be an optimal indicator given that a certain critical mass of plants must be accumulated and secured. Considering, however, that this certain level of plant supply must be consistently maintained, the provision of large accumulations at specific points in time will not satisfy the long-term requirements for plants. Plants must be grown at a constant rate. Accordingly, the damage indicator for primary production is shown by the change of NPP per unit of environmental load.

(4) Integrated DB for each safeguard subject

The marginal willingness to pay per unit of environmental property was the unit converted for each unit of the four safeguard subjects considered, as presented in Tables 3 and 4. Economic values (KRW, Korea Won) in Table 5 form the integrated DB for each safeguard subject. For reference, the unit conversion assumed that the loss of social assets (national economy) and economic loss are identical.

Table 3. Example of global warming potential (GWP).

GWP	Human Health (DALY/kg)	Social Assets (KRW/kg)	Biodiversity (EINES/kg)	Primary Production (kg/kg)
Ordinary Portland Cement	1.17×10^{-7}	2.58	0	0
Coarse aggregate	5.45×10^{-9}	1.20×10^{-1}	0	0
Blast furnace slag	3.81×10^{-9}	8.43×10^{-2}	0	0
Electric	7.91×10^{-9}	1.75×10^{-1}	0	0
Diesel	4.71×10^{-5}	1.04×10^{3}	0	0

Table 4. Example of abiotic depletion potential (ADP).

ADP	Human Health (DALY/kg)	Social Assets (KRW/kg)	Biodiversity (EINES/kg)	Primary Production (kg/kg)
Ordinary Portland Cement	0	2.72×10^{-2}	1.71×10^{-16}	2.93×10^{-4}
Coarse aggregate	0	2.58×10^{-1}	1.62×10^{-15}	2.78×10^{-3}
Blast furnace slag	0	3.17×10^{-3}	1.98×10^{-17}	3.41×10^{-5}
Electric	0	6.26×10^{-3}	3.92×10^{-17}	6.75×10^{-5}
Diesel	0	3.30×10^{2}	2.07×10^{-12}	3.55

Table 5. Economic value of safeguard [34].

Safeguard	Unit	Economic Value (KRW/unit)
Human health	DALY	2.82×10^{7}
Social assets	One thousand KRW	1.00×10^{3}
Biodiversity	Species	5.69×10^{5}
Primary production	Ton	4.93×10^{4}

3.4.2. Cost Performance (CP)

Economic performance assessment is performed on raw material production, transportation, and concrete manufacturing stages. The manufacturing cost per 1 m^3 of concrete was calculated using the unit price (Korean price information DB), based on the amount of materials and energy sources used in each step (Equation (12)). As shown in Table 6, the standard price information DB of Korea was investigated and applied to calculate the manufacturing costs for the raw materials and energy sources used in the production of concrete. Manufacturing cost (MC) was calculated by multiplying the unit price (KRW/kg) of raw materials, such as normal cement, slag cement, aggregate, admixtures (GGBS and fly ash), and chemical admixtures used to manufacture 1 m^3 of concrete, with the unit price (KRW/unit) of energy sources such as diesel, kerosene, electricity, and LNG used in the production of concrete by transport equipment and manufacturing facilities.

$$COST_{pf} = \sum(Input_{i,j} \times UC_i) \quad (12)$$

Here, $Input_{i,j}$ is (i) step (j) material and energy and UC_j is (j) unit price of material production (KRW). (i) Step: Raw material, Transportation, Manufacture. (j) Material: Cement; aggregate, admixture, etc.; electricity; and oil (diesel and kerosene) used by trains, trucks and batch plants.

Table 6. Unit price of raw materials and energy used in the production of concrete (example).

Substance	Unit Price	Unit	Reference
Ordinary Portland Cement	3410 KRW	40 kg	National information
Aggregate	13,000 KRW	ton	National information
Electric	110 KRW	Kwh	National information
Diesel	1700 KRW	L	National information

3.4.3. Durability Performance (DP)

Durability performance assessment of concrete was assumed to assess the service life of concrete used on a reinforced concrete (RC) structure based on carbonation. In addition, the covering thickness was 30 mm. For our purposes, service life was defined as the duration of utility until the carbonation depth of concrete reaches the outermost reinforcement. The most appropriate equation was selected as the standard specification for concrete to indicate that equations other than the proposed equation should be used. For this, service life was assessed on covering thicknesses of 30 and 40 mm with a water-binder ratio of 60%.

(1) Selection of prediction equation for the service life of concrete

Service life was assessed for covering thicknesses of 30 mm and 40 mm with a water-binder ratio of 60%. Although conditions for the calculation of service life differ, concrete comprised of ordinary Portland cement was assessed under outdoor conditions in order to apply identical parameters.

As shown in Table 7, an equation from the Architectural Institute of Japan (AIJ) [37] was used to deduce the results of assessment. The AIJ formula is appropriate because it yields results that are closest to the average service life of concrete.

Table 7. Service life evaluation results according to carbonation prediction equations (W/C 60%).

Prediction Formula	Coating Thickness	
	30 mm	40 mm
Kishitani [38]	42 years	76 years
Hamada [39]	60 years	109 years
Shirayama [40]	43 years	78 years
KCI [41], JSCE [42]	63 years	114 years
This study (AIJ)	43 years	87 years

(2) Prediction factors of the service life of concrete

Among existing carbonation prediction equations, Equations (13) and (14) (originating from the Architectural Institute of Japan) were used to derive the service life of concrete in order to assess the durability performance of concrete. However, because conditions of temperature and humidity in Japan (as presented in the values of the equations) are different from the values in Korea for an average year, more apt coefficients for the conditions of Korea were derived with annual average values provided by the Korea Meteorological Administration.

$$Durability_{pf}\ T = (C/A)^2 \tag{13}$$

$$A = K \times \alpha 1 \times \alpha 2 \times \alpha 3 \times \beta 1 \times \beta 2 \times \beta 3 \tag{14}$$

Here, T is the service life (year), C is the carbonation depth (mm), A is the carbonation velocity coefficient, K is the Kishitani coefficient, $\alpha 1$ is the coefficient by type of concrete, $\alpha 2$ is the coefficient by type of cement, $\alpha 3$ is the coefficient by the water-cement ratio, $\beta 1$ is the coefficient by temperature, $\beta 2$ is the coefficient by moisture, and $\beta 3$ is the coefficient by the concentration of carbon dioxide.

(3) Selection of covering thickness

Korean structural design standards for reinforced concrete have similar criteria to the code of the American Concrete Institute (ACI) [43]. Item 7.7.1 of ACI 318 suggests a covering thickness of 1.5 inches (38 mm) for reinforcement No. 14 and No. 18, and 3/4 inches (19 mm) for reinforcement No. 11 and below in the case of slabs, walls, and joist structures. The Korean standards suggest a covering thickness of 1.5 inches (38 mm) for main reinforcement, hoop reinforcement, and stirrup spiral reinforcement of beams and columns. In addition, the Japanese Architectural Standard Specification (JASS) [44] suggests covering thicknesses of 30 mm for slabs and 40 mm for column and bearing walls under the same conditions (i.e., when the structures do not directly contact soil and air).

Japan does not apply different criteria according to varying thicknesses of reinforcement. This study applied version 8.1.1 of the durability assessment criteria of green building certification standards (2013) [45]. The criteria of green building certification (2013) were selected based on the "Korean Building Code (KBC-2009) [46] and Concrete Structure Design Standards (2012)" [47]. Green building certification presents the legal minimum covering thickness to ensure the safety of structures by reflecting errors in covering thickness during actual construction. Accordingly, the covering thickness of reinforced concrete was defined as 30 mm to assess the service life of concrete based on carbonation according to the standards of green building certification.

4. Case Analysis

Eco-efficiency was assessed according to compressive strengths of concrete and mixing rates of admixtures. Changes in eco-efficiency according to increasing compressive strengths were assessed, and the major causes of any changes were analyzed.

4.1. Results of Eco-Efficiency Assessment for Different Compressive Strengths

As shown in Table 8, average eco-efficiency increased from 5700 (KRW/year) to 2200 (KRW/year) with an increase of compressive strength from 21 MPa to 40 MPa. In the case of 24 MPa and 40 MPa concretes, average eco-efficiency was 4700 (KRW/year) and 2200 (KRW/year), respectively (Figure 7).

Table 8. Eco-efficiency according to compressive strength.

Strength (MPa)	Eco-Efficiency (KRW/year)	Cost (KRW/m^3)	Service Life (year/m^3)
21	5700	308,000	54
24	4700	318,000	68
27	3800	333,000	89
30	3300	347,000	110
40	2200	387,000	190

The eco-efficiency of 40 MPa concrete was increased by about 50% in comparison to the eco-efficiency of 24 MPa concrete.

As shown in Figure 7, cost (comprising environmental and manufacturing costs) increased by about 21% from 318,000 (KRW/m^3) to 387,000 (KRW/m^3) according to increasing compressive strengths from 24 MPa to 40 MPa. Importantly, however, the service life of concrete based on carbonation was greatly increased from 68 (year/m^3) to 190 (year/m^3) with rising compressive strengths as well. As such, eco-efficiency was found to increase with the increasing compressive strength of concrete, because the rate of increase in service life is larger than the rate of increase in cost [48].

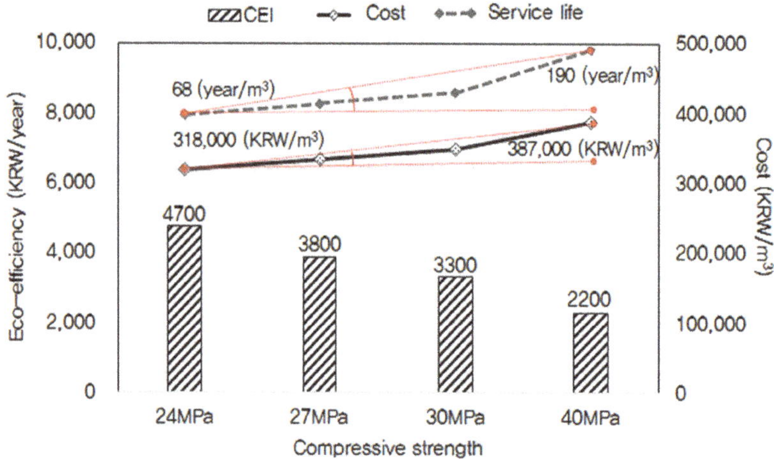

Figure 7. Eco-efficiency according to increasing compressive strength.

4.2. Results of Eco-Efficiency Assessment for Different Mixing Rates of Admixtures

As shown in Table 9, eco-efficiency (KRW/year) was shown to increase at all concrete strengths as mixing rates of admixtures (GGBS) increased to 30% during concrete mix design. As representative examples, the eco-efficiency of 24 MPa and 40 MPa concretes was increased by 10%–23% in comparison to OPC when the mixing rates of admixtures increased from 10% to 30% [49].

Table 9. Eco-efficiency according to admixture ratio.

Strength (MPa)	**Admixture Ratio**						
	0%	**10%**	**20%**	**30%**	**40%**	**50%**	**60%**
21	5157	5092	4636	4195	4503	3973	5739
24	4066	3564	3407	3101	3324	2926	3450
27	3874	3860	3527	3179	3398	2980	4272
30	2801	2512	2401	2168	2322	2043	2938
40	2014	1900	1792	1614	1724	1510	2162

As shown in Figure 8, the cost of 24 MPa concrete dropped from 362,000 (KRW/m^3) to 274,000 (KRW/m^3) according to increased mixing rates of admixtures due to reduced environmental impacts and manufacturing costs.

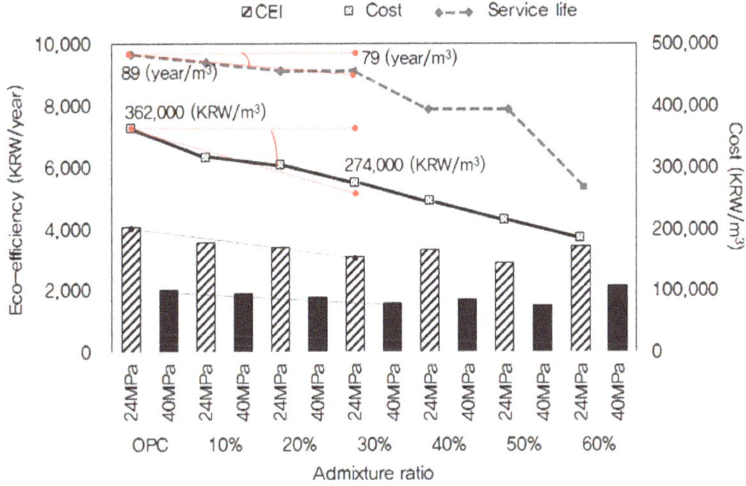

Figure 8. Analysis of eco-efficiency by admixture ratio (30%).

In addition, the service life of concrete based on carbonation was reduced from 89 (year/m^3) to 79 (year/m^3) according to increased mixing rates of admixtures. However, the rate of decrease in the service life was lower than the rate of decrease in the costs.

When mixing rates of admixtures were increased to 40% and 60%, however, the eco-efficiency of concrete samples dropped due to the rapid reduction of service life to 74 (year/m^3) and 44 (year/m^3), respectively (Figure 9).

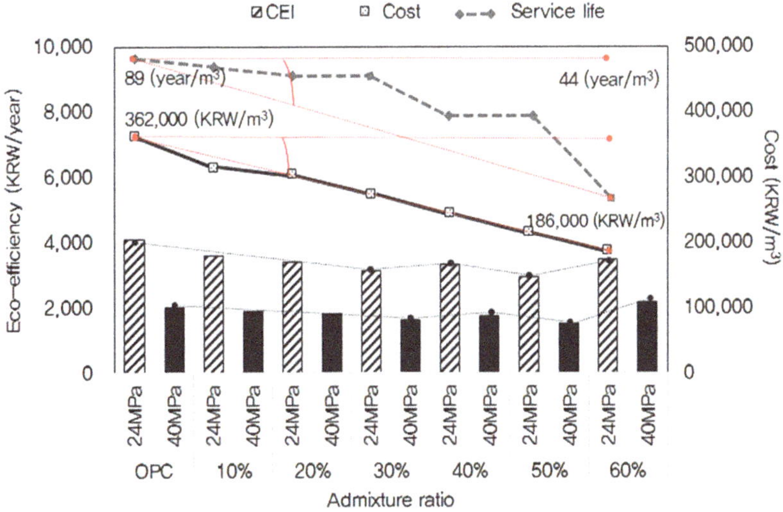

Figure 9. Analysis of eco-efficiency by admixture ratio (60%).

This tendency was verified in 21 MPa, 27 MPa, 30 MPa, and 40 MPa concretes in addition to 24 MPa concrete.

5. Discussion and Study Limitations

Two limitations of this study are a need to expand the assessment factors for durability performance (service life) and to develop an eco-efficiency assessment method for high-strength concrete (above 40 MPa).

(1) Expansion of assessment factors for durability performance (service life)

Among various deterioration phenomena, this study considers only neutralization as an assessment factor for durability performance. The assessment model for service life is based only on the neutralization of concrete, and the model fails to reflect durability performance for salt damage and freeze–thaw processes. Because the service life of concrete differs according to deterioration phenomena, differences may arise during eco-efficiency assessment of the same type of concrete. When assessment is conducted according to neutralization only, the service life of concrete is slightly reduced by the mixing of admixtures (GGBS). On the contrary, salt damage tends to show an increase in the service life of concrete due to the mixing of admixtures. Therefore, the measure of eco-efficiency will change according to differing service life values of concrete with the same mix design depending on the type of deterioration phenomenon. In future studies, the service life of concrete should be assessed according to various deterioration phenomena, such as neutralization and salt damage, simultaneously.

(2) Development of an eco-efficiency assessment method for high-strength concrete (above 40 MPa)

The equation for calculating service life applied herein for the assessment of durability showed mostly effective results for concrete with a water-binder ratio between 40% and 60%. The equation was inappropriate for high-strength concrete with a water-binder ratio below 40%, however.

It is necessary to develop an equation that can calculate the service life of high-strength concrete against compound deterioration.

6. Conclusions

(1) The concrete eco-efficiency index (CEI), an indicator for the eco-friendly concrete that accounts for environmental emissions, manufacturing costs, and durability performance in the production and use of concrete, is proposed as an advanced concept to assess the eco-friendliness of concrete.
(2) An assessment method for concrete eco-efficiency is developed as a technique to ensure the manufacture of highly durable and eco-friendly concrete,

while minimizing both the environmental load and manufacturing costs of concrete. The mechanism of eco-efficiency is based on the assessment results of environmental impact, manufacturing costs, and the service life of concrete.

(3) According to our findings, eco-efficiency increased when the compressive strength of concrete increased from 21 MPa to 40 MPa. This is because the rate of increase in the service life due to increasing compressive strength is higher than the rate of increase in costs.

(4) The eco-efficiency of 40 MPa concrete increased by about 50% in comparison to the eco-efficiency of 24 MPa concrete. While cost (comprising environmental and manufacturing costs) increased by about 21% from 318,000 (KRW/m^3) to 387,000 (KRW/m^3) according to the increase in compressive strength from 24 MPa to 40 MPa, the service life of concrete based on carbonation greatly increased from 68 (year/m^3) to 190 (year/m^3) between concretes with increasing compressive strength. Accordingly, eco-efficiency is found to increase with increasing compressive strength of concrete, because the rate of increase in the service life of concrete is larger than the rate of increase in costs.

(5) Eco-efficiency (KRW/year) was shown to increase for all concrete strengths as mixing rates of admixtures (GGBS) increased to 30% during concrete mix design. As representative examples, the eco-efficiency of 24 MPa and 40 MPa concretes increased by 10 to 23% in comparison to OPC when the mixing rates of admixtures increased from 10% to 30%. However, when the mixing rates of admixtures were increased to 40% and 60%, the eco-efficiency dropped due to the rapid reduction of service life values to 74 (year/m^3) and 44 (year/m^3), respectively.

(6) Environmental costs according to increasing compressive strengths of concrete were assessed, and the safeguard subjects and environmental impact categories responsible for the greatest part of overall environmental costs were analyzed. Environmental impact assessment according to the mixing of admixtures (GGBS) and recycled aggregate was performed to deduce a method of reducing environmental costs. In addition, manufacturing costs according to the compressive strength of concrete and the mixing of admixtures were assessed to analyze which material comprises the greatest proportion of manufacturing costs and to find a corresponding method of reduction.

Acknowledgments: This research was supported by Korea Institute of Civil Engineering and Building Technology (KICT) (No. 20160420-001).

Author Contributions: The paper was written by Taehyoung Kim and revised by Sungho Tae. Chang U. Chae and Kanghee Lee conducted the experimental and analytical work. All authors contributed to the analysis and approved the final manuscript.

Conflicts of Interest: The authors declare no conflict of interest.

Appendix A.

(1) Example of environmental cost assessment for ordinary cement)= CO_2 emission ($kg\text{-}CO_2$) * Damage factor ($DALY/kg\text{-}CO_2$)) * (Economic value (KRW/DARY)= 100 $kg\text{-}CO_2/kg$ (Input) * 1.17×10^{-7} $DALY/kg\text{-}CO_2$ (Table 3) * 28,200,000 KRW/DARY (Table 5)= 330 KRW (Korea Won)

(2) Example of eco-efficiency assessment for concrete)

(a) Cost = (Environmental Cost + Manufacture Cost)

* Environmental cost

Cement = (240 kg/m^3 * 0.9 CO_2 emission factor ($kg\text{-}CO_2/kg$)) * 1.17×10^{-7} $DALY/kg\text{-}CO_2$ (Table 3) * 28,200,000 KRW/DARY (Table 5) = 712 KRW (Korea Won)

Water = (160 kg/m^3 * 0.001 CO_2 emission factor ($kg\text{-}CO_2/kg$)) * 1.17×10^{-7} $DALY/kg\text{-}CO_2$ (Table 3) * 28,200,000 KRW/DARY (Table 5) = 0.5 KRW (Korea Won)

Coarse aggregate = (900 kg/m^3 * 0.015 CO_2 emission factor ($kg\text{-}CO_2/kg$)) * 1.17×10^{-7} $DALY/kg\text{-}CO_2$ (Table 3) * 28,200,000 KRW/DARY (Table 5) = 45 KRW (Korea Won)

* Manufacture cost

Cement = 240 kg/m^3 * 85 KRW/kg (Table 6) = 20,400 KRW (Korea Won)

Water = 160 kg/m^3 * 15 KRW/kg = 2,400 KRW (Korea Won)

Coarse aggregate = 900 kg/m^3 * 13 KRW/kg (Table 6) = 11,700 KRW (Korea Won)

(b) Service life = 54 year (using Equations (11) and (12))

(c) Eco-efficiency = (Cost/Service life) = (35260/54) = 652 (KRW/year)

References

1. Ministry of Environment. *The UNFCCC COP21 for a New Climate Regime to Take Place in Paris*; Ministry of Environment: Paris, France, 2015.
2. Roh, S.J.; Tae, S.H.; Shin, S.W.; Woo, J.H. A Study on the Comparison of Characterization of Environmental Impact of Major Building Material for Building Life Cycle Assessment. *Archit. Inst. Korea* **2013**, *29*, 93–100.
3. Taehyoung, K.; Sungho, T.; Seungjun, R.; Rakhyun, K. Life Cycle Assessment for Carbon Emission Impact Analysis of Concrete Mixing Ground Granulated Blast-furnace Slag(GGBS). *Archit. Inst. Korea* **2013**, *29*, 75–84.
4. Taehyoung, K.; Sungho, T.; KRWyoung, C.; Keunhyuk, Y. Development of the Eco-efficiency evaluation method for Concrete. *Korea Concr. Inst.* **2015**, *35*, 437–438.
5. WBCSD. *Eco-Efficiency: Creating More Value with Less Impact*; WBCSD Research Report; WBCSD: Geneva, Switzerland, 2000.

6. ISO 14045. *Environmental Management-Eco-Efficiency Assessment of Product Systems-Principles, Requirements and Guidelines*; ISO: Geneva, Switzerland, 2012.
7. Ministry of Environment. *Development & Diffusion of Eco-Efficiency Indicators and Software*; Ministry of Environment: Sejong-si, Korea, 2007.
8. Pacheco-Torgal, F. Introduction to biopolymers and biotech admixtures for eco-efficient construction materials. In *Biopolymers and Biotech Admixtures for Eco-Efficient Construction Materials*; Woodhead Publishing: Cambridge, UK, 2016; pp. 1–10.
9. Zuoren, N. Chapter 3—Eco-Materials and Life-Cycle Assessment. In *Green and Sustainable Manufacturing of Advanced Material*; Elsevier: Amsterdam, The Netherlands, 2016; pp. 31–76.
10. Agustín, P.G.; Arianna, G.V.; Guillermo, G.P. Building's eco-efficiency improvements based on reinforced concrete multilayer structural panels. *Energy Build.* **2014**, *85*, 1–11.
11. Jiaying, T.; Xianguo, W. Eco-footprint-based life-cycle eco-efficiency assessment of building projects. *Ecol. Indic.* **2014**, *39*, 160–168.
12. Lia, D.; Zhua, J.; Huib, E.C.M.; Leungb, B.Y.P.; Qiming Li, Q. An emergy analysis-based methodology for eco-efficiency evaluation of building manufacturing. *Ecol. Indic.* **2011**, *11*, 1419–1425.
13. Taehyoung, K.; Sungho, T. Development of Environmental Performance Index in Concrete. *Korea Concr. Inst.* **2014**, *26*, 651–652.
14. Taehyoung, K.; Sungho, T.; Sungjoon, S.; George, F.; Keunhyek, Y. An optimization System for Concrete Life Cycle Cost and Related CO_2 Emissions. *Sustainability* **2016**, *8*, 361.
15. Korea Environmental Industry Technology Institute. *Development of Integrated Evaluation Technology on Product Value for Dissemination of Environmentally Preferable Products*; Korea Environmental Industry Technology Institute (KEITI): Seoul, Korea, 2009.
16. Korea Price Information. Available online: http://www.kpi.or.kr/ (accessed on 14 March 2016).
17. Taehyoung, K.; Sungho, T.; Changu, C. Analysis of Environmental Impact for Concrete Using LCA by Varying the Recycling Components, the Compressive Strength and the Admixture Material Mixing. *Sustainability* **2016**, *8*, 361.
18. Organización Internacional de Normalizaci6n. *ISO 14044: Life Cycle Assessment-Requirements and Guidelines*; ISO: Geneva, Switzerland, 2006.
19. MOLIT. *National Database for Environmental Information of Building Products*; Ministry of Land, Transport and Maritime Affairs of the Korean Government: Sejong-si, Korea, 2008.
20. National Life Cycle Index Database Information Network. Available online: http://www.edp.or.kr (accessed on 11 August 2015).
21. The Ecoinvent Database. Available online: http://www.ecoinvent.org/database (accessed on 11 August 2015).
22. IPCC Guidelines for National Greenhouse Gas Inventories, 2006. Available online: http://www.ipcc-nggip.iges.or.jp/public/2006gl/ (accessed on 13 April 2016).

23. Guinee, J.B. Development of a Methodology for the Environmental Life Cycle Assessment of Products: With a Case Study on Margarines. Ph.D. Thesis, Leiden University, Leiden, The Netherlands, 1995.
24. Heijungs, R.; Guinée, J.B.; Huppes, G.; Lamkreijer, R.M.; Udo de Haes, H.A.; Wegener Sleeswijk, A.; Ansems, A.M.M.; Eggels, P.G.; van Duin, R.; de Goede, H.P. *Environmental Life Cycle Assessment of Products. Guide (Part1) and Background (Part 2)*; CML Leiden University: Leiden, The Netherlands, 1992.
25. World Metrological Organization (WMO). *Scientific Assessment of Ozone Depletion: Global Ozone Research and Monitoring Project*; WHO: Geneva, Switzerland, 1991; p. 25.
26. Derwent, R.G.; Jenkin, M.E.; Saunders, S.M.; Piling, M.J. Photochemical ozone creation potentials for organic compounds in Northwest Europe calculated with a master chemical mechanism. *Atmos. Environ.* **1998**, *32*, 2429–2441.
27. Jenkin, M.; Hayman, G. Photochemical Ozone Creation Potentials for oxygenated volatile organic compounds: Sensitivity to variation is in kinetic and mechanistic parameters. *Atmos. Environ.* **1999**, *33*, 1275–1293.
28. Goedkoop, M.; Spriensma, R. *The Eco-Indicator 99-A Damaged Oriented Method for Life Cycle Impact Assessment*; PRé: Amerfoort, The Netherlands, 1999.
29. Pilju, P.; Mannyoung, K. *The Development of Korean Life Cycle Impact Assessment Index Based on a Damage Oriented Modeling*; Korea Environmental Industry & Technology Institute: Seoul, Korea, 2010; pp. 499–508.
30. Itsubo, N.; Inaba, A. A new LCIA Method: LIME Has Been Completed. *Int. J. Life Cycle Assess.* **2003**, *8*, 305.
31. Ko, J.Y. Generation of DALYs for the Analysis of Damage to Human Health. Master's Thesis, Konkuk University, Seoul, Korea, 2005.
32. Murray, C.J.L. Quantifying the Burden of Disease, the Technical Basis for Disability Adjusted Life Years. *Bull. World Health Organ.* **1994**, *72*, 429–445.
33. Lane, J.L. *Life Cycle Impact Assessment—Chapter 4.6 Endpoint Assessment Methodologies for LCA*; Wiley: Hoboken, NJ, USA, 2015; pp. 64–65.
34. Jonathan, E.M.B.; Leon, A.B.; Thomas, M.B.; Stuart, H.M.B.; Janice, S.C.; Zoe, C.; Craig, H.T.; Michael, H.; Georgina, M.M.; Sue, A.M.; et al. *2004 IUCN Red List of Threatened Species: A Global Species Assessment*; IUCN: Gland, Switzerland; Cambridge, UK, 2004.
35. Robles Gil, P.; Pérez Gil, R.; Bolívar, A.; Bräutigam, A.; Jenkins, M.; Rabb, G.; Ceballos, G.; Ehrlich, P.; Dublin, H.T. The Red Book: The Extinction Crisis Face to Face. Available online: https://portals.iucn.org/library/node/7965 (accessed on 17 July 2016).
36. Itsubo, N.; Inaba, A. Lifecycle Impact Assessment Method Based on Endpoint Modeling (LIME2)—Chapter 2. Characterization and Damage Evaluation Methods. Available online: http://lca-forum.org/english/pdf/No15_Chapter2.1-2.3.pdf (accessed on 16 July 2016).
37. Architectural Institute of Japan (AIJ). Available online: https://www.aij.or.jp/ (accessed on 14 March 2016).
38. Kishitany, K. Consideration on durability of reinforced concrete. *Trans. Arch. Inst. Jpn.* **1963**, *65*, 9–16.

39. Hamada, M. Carbonation of concrete. In Proceedings of the Fifth International Symposium on the Chemistry of Cement, Tokyo, Japan, 31 December 1969; Volume 3, pp. 343–384.
40. Shirayama, K. Research activities and administrative measures on durability of buildings. The state of the art in Japan. *Mater. Constr. Mater. Struct.* **1985**, *18*, 215–221.
41. Korea Concrete Institute (KCI). Available online: http://www.kci.or.kr/eng/introduce1.asp/ (accessed on 16 March 2016).
42. Japan Society of Civil Engineers (JSCE). Available online: http://www.jsce-int.org/ (accessed on 16 March 2016).
43. American Concrete Institute (ACI). Available online: https://www.concrete.org/ (accessed on 16 March 2016).
44. Architectural Institute of Japan. *Japanese Architectural Standard Specification (JASS)*; Architectural Institute of Japan: Tokyo, Japan, 2012.
45. Ministry of Land, Infrastructure and Transport. *Green Standard for Energy and Environmental Design (G-SEED)*; Ministry of Land, Infrastructure and Transport: Sejong-si, Korea, 2013.
46. Ministry of Land, Infrastructure and Transport. *Korean Building Code (KBC)*; Ministry of Land, Infrastructure and Transport: Sejong-si, Korea, 2012.
47. Korea Concrete Institute. *Concrete Structure Design Standard*; Korea Concrete Institute: Seoul, Korea, 2012.
48. Junghoon, P.; Sungho, T.; Taehyoung, K. Life Cycle CO_2 Assessment of Concrete by compressive strength on Construction site in Korea. *Renew. Sustain. Energy Rev.* **2012**, *16*, 2940–2946.
49. Taehyoung, K.; Sungho, T.; Seongjun, R. Assessment of the CO_2 emission and cost reduction performance of a low-carbon-emission concrete mix design using an optimal mix design system. *Renew. Sustain. Energy Rev.* **2013**, *25*, 729–741.

Life Cycle CO_2 Assessment by Block Type Changes of Apartment Housing

Cheonghoon Baek, Sungho Tae, Rakhyun Kim and Sungwoo Shin

Abstract: The block type and structural systems in buildings affect the amount of building materials required as well as the CO_2 emissions that occur throughout the building life cycle ($LCCO_2$). The purpose of this study was to assess the life cycle CO_2 emissions when an apartment housing with 'flat-type' blocks (the reference case) was replaced with more sustainable 'T-type' blocks with fewer CO_2 emissions (the alternative case) maintaining the same total floor area. The quantity of building materials used and building energy simulations were analyzed for each block type using building information modeling techniques, and improvements in $LCCO_2$ emission were calculated by considering high-strength concrete alternatives. By changing the bearing wall system of the 'flat-type' block to the 'column and beam' system of the 'T-type' block, $LCCO_2$ emissions of the alternative case were 4299 kg-CO_2/m^2, of which 26% was at the construction stage, 73% was as the operational stage and 1% was at the dismantling and disposal stage. These total $LCCO_2$ emissions were 30% less than the reference case.

Reprinted from *Sustainability*. Cite as: Baek, C.; Tae, S.; Kim, R.; Shin, S. Life Cycle CO_2 Assessment by Block Type Changes of Apartment Housing. *Sustainability* **2016**, *8*, 752.

1. Introduction

Internationally, greenhouse gases are arguably the most prevalent global environmental problem. According to International Energy Agency (IEA), buildings account for almost 30% of greenhouse gas emissions [1–3]. Korea established the national Greenhouse Gas Reduction Roadmap to reduce greenhouse gas emissions by 37% for Business-As-Usual (BAU) levels by 2030 [4]. In Korea, the construction industry accounts for 40% of all material consumption, 24% of energy consumption and 42% of CO_2 emissions. Thus, reduction of the construction industry's CO_2 emissions is required to reach greenhouse gas reduction goals [5].

Apartment housing is the major type of the residential sector in Korea, making up 52.4% of residential building stock. The most common apartment building in Korea is the 'flat-type' block, which consists of two rectangular units side by side like a wide box [6]. Thus far, building types, building forms and structural systems have not been heavily studied in regard to Life Cycle CO_2 ($LCCO_2$) emission. However, for the majority of apartment housing blocks, it has been shown that a significant portion of the CO_2 emissions can be reduced by using a more sustainable block type instead of the 'flat-type' block [7–13].

The purpose of this study was to assess the $LCCO_2$ emissions when 'flat-type' blocks in an apartment building (the reference case) were replaced with more sustainable 'T-type' blocks with fewer CO_2 emissions (the alternative case) while maintaining the same total floor area. Therefore, this study focuses on changing the building type rather than by improving the insulation or the heating, ventilation and air conditioning (HVAC) equipment. The quantity of building materials used and energy simulations were analyzed on each block type using Building Information Modeling (BIM) techniques, and the $LCCO_2$ emissions were calculated.

The study results indicate that different block types have significantly different CO_2 emissions over the building life cycle, and 'T-type' blocks have the potential to significantly reduce greenhouse gas emission in the residential sector.

2. Literature Review

Life Cycle Assessment (LCA) that quantifies the consumption of resource and the occurrence of emissions throughout the entire process of products system is an environmental impact assessment scheme that evaluates their overall effects and is defined as ISO14040 [14]. The research on the construction sector began with the reference to the Product Life Cycle Assessment targeting materials and products. In order to apply to building structures, in consideration of the characteristics of having a complicated structure and a long lifetime, the process of establishing the evaluation subject's list of analysis and evaluation stages should be prioritized by setting a life cycle phase and range and by separating the inputs and outputs [15–17]. The building's previous LCA includes all processes and activities, during the life cycle of the building, that are divided into construction, operation, maintenance, management, dismantling and disposal phases [18]. It is used as a tool to calculate the environmental load of a quantitative structure. In addition, the previous LCA's ultimate purpose is to drive improvements that minimize the resource, energy consumption, CO_2 emissions, etc. at each step for a sustainable development. In consideration of the building's previous life cycle for an environmental impact assessment, European countries were undertaken in the development of national level since the early 1990s and the studies on building's environmental performance evaluation is being conducted in various fields using the LCA method [19]. Eco-Quantum is the world's first building LCA-based computer program and was developed by the IVAM Environmental Research Institute in Netherlands; it evaluates various aspects, such as the effects of energy consumption during the building life cycle, maintenance during the operational phase, differences in the durability of building-related parts, and recycling rates [20]. Becost, developed by the VTT Technical Research Centre of Finland Ltd., is a web-based program that is utilized in marketing and system management, and uses data relating to environmental effects throughout the building life cycle, including during building material production, transportation, construction, maintenance, and disposal [21]. Envest, developed by BRE in UK, is used to evaluate the LCA of building materials from the early phase of building design. Web-based Envest2 was developed in 2003. The system boundary of this system includes material extraction and manufacturing, related transport, on-site construction of assemblies, operation, maintenance and replacement, and demolition. It can evaluate greenhouse gas (GHG) emissions, acid deposition, ozone depletion, eutrophication, human toxicity, eco toxicity, waste disposal, etc. using the Ecoinvent database [22]. The analysis results produced by Envest provide information relating to both environmental performance and economic feasibility, through mean measured values of environmental effects (referred to as Eco-point) and whole-life cost analysis results [23,24]. Athena EcoCalculator is a spreadsheet-based LCA tool developed by the ATHENA Institute in Canada. Architects, engineers and other design professionals can have instant access to instant life cycle assessment results for hundreds of common building assemblies using Athena EcoCalculator for Assemblies [25,26]. The tool was commissioned by the Green Building Initiative (GBI) for use with the Green Globes environmental certification system. The boundary of this system includes material extraction and manufacturing, related transport, on-site construction of assemblies, maintenance and replacement, demolition, and transport to landfill. It can evaluate GHG emission, embodied primary energy, pollution to air, pollution to water, weighted resource use using the ATHENA database (cradle-to-grave) and US life cycle inventory database [27]. This system makes it easy to obtain the environmental impact result in real time and compare each other assemblies. However, it is only available custom assembly options. Column and beam sizes are fixed [25]. LISA, developed in Australia, offers advantages in terms of ease of analysis of environmental performance during the building material production phase by utilizing life cycle inventory (LCI) databases (DBs) for various materials; it also uses simple input methods, thereby reducing evaluation time and effort [24]. In Korea, SUSB–LCA was developed by the Sustainable Building Research Center. SUSB–LCA employs direct input of building materials and energy usage, together with an estimation model. SUSB–LCA can evaluate life-cycle energy, carbon emissions, and cost. It is also an evaluation program that allows a case comparison between target and alternative buildings [28].

Compared with the above LCA Method, SUSB–LCA enables easy assessment of building life cycle CO_2, offering outstanding performance of data renewal, and Becost and Envest2 offer users easy access and various analysis results but require many hours of in calculating CO_2 due to many input items. Moreover, Eco-Quantum also can perform various comprehensive assessments by the stages based on life cycle, but, due to many direct input items, require many hours in the assessment [29]. On the other hand, Athena EcoCalculator and LISA have outstanding capability to analyze construction material production by utilizing the LCI database of many materials and require relatively less assessment hours in the assessment owing to simple input method, but has limitation in detailed analysis of CO_2.

3. Assessment Method

Theory of the Building LCA Assessment Method

To compare the $LCCO_2$ of the different block types, an existing apartment housing project that consisted of all 'flat-type' apartment blocks was selected as the reference case, and CAD drawings (e.g., plans, sections, elevations and details) were obtained. To compare a more sustainable block type with the reference case, a 'T-type' block was proposed with the same levels of insulation and HVAC equipment as well as with the same total floor area in the block lay-out plan of the existing project. The 'T-type' block was developed based on the concepts of less building material used during the construction stage, and less energy used during the operation stage in the project life cycle (the alternative case). Second, each representative block for the base and alternative cases was composed using BIM software (ArchiCAD ver. 13, Graphisoft, Budapest, Hungary) based on 2D CAD drawings. After developing a 3D model, the cost of the building materials was assessed with SUSB–LCA, a software program developed by Sustainable Building Research Center at Hanyang University, Ansan, Korea. This information was used to quantitatively assess CO_2 emissions and calculate the cost and energy usage from the building's entire life cycle (construction, operation, maintenance and demolition and disposal) [28]. Third, the effect of CO_2 reduction was assessed by the application of high-strength concrete on the alternative cases only. This was performed by measuring the reduction in the quantity of materials used for construction and the life cycle extension of the structural system due to the use of high-strength concrete [30]. Fourth, CO_2 emissions during the operation stage were assessed by measuring energy consumption of each case using EcoDesigner, a building energy simulation software compatible with BIM (Graphisoft, Budapest, Hungary). This program has been validated for fast analysis results by international standards including IEA-BESTEST, ASHRAE-BESTEST and CEN-15265.

Finally, $LCCO_2$ emissions for the entire building life cycle were assessed: all CO_2 emissions were summed from the construction, the operation and the dismantling and disposal stages. In this study, for the evaluation of manufacturing process considering the construction materials' practical aspects and properties, the mixed analysis method, which the individual integration and the input-output analysis are complexly used, was applied. Especially, for the concrete CO_2 emission intensity that is different depending on the strength, since the input–output analysis and the individual integration currently indicate just the individual or partial intensity, the CO_2 basic unit through the database of concrete strength and CO_2 emissions for each admixtures, which were analyzed by the individual integration, was applied in the initial research and, in the case of materials other than concrete, the input–output analysis derived from the direct and indirect parts of the input–output relations table of the Bank of Korea was applied for consistency in the per-unit range analysis and evaluation results. In addition, the supply quantity table by specific-items, which is an attached table of input–output relations table, was prioritized in applying each material unit price, while the energy consumption and CO_2 emission per unit of each material were calculated by using the price information data and the construction cost analysis data of the Korea Housing Corporation for the materials that are difficult to apply the specific item. Figure 1 shows the process of assessment for this study.

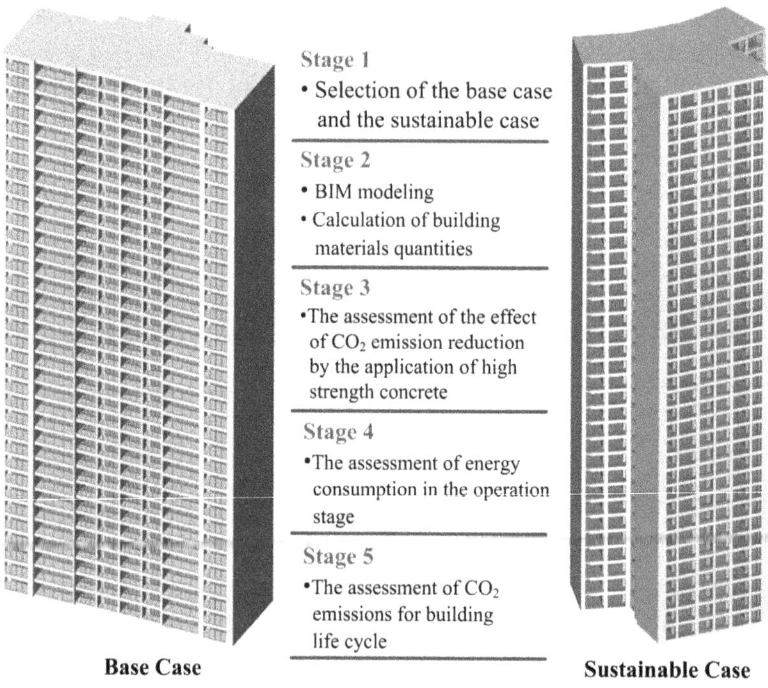

Figure 1. Process of life cycle CO_2 assessment.

4. Reference Case and Alternative Case Proposal

The apartment housing project selected as the reference case was composed of 14 'flat-type' blocks ranging from 30 to 35 stories. The land area was 99,744 m^2, and it had 1829 dwelling units with 249,951 m^2 of total floor area for residential use. The project was completed in 2004. A typical 35-story block in the housing project was selected as the reference block of the reference case (see Figure 2). The typical floor plan consisted of the same two units with one vertical circulation core; each unit area was 162.87 m^2, and the total floor area of the reference block was 11,400.9 m^2. The floor height was 2.9 m and the total height of the reference block was 104.8 m. The structural system was the bearing wall system and the concrete compressive strengths of the vertical members of the reference block were classified into four segments: 35 MPa from the ground floor to the 9th floor, 30 MPa from the 10th to 19th floors, 27 MPa from the 20th to 26th floors and 24 MPa from the 27th to 35th floors (See Table 1). The alternative case was designed to have the same level of insulation and HVAC equipment as well as with the same total floor area in the site level of the reference case. Therefore, the alternative case was composed of 14 'T-type' blocks ranging from 20 to 35 stories, which had 1820 dwelling units with 250,932 m^2 of total floor area for residential use.

A typical 35-story block was selected as the reference block in the alternative case. The floor height was 2.9 m and the total height of the reference block was 104.8 m; these dimensions were the same as the reference case. The typical floor plan of the alternative case was planned as the 'column and beam' structural system with a front 3–4 bay composition, and in the form of four units with one vertical circulation core to achieve spatial efficiency and openness in each unit (see Figure 3). Each unit area was 137.88 m^2, and the total floor area of the reference block was 19,302.5 m^2.

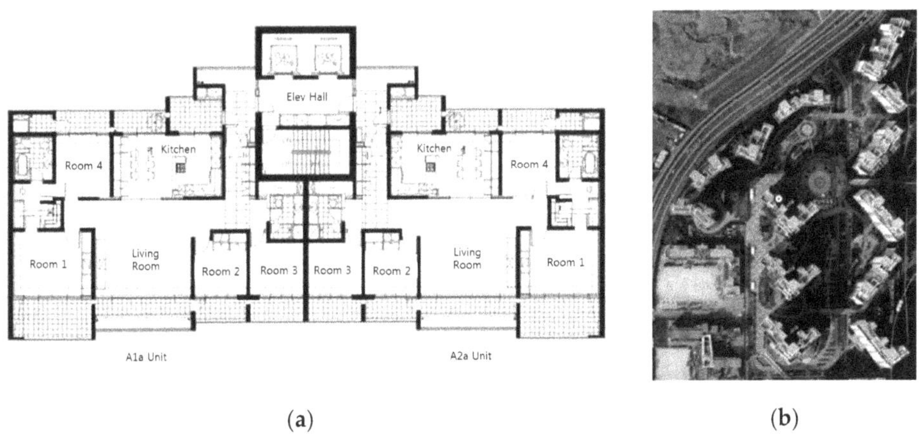

(a) (b)

Figure 2. The plan drawing and the bird's eye view of the reference floor of the reference case. (**a**) plan drawing of the reference floor; (**b**) bird's eye view.

Table 1. Building overview.

Category	Contents
Building Size	Above Ground 35 Stories, Basement 3 Stories
Structural system	Reference case: reinforced concrete, bearing wall structure
	Alternative case: reinforced concrete, column and beam structure
Concrete compressive strength	Reference case: Classified into 4 segments: 24, 27, 30, 35 MPa
	Alternative case: Classified into 4 segments: 24, 30, 40, 50 MPa
Others	Concrete and rebar quantities were reviewed comparatively based on the sum of the horizontal and vertical members.

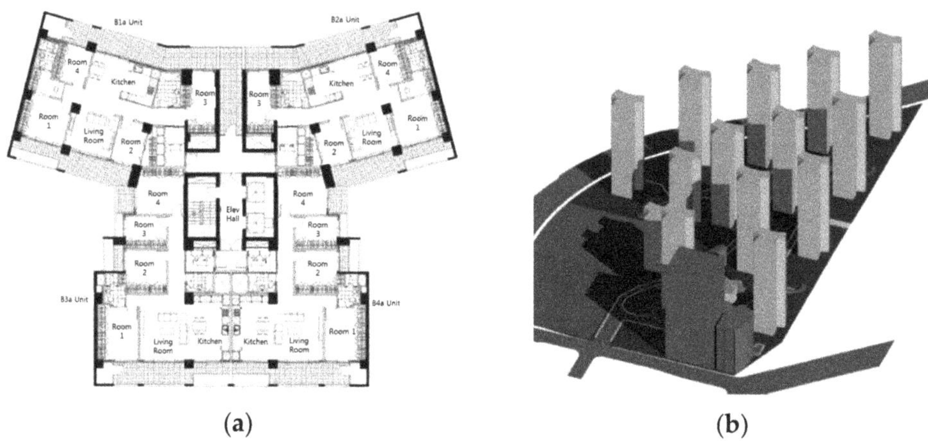

Figure 3. The plan drawing and bird's eye view of the reference floor of the alternative case. (**a**) reference floor plan drawing; (**b**) aerial view.

The structural system was the 'column and beam' system and the variable high-strength concrete was used as the structural material. These factors were selected in order to assess how many CO_2 emissions are reduced from each of these changes.

5. Assessment of CO_2 Emissions by Changes in Building Form

5.1. Comparison of the Amount of Major Materials and an Assessment of CO_2 Emissions

The bearing wall system of the reference case was changed to the 'column and beam' system and the changes in the amount of major materials were calculated using the quantity take-off function of the ArchiCAD BIM software. In addition, the quantity of materials per unit floor area was also assessed (Table 2). The materials used for each case were ready mixed concrete, rebar, cement bricks, tiles, expandable polystyrene, plasterboard, poly vinyl chloride (PVC) windows and glass. These elements make up 80% of CO_2 emissions in Korean apartments [31]. In the alternative case, the use of cement bricks increased by 236%, porcelain tile (wall) by 3%, expandable polystyrene by 53% and plasterboard by 59.3%. These increases were caused by changes to the different building types, including the 'column and beam' system in the structural system and the 'T-type' block in the building form. However, substituting load-bearing walls for columns and beams decreased the use of concrete and rebar by 11% and 36%, respectively. The CO_2 emissions for the reference case and the alternative case were calculated with SUSB–LCA (Table 3). The alternative case decreased CO_2 emissions per unit floor area by 9.45% (to 853.63 kg-CO_2/m^2) compared to the reference case. This change was mainly due to the decreased use of rebar and concrete in the alternative case.

Table 2. Comparisons of the major materials required for the reference case and the alternative case.

Materials	Reference Case (70 Households, Total Floor Area: 11,400.9 m^2)		Alternative Case (140 Households, Total Floor Area: 19,302.5 m^2)	
	Total	Quantity per Unit Area	Total	Quantity per Unit Area
Concrete				
Slab	3256.68 m^3	0.2856 m^3/m^2	5999.35 m^3	0.3108 m^3/m^2
Column	-	-	2207.45 m^3	0.1143 m^3/m^2
Beam	-	-	1388.93 m^3	0.0719 m^3/m^2
Wall	4706.8 m^3	0.4128 m^3/m^2	2350.6 m^3	0.1217 m^3/m^2
Total	7963.48 m^3	0.6984 m^3/m^2	11,946.33 m^3	0.6189 m^3/m^2
Rebar	944,656 kg	82.8 kg/m^2	1026.080 kg	53.1 kg/m^2
Cement brick	506,832 EA	44.455 EA/m^2	2,883,571 EA	149.388 EA/m^2
Tile				
Porcelain tile (floor)	64,680 kg	5.673 kg/m^2	91,272 kg	4.7285 kg/m^2
Porcelain tile (wall)	74,760 kg	6.557 kg/m^2	130,416 kg	6.7564 kg/m^2
Expandable polystyrene	7949.4 kg	0.6972 kg/m^2	20,599.5 kg	1.0671 kg/m^2
Plasterboard	154,918.4 kg	13.5882 kg/m^2	417,826.5 kg	21.6462 kg/m^2
PVC windows	11,945.85 kg	1.0477 kg/m^2	17,159.29 kg	0.8889 kg/m^2
Glass	6877.77 m^2	0.6032 m^2/m^2	7466.25 m^2	0.3868 m^2/m^2

5.2. Assessment of Changes in CO_2 Emissions Due to High Strength Concrete

The use of high-strength concrete may reduce $LCCO_2$ emissions by both extending the building life as well as reducing the amount of concrete and rebar used in the structural members. In this chapter, we specifically analyzed the decrease in CO_2 emissions due to life cycle extension, from 40 years in the reference case to 80 years in the alternative case.

5.2.1. Consideration of the Building Life Cycle

To extend the building life cycle to 80 years, we considered the carbonation phenomenon. Carbonation is where CO_2 in the atmosphere leaches into concrete and reacts with calcium hydroxide to form calcium carbonate, reducing the pH of the concrete pore solution down to 8.3–10.0. Once the pH inside the concrete is low, the rebar buried inside the concrete rusts thus decreasing its stability, and corrosion begins. Corrosion in rebar by carbonation is a representative deterioration phenomenon of reinforced concrete structures [31–34].

Table 3. Comparisons of the CO_2 emissions of the reference case and the alternative case.

Materials	Unit	CO_2 Emissions Unit (kg-CO_2/Unit)	Reference Case		Alternative Case	
			CO_2 Emissions (kg-CO_2)	CO_2 Emissions (kg-CO_2/m^2)	CO_2 Emission (kg-CO_2)	CO_2 Emissions (kg-CO_2/m^2)
Concrete						
24 MPa	m^3	329.37	1,471,295.79	129.05	2,433,385.56	126.07
27 MPa	m^3	353.02	332,191.82	29.14	0.00	0.00
30 MPa	m^3	383.77	516,170.65	45.27	749,502.81	38.83
35 MPa	m^3	406.71	492,119.10	43.16	0.00	0.00
40 MPa	m^3	429.65	0.00	0.00	559,404.30	28.98
50 MPa	m^3	508.39	0.00	0.00	661,923.78	34.29
Rebar	kg	3.84	3,627,479.04	318.17	3,940,147.20	204.13
Cement Block	EA	0.27	136,844.64	12.00	778,564.17	40.33
Tile	kg	13.80	1,924,272.00	168.78	3,059,294.40	158.49
Expandable Polystyrene	kg	12.73	101,195.86	8.88	262,231.64	13.59
Plasterboard	kg	4.45	689,386.88	60.47	1,859,327.93	96.33
PVC Windows	kg	12.10	144,544.79	12.68	207,627.41	10.76
Glass	m^2	27.33	187,969.45	16.49	204,052.61	10.57
	Total		10,748,180.97	942.75	15,988,617.15	853.63

The infiltration rate of CO_2 into concrete must be computed in order to compute the life cycle of the reinforced concrete in a carbonation environment. In general, it can be expressed as the square root of time, as shown in Equation (1). In addition, the velocity coefficient A used in Equation (1) is calculated from Equation (2), where A depends on: (1) the type of concrete; (2) the type of cement; (3) the water-cement ratio and (4) the temperature and humidity. The coefficient A for this study was determined using methods proposed by the Architectural Institute of Japan [35], and carbonation depth versus time was computed. Table 4 shows the values of the variables that determine the velocity coefficient of carbonation. We used the values shown in Table 4 to compute the carbonation velocity:

$$C = A\sqrt{t} \tag{1}$$

$$A = \alpha_1 \times \alpha_2 \times \alpha_3 \times \beta_1 \times \beta_2 \times \beta_3 \quad (2)$$

where C: Carbonation Depth (cm), A: Carbonation Velocity Coefficient, and t: Time (year).

Table 4. Variables of carbonation velocity coefficient A.

Variable	Details	Applied Value
$\alpha 1$	Concrete type	Normal concrete → 1
$\alpha 2$	Cement type	Normal concrete → 1
$\alpha 3$	Water to binder ratio	W/B = 0.6 → 0.22
$\beta 1$	Temperature	Annual average temperature 15.9 °C → 1
$\beta 2$	Humidity	Annual average humidity 63% → 1
$\beta 3$	Carbon dioxide concentration	CO_2 concentration 0.05% → 1

Figure 4 shows an estimation of the carbonation velocity. Figure 4 illustrates that concrete with a strength of 30 MPa or less may suffer from steel corrosion as carbonation may occur in the rebar inside the concrete within 80 years (the target service life). Therefore, in order to rule out the necessity of structure repair within 80 years, concrete with a minimum strength of 35 MPa should be used.

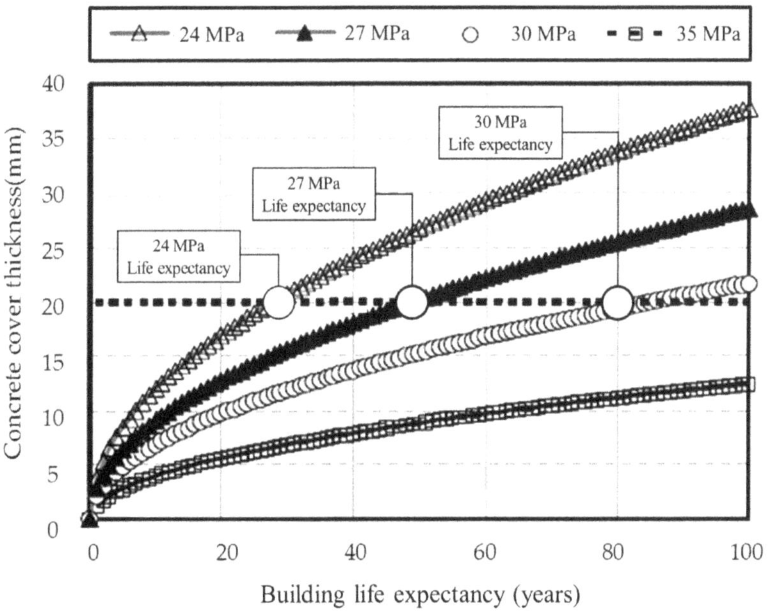

Figure 4. Results of carbonation velocity for each concrete strength.

5.2.2. Quantifying the Reduction of CO_2 Emissions by Using High-Strength Concrete

Based on the results of the above rate of carbonation analysis, the effects of high-strength concrete were assessed with both cases (Table 5).

Table 5. Overview of the applications of high-strength concrete.

	Case-1 (Reference Case)	Case-2 (Alternative Case 1)	Case-3 (Alternative Case 2)	Case-4 (Alternative Case 3)
Structural system	bearing wall	columns and beams	columns and beams	columns and beams
If carbonation is considered or not	Not considered	Not considered	Considered	Considered
Concrete strength	24, 27, 30, 35 MPa	24, 30, 40, 50 MPa	35, 40, 50 MPa	35, 40, 50 MPa
Whole repair	Once	Once	Unnecessary	Unnecessary
Blast furnace slag	Not used	Not used	Not used	Used (substitution rate 20%)

Case 1 represents the reference case and Case 2 represents the alternative case with repairs once every 40 years. Case 3 shows a situation in which no repair is required over the target life cycle (80 years) with the use of 35 MPa high-strength concrete. Case 4 shows a situation in which 20% blast furnace slag is substituted for the high-strength concrete in Case 3. Relatively more CO_2 is emitted when high-strength concrete is used because the amount of cement used is increased compared to normal strength concrete. In order to solve this problem, methods such as substitution of a portion of the cement with industrial waste such as blast furnace slag have been proposed [36,37]. This study assumed a mixture with 20% blast furnace slag in the cement.

Based on the actual structural calculations on each case and the quantities of concrete and rebar required, the CO_2 emissions were computed and compared (Table 6).

As for structural repair, partial repairs were assessed assuming that the entire repair is done in consideration of inefficiency of construction, and according to the Japan Society of Civil Engineers [38] research results, the CO_2 emissions of materials consumed for one session of repair were set to 40% of CO_2 emitted from the materials related to the structure for one session of new construction.

When high-strength concrete was used for the alternative cases (Cases 2 to 4), CO_2 emissions of concrete and rebar were reduced by 21.08% compared to the reference case in Case 2 (structural repairs at 40 years), 35.39% in Case 3 (without structural repair) and 37.99% in Case 4 (when blast furnace slag is substituted at 20%).

Table 6. CO_2 emissions of concrete and rebar by whether high-strength concrete is applied or not.

		Unit	Case-1 (Reference Case)		Case-2 (Alternative Case 1)		Case-3 (Alternative Case 2)		Case-4 (Alternative Case 3)	
			Volume	CO_2 Emission (kg-CO_2/m^2)	Volume	CO_2 Emission (kg-CO_2/m^2)	Volume	CO_2 Emission (kg-CO_2/m^2)	Volume	CO_2 Emission (kg-CO_2/m^2)
Concrete	24 MPa	m^3	4467	129.05	7388	126.07	-	-	-	-
	27 MPa	m^3	941	29.14	-	-	-	-	-	-
	30 MPa	m^3	1345	45.27	1953	33.83	-	-	-	-
	35 MPa	m^3	1210	43.16	-	-	9301	195.98	9301	183.00
	40 MPa	m^3	-	-	1302	28.98	1302	28.98	1302	27.00
	50 MPa	m^3	-	-	1302	34.29	1302	34.29	1302	32.00
	Repair	-	-	98.65	-	91.27	0	-	0	-
	Sub-total	m^3	7963	345.28	11,945	319.44	11,905	259.25	11,905	242.00
Rebar		kg	944,656	318.17	1,026,080	204.13	851,646	169.42	851,646	169.42
Total			-	663.45	-	523.57	-	428.68	-	411.42
Ratio of reduction over case 1					21.08%		35.39%		37.99%	

5.2.3. CO_2 Emission for the Construction Stage

Based on our assessment of changes in CO_2 emissions due to high-strength concrete, Figure 5 shows CO_2 emissions of the construction stage, which is composed of emissions from construction works on the site ("Construction"), building material transportation to the site ("Transportation"), and building material production off the site ("Production"). As shown in Figure 5, Case 4 produced the least amount of emissions and Case 1 produced the most. In all cases, Production was the stage that produced the highest percentage of emissions.

Specifically, Case 2, which had the "column and beam" structural system, showed 11.98% fewer emissions than Case 1 (the reference case), which had the bearing wall structural system and less concrete and rebar. Therefore, the 'column and beam' structural system was effective at reducing CO_2 emissions in the construction stage if the building blocks are in similar conditions. By design, Cases 3 and 4 used high-strength concrete and had twice the building life cycle than Case 1 and 2 (80 years vs. a repair at 40 years). Despite the shorter life cycle, Case 1 had 26.7% more emissions and Case 2 had 11.5% more CO_2 emissions than Case 4.

We found that the effects of the structural systems on CO_2 emissions were relatively large and that the 'column and beam' system was very effective at reducing CO_2 emissions compared to the load-bearing wall system during the construction stage. In addition, applying high-strength concrete to apartment housing is advantageous for reducing not only building material amounts but also reducing the requirement for repairs due to extending the building's life cycle.

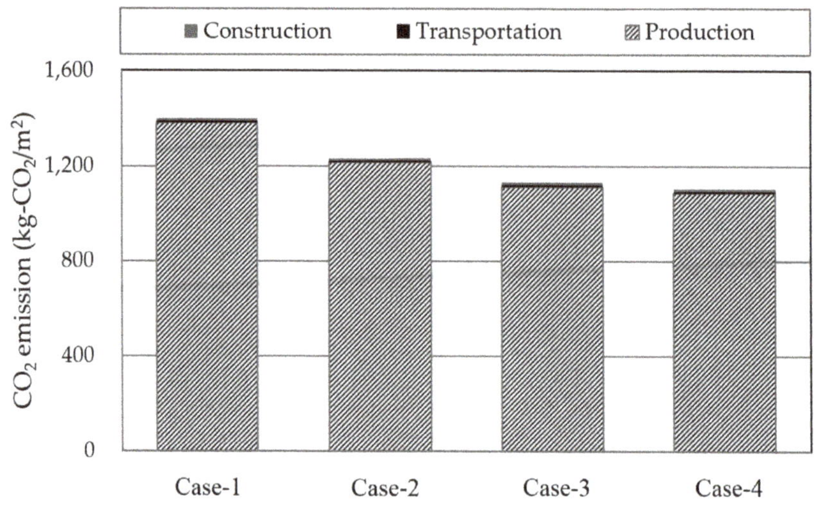

Figure 5. CO_2 emissions for the construction stage.

5.3. Assessment of Energy Consumption and CO_2 Emissions in the Operation Stage

ArchiCAD modeling files and the EcoDesigner add-on energy simulation program were used to assess changes in energy consumption in the operation stage. This study did not consider the reduction rate of operational energy effectiveness [39]. Both cases were simulated under the same conditions (see Table 7) in order to assess the energy consumption due to the different building forms, i.e., "flat-type" blocks and "T-type" blocks.

Table 7. EcoDesigner input data for each case.

	Mechanical Electrical and Plumbing System	Value
Heating and Cooling	Hot Water Generation	60 °C
	Cooling Type	Natural
Ventilation	Ventilation Type	Natural
	Air Change per Hour	0.7 times/hour
Energy Source	Heating	Natural Gas
	Other energy use	Electricity

As for the heat transfer coefficient of the wall parts, both cases were set based on the regional energy code in Korea. The glass used in the windows was 6 mm thick double glazing with a heat transfer coefficient of 3.1 W/m^2·K, solar heat gain coefficient (SHGC) of 0.66 and infiltration of 3.06 L/m^2.

Table 8 shows the calculated annual energy consumption and annual CO_2 emissions depending on what direction the block is facing. These results were generated with the EcoDesigner energy simulation software.

Table 8. Annual energy consumption and CO_2 emissions per unit area.

	Facing Direction	Annual Energy Consumption	Annual CO_2 Emission
Reference case	South	130.848 kWh/m^2	26.43 kgCO$_2$/m^2
	Southeast	134.234 kWh/m^2	27.11 kg CO$_2$/m^2
	Southwest	137.592 kWh/m^2	27.79 kg CO$_2$/m^2
Alternative case	South	87.547 kWh/m^2	17.68 kg CO$_2$/m^2

When facing south, the energy used by the alternative case decreased 33.09% from the reference case. This was mainly due to the 24.9% reduction in the Surface to Volume ratio (S/V ratio), which is attributed to the 25.3% decrease in the envelope area by efficient design of four units per floor and one vertical circulation core of the alternative case. The wall area ratio was also raised from 61.30% to 67.49%.

5.4. Discussion about Assessment of CO_2 Emissions for Building Life Cycle

To assess CO_2 emissions during the whole life cycle of the building, all CO_2 emissions should be totaled from the construction, operational and final stages of the building life cycle, with the final stage consisting of dismantling and disposal. In this chapter, we summarize all previous assessments of CO_2 emissions on the reference case and the alternative cases including high-strength concrete alternatives.

Based on the assessments discussed in 5.2 and 5.3 as well as an additional Dismantling and Disposal Assessment, Figure 6 and Table 9 show LCCO$_2$ emissions of all the test cases. LCCO$_2$ emission of Case 4 was 4299 kg-CO$_2$/m^2, which consisted of 26% in the construction stage, 73% in the operational stage, and 1% in the dismantling and disposal stage. The total amount of emissions for Case 4 was 30% less than Case 1.

As shown in Figure 6, "Production" in the construction stage and "Occupancy" in the operational stage are the most significant contributors to LCCO$_2$ emission. Therefore, applying effective CO_2 emission-reducing technologies to these two sub stages will substantially reduce total LCCO$_2$ emissions. In addition, CO_2 emissions from heating the building and the electrical energy required for operation, both in the operation stage, and from "Production" in the construction stage also contribute a fair amount of LCCO$_2$ emissions. The proportion of LCCO$_2$ emissions from each stage of the life cycle is similar in all four cases.

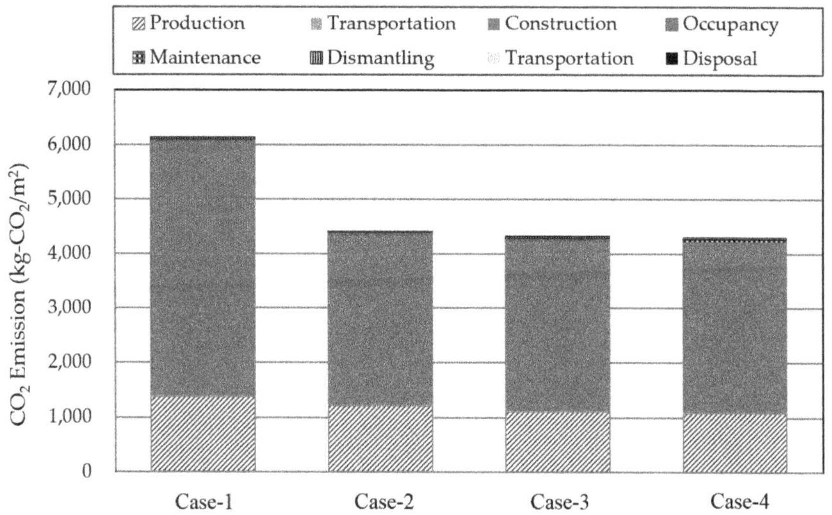

Figure 6. CO_2 emissions during the building life cycle.

Table 9. Life cycle CO_2 emissions.

CO_2 Assessment Stage		**$LCCO_2$ Emissions (kg-CO_2/m^2)**			
		Case-1	**Case-2**	**Case-3**	**Case-4**
Construction	Production	1381.52	1213.83	1111.87	1086.59
	Transportation	7.39	7.39	7.39	7.39
	Construction	10.96	10.96	10.96	10.96
	Sub-total	1399.87	1232.19	1130.22	1104.95
Operation	Occupancy	4656.10	3115.27	3115.27	3115.27
	Maintenance	42.40	42.40	42.40	42.40
	Sub-total	4698.50	3157.67	3157.67	3157.67
Dismantling and Disposal	Dismantling	32.40	32.40	32.40	32.40
	Transportation	3.60	3.60	3.60	3.60
	Disposal	0.58	0.58	0.58	0.58
	Sub-total	36.58	36.58	36.58	36.58
TOTAL		**6134.95**	**4426.44**	**4324.47**	**4299.20**

As shown in Table 9, $LCCO_2$ emissions in Case 2 were 27.8% less than that of Case 1, mainly because the operation stage produced 32.8% fewer emissions than Case 1. As we discussed in 5.3, the fewer emissions in the alternative cases stemmed from the different building forms: "flat-type" blocks vs. "T-type" blocks. Therefore, we recommend energy-efficient design strategies that optimize the S/V ratio and the wall area ratio in order to minimize the operational energy requirements and $LCCO_2$ emissions.

Applying high-strength concrete as well as a "column and beam" system led to only a 2.9% decrease in total $LCCO_2$ emission, but a 26.7% reduction in the construction stage is not a small portion of $LCCO_2$ emission. There is a reason that this amount is usually ignored in the construction process of apartment buildings. When apartment housing is planned and constructed initially, developers generally try to reduce the initial construction costs and do not consider $LCCO_2$ emissions. However, when the building is constructed with normal concrete rather than high-strength concrete, the building generally requires normal repairs after approximately 40 years. Although initially cheaper, normal concrete will lead to more CO_2 emissions through the whole building life cycle and lower the quality of the structure.

Based on the comparison of block types, we highly recommend the combination of an effective structural system such as the 'column and beam' system with a long life cycle technology such as high-strength concrete to help reduce $LCCO_2$ emissions in apartment housing projects. Our results indicate that the block type and system structure have significant impacts on building environmental load over its lifecycle, and significantly contribute to optimal greenhouse gas reduction. Therefore, it is expected that the assessment process of CO_2 emission based on the change in shapes of multi-unit dwellings that are examined in this research would be applicable in other countries, including Korea, as an alternative technique for estimating and assessing the environment performance of apartment houses. However, the regional applicability range could be comparatively limited as the established database of the research is based on the actual data of multi-unit dwellings that are built in Korea.

6. Conclusions

This paper assessed $LCCO_2$ emissions when an apartment building in Korea with 'flat-type' blocks (the reference case) was changed to a more sustainable 'T-type' block structure with fewer CO_2 emissions (the alternative case) while maintaining the same total floor area. The quantity of building materials used and building energy simulations were analyzed with each block type using BIM techniques, and the $LCCO_2$ was calculated with high-strength concrete alternatives. The conclusions are as follows:

1. By changing the bearing wall system of the 'flat-type' block to the 'column and beam' system of the 'T-type' block, the alternative case decreased the concrete and rebar used by 11% and 36%, respectively, compared with the ba3se case and as a result, CO_2 emission decreased by 9.45%.
2. When concrete strength was raised in order to decrease carbonation and increase durability in the 'T-type' block, CO_2 emissions of the concrete and rebar in the alternative case decreased by 35.39% compared with the reference case.

Moreover, there was an additional 2.6% reduction when the blast furnace slag was substituted at 20%.

3. By changing the building forms, the envelope volume ratio of the 'T-type' block decreased by 24% compared with the 'flat-type' block and, as a result, the CO_2 emissions of the alternative case during the operation stage decreased by 33.1%.
4. $LCCO_2$ emission of Case 4 was 4299 kg-CO_2/m^2, which consisted of 26% of the construction stage, 73% of the operational stage, and 1% of the dismantling and disposal stage. The total emissions were 30% less than Case 1.

Acknowledgments: This research was supported by Basic Science Research Program through the National Research Foundation of Korea (NRF) funded by the Ministry of Science, ICT & Future Planning (No. 2015R1A5A1037548, No. 2015R1D1A1A01057925).

Author Contributions: The paper was written by Cheonghoon Baek and revised by Sungho Tae. Rakhyun Kim and Sungwoo Shin conducted the experimental and analytical work. All authors contributed to the analysis and approved the final manuscript.

Conflicts of Interest: The authors declare no conflict of interest.

References

1. Cabeza, L.; Rincon, L.; Vilarino, V.; Perez, G.; Castell, A. Life cycle assessment (LCA) and life cycle energy analysis (LCEA) of buildings and the building sector: A review. *Renew. Sustain. Energy Rev.* **2014**, *29*, 394–416.
2. Dixit, M.; Culp, C.; Fernandez-Solis, J. System boundary for embodied energy in buildings: A conceptual model for definition. *Renew. Sustain. Energy Rev.* **2013**, *21*, 153–164.
3. International Energy Agency (IEA). *Energy Technology Perspectives 2010*; OECD/IEA: Washington, DC, USA, 2010.
4. UNEP. Overview of the Republic of Korea's National Strategy for Green Growth. 2010. Available online: http://www.unep.org/PDF/PressReleases/201004_UNEP_NATIONAL_STRATEGY.pdf (accessed on 25 July 2016).
5. Tae, S.; Shin, S. Current Work & Future Trends for Sustainable Buildings in South Korea. *Renew. Sustain. Energy Rev.* **2009**, *13*, 1910–1921.
6. Yi, I.S.; Seo, K.S. *Social Indicators in Korea*; National Statistical Office: Seoul, Korea, 2010.
7. Federica, C.; Idiano, D.; Massimo, G. Sustainable management of waste-to-energy facilities. *Renew. Sustain. Energy Rev.* **2014**, *33*, 719–728.
8. March, L.; Trace, M. *The Land-Use Performances of Selected Arrays of Built Forms, Land Use and Built form Studies*; University of Cambridge: Cambridge, UK, 1968.
9. Fathy, H. *Natural Energy and Vernacular Architecture*; The University of Chicago Press: Chicago, IL, USA, 1986.
10. Berköz, E. *Energy Efficient Building and Settlement Design*; Research Report, INTAG 201; The Scientific and Technical Research Council of Turkey: Istanbul, Turkey, 1995; pp. 98–138.

11. Givoni, B. *Climate Considerations in Buildings*; Van Nostrand Reinhold: Hoboken, NJ, USA, 1998.
12. Ratti, C.; Raydan, D.; Steemers, K. Building form and environmental performance: Archetypes, and arid climate. *Energy Build.* **2003**, *35*, 49–59.
13. Ba, G.; Gong, G. Research on passive solar ventilation and building self-sunshading. In Proceedings of the International Conference on Computer Distributed Control and Intelligent Environmental Monitoring, CDCIEM, Changsha, China, 19–20 February 2011; pp. 950–953.
14. International Organization for Standardization. *ISO 14040: Environmental Management-Life Cycle Assessment-Principles and Framework*; ISO: Geneva, Switzerland, 2006.
15. Ali, H.H.; Nsairat, S.A. Developing a green building assessment tool for developing countries—Case of Jordan. *Build. Environ.* **2009**, *44*, 1053–1064.
16. Haapio, A.; Viitaniemi, P. A critical review of building environmental assessment tools. *Environ. Impact Assess. Rev.* **2008**, *28*, 469–482.
17. Sharifi, A.; Murayama, A. A critical review of seven selected neighborhood sustainability assessment tools. *Environ. Impact Assess. Rev.* **2013**, *38*, 73–87.
18. Shin, S.; Tae, S.; Woo, J.; Roh, S. The development of apartment house life cycle CO_2 simple assessment system using standard apartment houses of South Korea. *Renew. Sustain. Energy Rev.* **2011**, *15*, 1454–1467.
19. Jukka, H.; Antti, S.; Juha-Matti, J.; Amalia, P.; Seppo, J. Pre-use phase LCA of a multi-story residential building: Can greenhouse gas emissions be used as a more general environmental performance indicator? *Build. Environ.* **2016**, *95*, 116–125.
20. Saxena, A.; Sethi, M.; Varun, V. Life cycle assessment of buildings: A review. *Renew. Sustain. Energy Rev.* **2011**, *15*, 871–875.
21. Hakkinen, T.; Huovila, P.; Tattari, K. Eco-efficient Building Process, VTT ProP Systematics on Building Headings. 2001. Available online: http://virtual.vtt.fi/virtual/environ/sb02-eco-efficient%20b_process2.pdf (accessed on 25 July 2016).
22. Roh, S.; Tae, S.; Shin, S.; Woo, J. Development of an optimum design program (SUSB-OPTIMUM) for the life cycle CO_2 assessment of an apartment house in Korea. *Build. Environ.* **2014**, *73*, 40–54.
23. Watson, P.; Mitchell, P.; Jones, D. Environmental Assessment for Commercial Buildings: Stakeholder Requirements and Tool Characteristics, Environmental Assessment Systems for Commercial Buildings. 2004. Available online: http://www.sbenrc.com.au/wp-content/uploads/2013/10/20-environmentalassessmentforcommercialbuildings.pdf (accessed on 25 July 2016).
24. Baek, C.; Park, S.; Suzuki, M.; Lee, S. Life cycle carbon dioxide assessment tool for buildings in the schematic design phase. *Energy Build.* **2013**, *611*, 75–87.
25. AIA. *AIA Guide to Building Life Cycle Assessment in Practice*; The American Institute of Architects: Washington, DC, USA, 2010.
26. ATHENA Sustainable Materials Institute—Athena EcoCalcuator. Available online: http://calculatelca.com/software/ecocalculator (accessed on 25 July 2016).

27. US Life Cycle Inventory Database. Available online: http://www.nrel.gov/lci (accessed on 25 July 2016).
28. Lee, K.; Tae, S.; Shin, S. Development of a Life Cycle Assessment Program for building (SUSB-LCA) in South Korea. *Renew. Sustain. Energy Rev.* **2009**, *13*, 1994–2002.
29. Lombera, J.S.; Rojo, J.C. Industrial building design stage based on a system approach to their environmental sustainability. *Constr. Build. Mater.* **2010**, *24*, 438–447.
30. Tae, S.; Baek, C.; Shin, S. Life cycle CO_2 evaluation on reinforced concrete structure with high-strength concrete. *Environ. Impact Assess. Rev.* **2011**, *31*, 253–260.
31. Shin, S.; Tae, S.; Woo, J.; Roh, S. The development of environmental load evaluation system of a standard Korean apartment house. *Renew. Sustain. Energy Rev.* **2011**, *15*, 1239–1249.
32. Sisomphon, K.; Franke, L. Carbonation rates of concrete containing high volume of pozzolanic materials. *Cem. Concr. Res.* **2007**, *37*, 1647–1653.
33. Chang, C.F.; Chen, J.W. The experimental investigation of concrete carbonation depth. *Cem. Concr. Res.* **2006**, *36*, 1760–1767.
34. Tae, S.; Ujiro, T. A study on the corrosion resistance of Cr-bearing rebar in mortar in corrosive environments involving chloride attack and carbonation. *ISIJ Int.* **2007**, *47*, 715–722.
35. AIJ (Architectural Institute of Japan). *Recommendations for Durability Design and Construction Practice of Reinforced Concrete*; Architectural Institute of Japan: Tokyo, Japan, 2004. (In Japanese)
36. Yuksel, I.; Bilir, T.; Ozkan, O. Durability of concrete incorporating non-ground blast furnace slag and bottom ash as fine aggregate. *Build. Environ.* **2007**, *42*, 2676–2685.
37. Robeyst, N.; Gruyaert, E.; Grosse, C.U.; Belie, N.D. Monitoring the setting of concrete containing blast-furnace slag by measuring the ultrasonic p-wave velocity. *Cem. Concr. Res.* **2008**, *38*, 1169–1176.
38. JSCE (Japan Society of Civil Engineers). *Standard Specifications of Concrete—Part Maintenance*; Japan Society of Civil Engineers: Tokyo, Japan, 2004. (In Japanese)
39. The Energy Information Administration (EIA). *Annual Energy Outlook—Fossil Fuels Remain Dominant through 2040*; EIA: Washington, DC, USA, 2014.

MDPI AG
St. Alban-Anlage 66
4052 Basel, Switzerland
Tel. +41 61 683 77 34
Fax +41 61 302 89 18
http://www.mdpi.com

Sustainability Editorial Office
E-mail: sustainability@mdpi.com
http://www.mdpi.com/journal/sustainability

www.ingramcontent.com/pod-product-compliance
Lightning Source LLC
LaVergne TN
LVHW061954050326
832904LV00011B/304

* 9 7 8 3 0 3 8 4 2 2 5 6 3 *